A·N·N·U·A·L ED·I·T·I

Developing World *03/04*

Thirteenth Edition

EDITOR

Robert J. Griffiths

University of North Carolina at Greensboro

Robert J. Griffiths is associate professor of political science and director of the International Studies Program at the University of North Carolina at Greensboro. His teaching and research interests are in the fields of comparative and international politics, and he teaches courses on the politics of development, African politics, international law and organization, and international political economy. His publications include articles on South African civil/military relations, democratic consolidation in South Africa, and developing countries and global commons negotiations.

McGraw-Hill/Dushkin

530 Old Whitfield Street, Guilford, Connecticut 06437

Visit us on the Internet
http://www.dushkin.com

Credits

1. **Understanding the Developing World**
   Unit photo—© 2003 by PhotoDisc, Inc.
2. **Political Economy and the Developing World**
   Unit photo—© 2003 by PhotoDisc, Inc.
3. **Conflict and Instability**
   Unit photo—AP/Wide World Photo.
4. **Political Change in the Developing World**
   Unit photo—United Nations photo by John Isaac.
5. **Population, Development, Environment, and Health**
   Unit photo—United Nations photo by J. K. Isaac.
6. **Women and Development**
   Unit photo—World Bank photo.

Copyright

Cataloging in Publication Data
Main entry under title: Annual Editions: Developing World. 2003/2004.
1. Underdeveloped areas—Periodicals. I. Griffiths, Robert J., *comp.* II. Title: Developing World.
ISBN 0–07–283855–8   303.4'5   ISSN 1096–4215

Thirteenth Edition

Cover image © 2003 PhotoDisc, Inc.
Printed in the United States of America   1234567890BAHBAH543   Printed on Recycled Paper

Editors/Advisory Board

Members of the Advisory Board are instrumental in the final selection of articles for each edition of ANNUAL EDITIONS. Their review of articles for content, level, currentness, and appropriateness provides critical direction to the editor and staff. We think that you will find their careful consideration well reflected in this volume.

To the Reader

In publishing ANNUAL EDITIONS we recognize the enormous role played by the magazines, newspapers, and journals of the public press in providing current, first-rate educational information in a broad spectrum of interest areas. Many of these articles are appropriate for students, researchers, and professionals seeking accurate, current material to help bridge the gap between principles and theories and the real world. These articles, however, become more useful for study when those of lasting value are carefully collected, organized, indexed, and reproduced in a low-cost format, which provides easy and permanent access when the material is needed. That is the role played by ANNUAL EDITIONS.

The developing world is home to the vast majority of the world's population. Because of its large population, increasing role in the international economy, frequent conflicts and humanitarian crises, and importance to environmental preservation, the developing world is a growing focus of international concern. The developing world has also figured prominently in the recent protests over globalization, and the September 11, 2001, attacks against the United States are tied to certain circumstances in developing countries.

Developing countries demonstrate considerable ethnic, cultural, political, and economic diversity, making generalizations about them difficult. Increasing differentiation further complicates our ability to understand developing countries and to comprehend the challenges of modernization, development, and globalization that they face. Confronting these challenges must take into account the combination of internal and external factors that contribute to the current circumstances. Issues of peace and security, international trade and finance, debt, poverty, and the environment illustrate the effects of growing interdependence and the need for a cooperative approach to dealing with these issues. There is significant debate regarding the best way to address the developing world's problems. Moreover, the developing world's needs compete for attention on an international agenda that is often dominated by relations between the industrialized nations and more recently by the war on terrorism. Domestic concerns within the industrial nations also continue to overshadow the plight of the developing world.

This thirteenth edition of *Annual Editions: Developing World* seeks to provide students with an understanding of the diversity and complexity of the developing world and to acquaint them with the challenges that these nations confront. I feel very strongly that there is a need for greater awareness of, and attention to, the problems of the developing world, especially in the aftermath of the September 11, 2001, terrorist attacks. I hope that this volume contributes to students' knowledge and understanding and serves as a catalyst for further discussion.

Approximately two-thirds of the articles in this edition are new. I have chosen articles that I hope are both interesting and informative and that can serve as a basis for further student research and discussion. The units deal with what I regard as the major issues facing the developing world. In addition, I have attempted to suggest similarities and differences between developing countries, the nature of their relationships with the industrialized nations, and the differences in perspective regarding the causes of and approaches to the issues.

I would again like to thank McGraw-Hill/Dushkin for the opportunity to put together a reader on a subject that is the focus of my teaching and research. I would also like to thank those who have sent in the response forms with their comments and suggestions. I have tried to take these into account in preparing the current volume.

No book on a topic as broad as the developing world can be comprehensive. There are certainly additional and alternative readings that might be included. Any suggestions for improvement are welcome. Please complete and return the postage-paid *article rating form* at the end of the book with your comments.

Robert J. Griffiths
Editor

Contents

UNIT 1
Understanding the Developing World

Five selections examine how the developing world's problems are interrelated with world order.

Unit Overview   xviii

The concepts in bold italics are developed in the article. For further expansion, please refer to the Topic Guide and the Index.

UNIT 2
Political Economy and the Developing World

Eleven articles discuss the impact that debt has on the developing world's politics and economy.

The concepts in bold italics are developed in the article. For further expansion, please refer to the Topic Guide and the Index.

UNIT 3
Conflict and Instability

Nine articles discuss the current state of ethnic conflicts throughout the developing world.

The concepts in bold italics are developed in the article. For further expansion, please refer to the Topic Guide and the Index.

The concepts in bold italics are developed in the article. For further expansion, please refer to the Topic Guide and the Index.

UNIT 4
Political Change in the Developing World

Six selections examine the innate problems faced by developing countries as they experience political change.

The concepts in bold italics are developed in the article. For further expansion, please refer to the Topic Guide and the Index.

UNIT 5
Population, Development, Environment, and Health

Eight articles examine some of the effects that the developing world's growth has on Earth's sustainability.

The concepts in bold italics are developed in the article. For further expansion, please refer to the Topic Guide and the Index.

UNIT 6
Women and Development

Four articles discuss the role of women in the developing world.

The concepts in bold italics are developed in the article. For further expansion, please refer to the Topic Guide and the Index.

Topic Guide

This topic guide suggests how the selections in this book relate to the subjects covered in your course. You may want to use the topics listed on these pages to search the Web more easily.

On the following pages a number of Web sites have been gathered specifically for this book. They are arranged to reflect the units of this *Annual Edition.* You can link to these sites by going to the DUSHKIN ONLINE support site at *http://www.dushkin.com/online/.*

ALL THE ARTICLES THAT RELATE TO EACH TOPIC ARE LISTED BELOW THE BOLD-FACED TERM.

World Wide Web Sites

The following World Wide Web sites have been carefully researched and selected to support the articles found in this reader. The easiest way to access these selected sites is to go to our DUSHKIN ONLINE support site at *http://www.dushkin.com/online/*.

AE: Developing World 03/04

The following sites were available at the time of publication. Visit our Web site—we update DUSHKIN ONLINE regularly to reflect any changes.

General Sources

Central Intelligence Agency—U.S.

http://www.odci.gov

This site includes publications of the CIA, such as the *World Fact Book* and CIA world and regional maps.

Foreign Policy in Focus (FPIF): Progressive Response Index

http://fpif.org/progresp/index_body.html

This index is produced weekly by FPIF, a "think tank without walls," which is an international network of analysts and activists dedicated to "making the U.S. a more responsible global leader and partner by advancing citizen movements and agendas." This index lists volume and issue numbers, dates, and topics covered by the articles.

Our Developing World

http://www.magiclink.net/~odw/

Dedicated to bringing the realities of the "third world" and the richness of diverse cultures to North Americans, Our Developing World provides hands-on as well as visual and print materials and training for teachers, and programs for community groups and classes, study tours, a tri-annual newsletter   Our Developing World's Voices   and a lending resource library free to local teachers.

People & Planet

http://www.peopleandplanet.org

People & Planet is an organization of student groups at universities and colleges across the United Kingdom. Organized in 1969 by students at Oxford University, it is now an independent pressure group campaigning on world poverty, human rights, and the environment.

United Nations System Web Locator

http://www.unsystem.org

This is the Web site for all the organizations in the United Nations family. According to its brief overview, the United Nations, an organization of sovereign nations, provides the machinery to help find solutions to international problems or disputes and to deal with pressing concerns that face people everywhere, including the problems of the developing world, through the UN Development Program at *http://www.undp.org* and UNAIDS at *http://www.unaids.org*.

United States Census Bureau: International Summary Demographic Data

http://www.census.gov/ipc/www/idbsum.html

The International Data Base (IDB)is a computerized data bank containing statistical tables of demographic and socioeconomic data for all countries of the world.

World Health Organization (WHO)

http://www.who.ch

The WHO's objective, according to its Web site, is the attainment by all peoples of the highest possible level of health. Health, as defined in the WHO constitution, is a state of complete physical, mental, and social well-being and not merely the absence of disease or infirmity.

UNIT 1: Understanding the Developing World

Africa Index on Africa

http://www.afrika.no/index/

A complete reference source on Africa is available on this Web site.

African Policy Information Center (APIC)

http://www.igc.apc.org/apic/index.shtml

The primary objective of the APIC is to widen the policy debate in the United States on African issues and the U.S. role in Africa, by providing accessible policy-relevant information and analysis usable by a wide range of groups and individuals.

African Studies WWW (U. Penn)

http://www.sas.upenn.edu/African_Studies/AS.html

The African Studies Center at the University of Pennsylvania supports this ongoing project that lists online resources related to African Studies.

UNIT 2: Political Economy and the Developing World

Center for Third World Organizing

http://www.ctwo.org/

The Center for Third World Organizing (CTWO, pronounced "C-2") is a racial justice organization dedicated to building a social justice movement led by people of color. CTWO is a 20-year-old training and resource center that promotes and sustains direct action organizing in communities of color in the United States.

ENTERWeb

http://www.enterweb.org

ENTERWeb is an annotated meta-index and information clearinghouse on enterprise development, business, finance, international trade, and the economy in this age of cyberspace and globalization. The main focus is on micro-, small-, and medium-scale enterprises, cooperatives, and community economic development both in developed and developing countries.

Grameen Bank (GB)

http://www.grameen-info.org

Grameen Bank has reversed covential banking practice by removing the need for collateral and creating a banking system based on mutual trust, accountability, participation, and creativity. GB provides credit to the poorest of the poor in rural Bangladesh without any collateral.

www.dushkin.com/online/

International Monetary Fund (IMF)
http://www.imf.org

The IMF was created to promote international monetary cooperation, to facilitate the expansion and balanced growth of international trade, to promote exchange stability, to assist in the establishment of a multilateral system of payments, to make its general resources temporarily available under adequate safeguards to its members experiencing balance-of-payments difficulties, and to shorten the duration and lessen the degree of disequilibrium in the international balances of payments of members.

TWN   Third World Network
http://www.twnside.org.sg/

The Third World Network is an independent non-profit international network of organizations and individuals involved in issues relating to development, the Third World and North-South issues.

U.S. Agency for International Development (USAID)
http://www.info.usaid.gov

USAID is an independent government agency that provides economic development and humanitarian assistance to advance U.S. economic and political interests overseas.

The World Bank
http://www.worldbank.org

The International Bank for Reconstruction and Development, frequently called the World Bank, was established in July 1944 at the UN Monetary and Financial Conference in Bretton Woods, New Hampshire. The World Bank's goal is to reduce poverty and improve living standards by promoting sustainable growth and investment in people. The bank provides loans, technical assistance, and policy guidance to developing country members to achieve this objective.

UNIT 3: Conflict and Instability

The Carter Center
http://www.cartercenter.org

The Carter Center is dedicated to fighting disease, hunger, poverty, conflict, and oppression through collaborative initiatives in the areas of democratization and development, global health, and urban revitalization.

Center for Strategic and International Studies (CSIS)
http://www.csis.org/

For four decades, the Center for Strategic and International Studies (CSIS) has been dedicated to providing world leaders with strategic insights on, and policy solutions to, current and emerging global issues.

Conflict Research Consortium
http://www.Colorado.EDU/conflict/

The site offers links to conflict- and peace-related Internet sites.

Institute for Security Studies
http://www.iss.co.za

This site is South Africa's premier source for information related to African security studes.

PeaceNet
http://www.igc.org/peacenet/

PeaceNet promotes dialogue and sharing of information to encourage appropriate dispute resolution, highlights the work of practitioners and organizations, and is a proving ground for ideas and proposals across the range of disciplines within the conflict-resolution field.

Refugees International
http://www.refintl.org

Refugees International provides early warning in crises of mass exodus. It seeks to serve as the advocate of the unrepresented—the refugee. In recent years, Refugees International has moved from its initial focus on Indochinese refugees to global coverage, conducting almost 30 emergency missions in the last 4 years.

UNIT 4: Political Change in the Developing World

Greater Horn Information Exchange (GHIE)
http://edcsnw3.cr.usgs.gov/ghai/ghai.html

The GHIE is a no-fee resource accessible via e-mail, Telnet, Gopher, and the World Wide Web. It expedites the sharing of site reports, fact sheets, activity summaries, data sets, scientific papers, and analyses on the Horn region of Africa.

Latin American Network Information Center—LANIC
http://www.lanic.utexas.edu

According to *Latin Trade*, LANIC is "a good clearing house for Internet-accessible information on Latin America."

ReliefWeb
http://www.reliefweb.int/w/rwb.nsf

ReliefWeb is the UN's Department of Humanitarian Affairs clearinghouse for international humanitarian emergencies.

World Trade Organization (WTO)
http://www.wto.org

The WTO is promoted as the only international body dealing with the rules of trade between nations. At its heart are the WTO agreements, the legal ground rules for international commerce and for trade policy.

UNIT 5: Population, Development, Environment, and Health

Earth Pledge Foundation
http://www.earthpledge.org

The Earth Pledge Foundation promotes the principles and practices of sustainable development—the need to balance the desire for economic growth with the necessity of environmental protection.

EnviroLink
http://envirolink.org

EnviroLink is committed to promoting a sustainable society by connecting individuals and organizations through the use of the World Wide Web.

Greenpeace
http://www.greenpeace.org

Greenpeace is an international NGO (nongovernmental organization) that is devoted to environmental protection.

Linkages on Environmental Issues and Development
http://www.iisd.ca/linkages/

Linkages is a site provided by the International Institute for Sustainable Development. It is designed to be an electronic clearinghouse for information on past and upcoming international meetings related to both environmental issues and economic development in the developing world.

Population Action International
http://www.populationaction.org

According to their mission statement, Population Action International is dedicated to advancing policies and programs that slow population growth in order to enhance the quality of life for all people.

The Worldwatch Institute

http://www.worldwatch.org

The Worldwatch Institute advocates environmental protection and sustainable development.

UNIT 6: Women and Development

African Women Global Network

http://www.osu.edu/org/awognet/main.html

African Women Global Network is a global organization that networks all men and women, organizations, institutions, and indigenous national organizations within Africa. The activities at this Web site target the improvement of the living conditions of women and children in Africa.

WIDNET: Women in Development NETwork

http://www.focusintl.com/widnet.htm

This site provides a wealth of information about women in development, including the Beijing '95 Conference, WIDNET statistics, and women's studies.

Women's Issues—3rd World

http://www.women3rdworld.about.com/newsissues/women3rdworld/ mbody.htm

Information on the economic, political, religious, and sociocultural issues facing women in the developing world is available here.

WomenWatch/Regional and Country Information

http://www.un.org/womenwatch/

The UN Internet Gateway on the Advancement and Empowerment of Women provides a rich mine of information.

We highly recommend that you review our Web site for expanded information and our other product lines. We are continually updating and adding links to our Web site in order to offer you the most usable and useful information that will support and expand the value of your Annual Editions. You can reach us at: *http://www.dushkin.com/annualeditions/*.

UNIT 1

Understanding the Developing World

Unit Selections

1. **The Great Divide in the Global Village**, Bruce R. Scott
2. **Prisoners of Geography**, Ricardo Hausmann
3. **The Poor Speak Up**, Rana Foroohar
4. **The Rich Should Not Forget the ROW (Rest of the World)**, Jose Ramos-Horta
5. **Putting a Human Face on Development**, Rubens Ricupero

Key Points to Consider

- What factors account for the economic disparity both within developing countries and between them and the world's industrialized countries?

- How do immigration and trade policies in the industrialized countries affect the developing world?

- What are the features of the Washington Consensus?

- How does geography affect development?

- In what ways have the developing countries cooperated to achieve their goals?

- What should a cooperative agenda for the developing world include?

- In what ways should the concept of development be broadened?

 **Links: www.dushkin.com/online/**
These sites are annotated in the World Wide Web pages.

Africa Index on Africa
http://www.afrika.no/index/
African Policy Information Center (APIC)
http://www.igc.apc.org/apic/index.shtml
African Studies WWW (U. Penn)
http://www.sas.upenn.edu/African_Studies/AS.html

Understanding the diverse countries that make up the developing world has never been an easy task and it has become even more difficult as further differentiation among these countries has occurred. "Developing world" is a catch-all term that lacks precision and explanatory power. It encompasses a wide range of societies from traditional to modernizing. There is also controversy over what actually constitutes development. For some, it is economic growth or progress towards democracy; for others it involves greater empowerment and dignity. There are also differing views on why progress toward development has been uneven. The West tends to see the problem as stemming from institutional weakness and failure to embrace free-market principles. Critics from the developing world cite the legacy of colonialism and the international political and economic structures as the reasons for a lack of development. Lumping together the 100-plus nations that make up the developing world obscures the disparities in size, population, resources, forms of government, industrialization, distribution of wealth, ethnic diversity, and a host of other indicators that make it difficult to understand this large, diverse group of countries.

Despite their diversity, most nations of the developing world share some characteristics. Developing countries often have large populations, with annual growth rates between 2 and 4 percent. Poverty is widespread in both rural and urban areas, with rural areas often containing the poorest of the poor. While the majority of the developing world's inhabitants continue to live in the countryside, there is a massive rural-to-urban migration under way, and cities are growing rapidly. Wealth is unevenly distributed, making education, employment opportunities, and access to health care luxuries that few can enjoy. Corruption and mismanagement are widespread. With very few exceptions, these nations share a colonial past that has affected them both politically and economically. Moreover, critics charge that the neocolonial structure of the international economy and the West's political, military, and cultural links with the developing world amount to continued domination.

Developing countries continue to struggle to improve their citizens' living standards. Despite the economic success in some areas, poverty remains widespread, and over a billion people live on less than a dollar a day. There is also growing economic inequality between the industrial countries and the developing world. This is especially true of the poorest countries, which have become further marginalized due to their fading strategic importance since the end of the cold war and their limited participation in the global economy. Inequality is also growing within developing countries where elite access to education, capital, and technology has significantly widened the gap between rich and poor.

Although the gap between rich and poor nations persists, some emerging markets saw significant growth during the 1990s. However, even these countries experienced the harsh realities of the global economy. The 1997 Asian financial crisis demonstrated the potential consequences of global finance and investment, and investors remain wary of investing in all but a few developing countries. Focus on emerging markets' prob-

lems further marginalizes the majority of developing countries that have not experienced economic growth. The reasons for poor economic performance are complex. Both internal and external factors play a role in the inability of some developing countries to make economic progress. Among the factors are the colonial legacy, continued reliance on the export of primary products, stagnating or declining terms of trade for those primary products, protectionism in the industrialized countries, meager foreign aid contributions, debt, and the array of domestic social problems. The economic slowdown in the United States, worsened by the September 2001 terrorist attacks, has also had international consequences that affect economic growth in the developing world.

As colonies gained their independence in the post–World War II era, a combination of thought and action that constituted a loosely defined third world movement emerged. Besides shaping the domestic goals of many leaders, this perspective emphasized the revolutionary aspirations of developing world peoples and contributed to the view of a world divided between the industrialized North and the developing South. Based on this view, which portrayed the North as continuing to exploit the South, the developing countries challenged the North, calling for the establishment of a New International Economic Order and seeking to influence the agenda of various international organizations. Although this effort faded during the 1980s, elements of this perspective are still evident in a radical, non-Western worldview. Emphasizing the North's continuing domination of the developing world, its proponents view international relations as a struggle of the oppressed against their oppressors. Echoing the earlier efforts at solidarity, developing countries coordinated efforts to extract concessions from the industrialized countries at the November 2001 Doha trade summit. Such cooperative efforts are likely to enhance the ability of the developing world to shape the international agenda. Whether the perspective is radical or mainstream, there is a divergence of opinion between the industrialized countries and the developed world on issues ranging from human rights to economic development.

1

The Great Divide
in the Global Village

Bruce R. Scott

INCOMES ARE DIVERGING

MAINSTREAM economic thought promises that globalization will lead to a widespread improvement in average incomes. Firms will reap increased economies of scale in a larger market, and incomes will converge as poor countries grow more rapidly than rich ones. In this "win–win" perspective, the importance of nation-states fades as the "global village" grows and market integration and prosperity take hold.

But the evidence paints a different picture. Average incomes have indeed been growing, but so has the income gap between rich and poor countries. Both trends have been evident for more than 200 years, but improved global communications have led to an increased awareness among the poor of income inequalities and heightened the pressure to emigrate to richer countries. In response, the industrialized nations have erected higher barriers against immigration, making the world economy seem more like a gated community than a global village. And although international markets for goods and capital have opened up since World War II and multilateral organizations now articulate rules and monitor the world economy, economic inequality among countries continues to increase. Some two billion people earn less than $2 per day.

At first glance, there are two causes of this divergence between economic theory and reality. First, the rich countries insist on barriers to immigration and agricultural imports. Second, most poor nations have been unable to attract much foreign capital due to their own government failings. These two issues are fundamentally linked: by forcing poor people to remain in badly governed states, immigration barriers deny those most in need the opportunity to "move up" by "moving out." In turn, that immobility eliminates a potential source of pressure on ineffective governments, thus facilitating their survival.

Since the rich countries are unlikely to lower their agricultural and immigration barriers significantly, they must recognize that politics is a key cause of economic inequality. And since most developing countries receive little foreign investment, the wealthy nations must also acknowledge that the "Washington consensus," which assumes that free markets will bring about economic convergence, is mistaken. If they at least admit these realities, they will abandon the notion that their own particular strategies are the best for all countries. In turn, they should allow poorer countries considerable freedom to tailor development strategies to their own circumstances. In this more pragmatic view, the role of the state becomes pivotal.

Why have economists and policymakers not come to these conclusions sooner?

Since the barriers erected by rich countries are seen as vital to political stability, leaders of those countries find it convenient to overlook them and focus instead on the part of the global economy that has been liberalized. The rich countries' political power in multilateral organizations makes it difficult for developing nations to challenge this self-serving world-view. And standard academic solutions may do as much harm as good, given their focus on economic stability and growth rather than on the institutions that underpin markets. Economic theory has ignored the political issues at stake in modernizing institutions, incorrectly assuming that market-based prices can allocate resources appropriately.

The fiasco of reform in Russia has forced a belated reappraisal of this blind trust in markets. Many observers now admit that the transition economies needed appropriate property rights and an effective state to enforce those rights as much as they needed the liberalization of prices. Indeed, liberalization without property rights turned out to be the path to gangsterism, not capitalism. China, with a more effective state, achieved much greater success in its transition than did Russia, even though Beijing proceeded much more slowly with liberalization and privatization.

Economic development requires the transformation of institutions as well as the freeing of prices, which in turn requires political and social modernization as well as economic reform. The state plays a key role in this process; without it, developmental strategies have little hope of succeeding. The creation of effective states in the developing world will not be driven by familiar market forces, even if pressures from capital markets can force fiscal and monetary discipline. And in a world still governed by "states rights," real progress in achieving accountable governments will require reforms beyond the mandates of multilateral institutions.

GO WITH THE FLOW

In THEORY, globalization provides an opportunity to raise incomes through increased specialization and trade. This opportunity is conditioned by the size of the markets in question, which in turn depends on geography, transportation costs, communication networks, and the institutions that underpin markets. Free trade increases both the size of the market and the pressure to improve economic performance. Those who are most competitive take advantage of the enhanced market opportunities to survive and prosper.

Neoclassical economic theory predicts that poor countries should grow faster than rich ones in a free global market. Capital from rich nations in search of cheaper labor should flow to poorer economies, and labor should migrate from low-income areas toward those with higher wages. As a result, labor and capital costs—and eventually income—in rich and poor areas should eventually converge.

The U.S. economy demonstrates how this theory can work in a free market with the appropriate institutions. Since the 1880s, a remarkable convergence of incomes among the country's regions has occurred. The European Union has witnessed a similar phenomenon, with the exceptions of Greece and Italy's southern half, the *Mezzogiorno*. What is important, however, is that both America and the EU enjoy labor and capital mobility as well as free internal trade.

But the rest of the world does not fit this pattern. The most recent *World Development Report* shows that real per capita incomes for the richest one-third of countries rose by an annual 1.9 percent between 1970 and 1995, whereas the middle third went up by only 0.7 percent and the bottom third showed no increase at all. In the

Western industrial nations and Japan alone, average real incomes have been rising about 2.5 percent annually since 1950—a fact that further accentuates the divergence of global income. These rich countries account for about 60 percent of world GDP but only 15 percent of world population.

Why is it that the poor countries continue to fall further behind? One key reason is that most rich countries have largely excluded the international flow of labor into their markets since the interwar period. As a result, low-skilled labor is not free to flow across international boundaries in search of more lucrative jobs. From an American or European perspective, immigration appears to have risen in recent years, even approaching its previous peak of a century ago in the United States. Although true, this comparison misses the central point. Billions of poor people could improve their standard of living by migrating to rich countries. But in 1997, the United States allowed in only 737,000 immigrants from developing nations, while Europe admitted about 665,000. Taken together, these flows are only 0.04 percent of all potential immigrants.

Global markets offer opportunities for all, but opportunities do not guarantee results

---

The point is not that the rich countries should permit unfettered immigration. A huge influx of cheap labor would no doubt be politically explosive; many European countries have already curtailed immigration from poor countries for fear of a severe backlash. But the more salient issue is that rich nations who laud liberalism and free markets are rejecting those very principles when they restrict freedom of movement. The same goes for agricultural imports. Both Europe and Japan have high trade barriers in agriculture, while the United States remains modestly protectionist.

Mainstream economic theory does provide a partial rationalization for rich-country protectionism: Immigration barriers need not be a major handicap to poor nations because they can be offset by capital flows from industrialized economies to developing ones. In other words, poor people need not demand space in rich countries because the rich will send their capital to help develop the poor countries. This was

indeed the case before World War I, but it has not been so since World War II.

But the question of direct investment, which typically brings technologies and know-how as well as financial capital, is more complicated than theories would predict. The total stock of foreign direct investment did rise almost sevenfold from 1980 to 1997, increasing from 4 percent to 12 percent of world GDP during that period. But very little has gone to the poorest countries. In 1997, about 70 percent went from one rich country to another, 8 developing countries received about 20 percent, and the remainder was divided among more than 100 poor nations. According to the World Bank, the truly poor countries received less than 7 percent of the foreign direct investment to all developing countries in 1992–98. At the same time, the unrestricted opening of capital markets in developing countries gives larger firms from rich countries the opportunity for takeovers that are reminiscent of colonialism. It is not accidental that rich countries insist on open markets where they have an advantage and barriers in agriculture and immigration, where they would be at a disadvantage.

As for the Asian "tigers," their strong growth is due largely to their high savings rate, not foreign capital. Singapore stands out because it has enjoyed a great deal of foreign investment, but it has also achieved one of the highest domestic-savings rates in the world, and its government has been a leading influence on the use of these funds. China is now repeating this pattern, with a savings rate of almost 40 percent of GDP. This factor, along with domestic credit creation, has been its key motor of economic growth. China now holds more than $100 billion in low-yielding foreign-exchange reserves, the second largest reserves in the world.

In short, global markets offer opportunities for all, but opportunities do not guarantee results. Most poor countries have been unable to avail themselves of much foreign capital or to take advantage of increased market access. True, these countries have raised their trade ratios (exports plus imports) from about 35 percent of their GDP in 1981 to almost 50 percent in 1997. But without the Asian tigers, developing-country exports remain less than 25 percent of world exports.

Part of the problem is that the traditional advantages of poor countries have been in primary commodities (agriculture and minerals), and these categories have shrunk from about 70 percent of world trade in 1900 to about 20 percent at the end of the century. Opportunities for growth in the world market have shifted from raw or semiprocessed commodities toward manufactured goods and services—and, within these categories, toward more knowledge-intensive segments. This trend obviously favors rich countries over poor ones, since most of the latter are still peripheral players in the knowledge economy. (Again, the Asian tigers are the exception. In 1995, they exported as much in high-technology goods as did France, Germany, Italy, and Britain combined—which together have three times the population of the tigers.)

ONE COUNTRY, TWO SYSTEMS

WHY is the performance of poor countries so uneven and out of sync with theoretical forecasts? Systemic barriers at home and abroad inhibit the economic potential of poorer nations, the most formidable of these obstacles being their own domestic political and administrative problems. These factors, of course, lie outside the framework of mainstream economic analysis. A useful analogy is the antebellum economy of the United States, which experienced a similar set of impediments.

Like today's "global village," the U.S. economy before the Civil War saw incomes diverge as the South fell behind the North. One reason for the Confederacy's secession and the resulting civil war was Southern recognition that it was falling behind in both economic and political power, while the richer and more populous North was attracting more immigrants. Half of the U.S. population lived in the North in 1780; by 1860, this share had climbed to two-thirds. In 1775, incomes in the five original Southern states equaled those in New England, even though wealth (including slaves) was disproportionately concentrated in the South. By 1840, incomes in the northeast were about 50 percent higher than those in the original Southern states; the North's railroad mileage was about 40 percent greater (and manufacturing investment four times higher) than the South's. As the economist Robert Fogel has pointed out, the South was not poor—in 1860 it was richer than all European states except England—but Northern incomes were still much higher and increasing.

Why had Southern incomes diverged from those in the North under the same government, laws, and economy? Almost from their inception, the Southern colonies followed a different path from the North—specializing in plantation agriculture rather than small farms with diversified crops—due to geography and slavery. Thanks to slave labor, Southerners were gaining economies of scale and building comparative advantage in agriculture, exporting their goods to world markets and the North. Gang labor outproduced "free" (paid) labor. But the North was building even greater advantages by developing a middle class, a manufacturing sector, and a more modern social and political culture. With plans to complete transcontinental railroads pending, the North was on the verge of achieving economic and political dominance and the capacity to shut off further expansion of slavery in the West. The South chose war over Northern domination—and modernization.

Although the Constitution guaranteed free trade and free movement of capital and labor, the institution of slavery meant that the South had much less factor mobility than the North. It also ensured less development of its human resources, a less equal distribution of income, a smaller market for manufactures, and a less dynamic economy. It was less attractive to both European immigrants and external capital. With stagnant incomes in the older states, it was falling behind. In these respects, it was a forerunner of many of today's poor countries, especially those in Latin America.

What finally put the South on the path to economic convergence? Four years of civil war with a total of 600,000 deaths and vast destruction of property were only a start. Three constitutional amendments and twelve years of military "reconstruction" were designed to bring equal rights and due process to the South. But the reestablishment of racial segregation following Reconstruction led to sharecropping as former slaves refused to return to the work gangs. Labor productivity dropped so much that Southern incomes fell to about half of the North's in 1880. In fact, income convergence did not take off until the 1940s, when a wartime boom in the North's industrial cities attracted Southern migrants in search of better jobs. At the same time, the South began drawing capital as firms sought lower wages, an anti-union environment, and military contracts in important congressional districts. But this process did not fully succeed until the 1960s, as new federal laws and federal troops brought full civil rights to the South

and ensured that the region could finally modernize.

THE GREAT DIVIDE

ALTHOUGH slavery is a rarity today, the traditional U.S. divide between North and South provides a good model for understanding contemporary circumstances in many developing countries. In the American South, voter intimidation, segregated housing, and very unequal schooling were the rule, not the exception—and such tactics are repeated today by the elites in today's poor countries. Brazil, Mexico, and Peru had abundant land relative to population when the Europeans arrived, and their incomes roughly approximated those in North America, at least until 1700. The economists Stanley Engerman and Kenneth Sokoloff have pointed out that these states, like the Confederacy, developed agricultural systems based on vast landholdings for the production of export crops such as sugar and coffee. Brazil and many Caribbean islands also adopted slavery, while Peru and Mexico relied on forced indigenous labor rather than African slaves.

History shows that the political development of North America and developing nations—most of which were colonized by Europeans at some point—was heavily influenced by mortality. In colonies with tolerable death rates (Australia, Canada, New Zealand, and the United States), the colonists soon exerted pressure for British-style protections of persons and property. But elsewhere (most of Africa, Latin America, Indonesia, and to a lesser degree, India), disease caused such high mortality rates that the few resident Europeans were permitted to exploit a disenfranchised laboring class, whether slave or free. When the colonial era ended in these regions, it was followed by "liberationist" regimes (often authoritarian and incompetent) that maintained the previous system of exploitation for the advantage of a small domestic elite. Existing inequalities within poor countries continued; policies and institutions rarely protected individual rights or private initiative for the bulk of the population and allowed elites to skim off rents from any sectors that could bear it. The economist Hernando de Soto has shown how governments in the developing world fail to recognize poor citizens' legal titles to their homes and businesses, thereby depriving them of the use of their assets for collateral. The losses in potential capital to these countries have dwarfed the cumulative capital inflows going to these economies in the last century.

The legacy of these colonial systems also tends to perpetuate the unequal distribution of income, wealth, and political power while limiting capital mobility. Thus major developing nations such as Brazil, China, India, Indonesia, and Mexico are experiencing a divergence of incomes by province within their economies, as labor and capital fail to find better opportunities. Even in recent times, local elites have fought to maintain oppressive conditions in Brazil, El Salvador, Guatemala, Mexico, Nicaragua, and Peru. Faced with violent intimidation, poor people in these countries have suffered from unjust law enforcement similar to what was once experienced by black sharecroppers in the American South.

Modernization and economic development inevitably threaten the existing distribution of power and income, and powerful elites continue to protect the status quo—even if it means that their society as a whole falls further behind. It takes more than a constitution, universal suffrage, and regular elections to achieve governmental accountability and the rule of law. It may well be that only the right of exit—emigration—can peacefully bring accountability to corrupt and repressive regimes. Unlike the U.S. federal government, multilateral institutions lack the legitimacy to intervene in the internal affairs of most countries. Europe's economic takeoff in the second half of the nineteenth century was aided by the emigration of 60 million people to North America, Argentina, Brazil, and Australia. This emigration—about 10 percent of the labor force—helped raise European wages while depressing inflated wages in labor-scarce areas such as Australia and the United States. A comparable out-migration of labor from today's poor countries would involve hundreds of millions of people.

Of course, Latin America has seen some success. Chile has received the most attention for its free market initiatives, but its reforms were implemented by a brutally repressive military regime—hardly a model for achieving economic reform through democratic processes. Costa Rica would seem to be a much better model for establishing accountability, but its economic performance has not been as striking as Chile's.

Italy, like the United States in an earlier era, is another good example of "one country, two systems." Italy's per capita in-come has largely caught up with that of its European neighbors over the past 20 years, even exceeding Britain's and equaling France's in 1990, but its *Mezzogiorno* has failed to keep up. Whereas overall Italian incomes have been converging toward those of the EU, *Mezzogiorno* incomes have been diverging from those in the north. Southern incomes fell from 65 percent of the northern average in 1975 to 56 percent 20 years later; in Calabria, they fell to 47 percent of the northern average. Southern unemployment rose from 8 percent in 1975 to 19 percent in 1995—almost three times the northern average. In short, 50 years of subsidies from Rome and the EU have failed to stop the *Mezzogiorno* from falling further behind. Instead, they have yielded local regimes characterized by greatly increased public-sector employment, patronage, dependency, and corruption—not unlike the results of foreign aid for developing countries. And the continuing existence of the Mafia further challenges modernization.

Democracy is not enough to ensure that the governed reap the gains of their own efforts.

Democracy, then, is not enough to ensure that the governed are allowed to reap the gains of their own efforts. An effective state requires good laws as well as law enforcement that is timely, evenhanded, and accessible to the poor. In many countries, achieving objective law enforcement means reducing the extralegal powers of vested interests. When this is not possible, the only recourse usually available is emigration. But if the educated elite manages to emigrate while the masses remain trapped in a society that is short of leaders, the latter will face even more formidable odds as they try to create effective institutions and policies. Although Italians still emigrate from south to north, the size of this flow is declining, thanks in part to generous transfer payments that allow them to consume almost as much as northerners. In addition, policymaking for the *Mezzogiorno* is still concentrated in Rome.

The immigration barriers in rich countries not only foreclose opportunities in the global village to billions of poor people, they help support repressive, pseudo-democratic governments by denying the citizens of these countries the right to vote against the regime with their feet. In effect, the strict dictates of sovereignty allow wealthy nations to continue to set the rules in their own favor while allowing badly governed poor nations to continue to abuse their own citizens and retard economic development. Hence the remedy for income divergence must be political as well as economic.

GETTING INSTITUTIONS RIGHT

ACCORDING TO ECONOMIC THEORY, developing nations will create and modernize the institutions needed to underpin their markets so that their markets and firms can gradually match the performance of rich countries. But reality is much more complex than theory. For example, de Soto's analysis makes clear that effectively mobilizing domestic resources offers a much more potent source of capital for most developing nations than foreign inflows do. Yet mainstream economists and their formal models largely ignore these resources. Western economic advisers in Russia were similarly blindsided by their reliance on an economic model that had no institutional context and no historical perspective. Economists have scrambled in recent years to correct some of these shortcomings, and the Washington consensus now requires the "right" institutions as well as the "right" prices. But little useful theory exists to guide policy when it comes to institutional analysis, and gaps in the institutional foundations in most developing countries leave economic models pursuing unrealistic solutions or worse.

The adjustment of institutions inevitably favors certain actors and disadvantages others. As a result, modernization causes conflict that must be resolved through politics as well as economics. At a minimum, successful development signifies that the forces for institutional change have won out over the status quo. Achieving a "level playing field" signifies that regulatory and political competition is well governed.

Economists who suggest that all countries must adopt Western institutions to achieve Western levels of income often fail to consider the changes and political risks involved. The experts who recommended that formerly communist countries apply "shock therapy" to markets and democracy disregarded the political and regulatory issues involved. Each change requires a victory in the "legislative market" and successful persuasion within the state bureaucracy for political approval. Countries with lower incomes and fewer educated people than Russia face even more significant developmental challenges just to achieve economic stability, let alone attract foreign investment or make effective use of it. Institutional deficiencies, not capital shortages, are the major impediment to development, and as such they must be addressed before foreign investors will be willing to send in capital.

Although price liberalization can be undertaken rapidly, no rapid process (aside from revolution) exists for an economy modernizing its institutions. Boris Yeltsin may be credited with a remarkable turnover, if not a coup d'état, but his erratic management style and the lack of parliamentary support ensured that his government would never be strong. In these circumstances, helping the new Russian regime improve law enforcement should have come ahead of mass privatization. Launching capitalism in a country where no one other than apparatchiks had access to significant amounts of capital was an open invitation to gangsterism and a discredited system. Naive economic models made for naive policy recommendations.

HOW THE WEST WON

THE STATE'S crucial role is evident in the West's economic development. European economic supremacy was forged not by actors who followed a "Washington consensus" model but by strong states. In the fifteenth century, European incomes were not much higher than those in China, India, or Japan. The nation-state was a European innovation that replaced feudalism and established the rule of law; in turn, a legal framework was formed for effective markets. Once these countries were in the lead, they were able to continuously increase their edge through technological advances. In addition, European settlers took their civilization with them to North America and the South Pacific, rapidly raising these areas to rich-country status as well. Thus Europe's early lead became the basis for accumulating further advantages with far-reaching implications.

Europe's rise to economic leadership was not rapid at first. According to the economist Angus Maddison, Europe's economy grew around 0.07 percent a year until 1700; only after 1820 did it reach one

percent. But the pace of technological and institutional innovation accelerated thereafter. Meanwhile, discovery of new markets in Africa, Asia, and the Americas created new economic opportunities. Secular political forces overthrew the hegemony of the Catholic Church. Feudalism was eroded by rising incomes and replaced by a system that financed government through taxes, freeing up land and labor to be traded in markets. Markets permitted a more efficient reallocation of land and labor, allowing further rises in incomes. Effective property rights allowed individuals to keep the fruits of their own labor, thereby encouraging additional work. And privatization of common land facilitated the clearing of additional acreage.

The nation-state helped forge all these improvements. It opened up markets by expanding territory; reduced transaction costs; standardized weights, measures, and monetary units; and cut transport costs by improving roads, harbors, and canals. In addition, it was the state that established effective property rights. The European state system thrived on flexible alliances, which constantly changed to maintain a balance of power. Military and economic rivalries prompted states to promote development in agriculture and commerce as well as technological innovation in areas such as shipping and weaponry. Absent the hegemony of a single church or state, technology was diffused and secularized. Clocks, for instance, transferred timekeeping from the monastery to the village clock tower; the printing press did much the same for the production and distribution of books.

Europe's development contrasts sharply with Asia's. In the early modern era, China saw itself as the center of the world, without real rivals. It had a much larger population than Europe and a far bigger market as well. But though the Chinese pioneered the development of clocks, the printing press, gunpowder, and iron, they did not have the external competitive stimulus to promote economic development. Meanwhile, Japan sealed itself off from external influences for more than 200 years, while India, which had continuous competition within the subcontinent, never developed an effective national state prior to the colonial era.

The Europeans also led in establishing accountable government, even though it was achieved neither easily nor peacefully. Most European states developed the notion that the sovereign (whether a monarch or a parliament) had a duty to protect subjects

and property in return for taxes and service in the army. Rulers in the Qing, Mughal, and Ottoman Empires, in contrast, never recognized a comparable responsibility to their subjects. During the Middle Ages, Italy produced a number of quasi-democratic city-states, and in the seventeenth century Holland created the first modern republic after a century of rebellion and warfare with Spain. Britain achieved constitutional monarchy in 1689, following two revolutions. After a bloody revolution and then dictatorship, France achieved accountable government in the nineteenth century.

Europe led in establishing accountable government, although it was not easy or peaceful.

Europe led the way in separating church and state—an essential precursor to free inquiry and adoption of the scientific method—after the Thirty Years' War. The secular state in turn paved the way for capitalism and its "creative destruction." Creative destruction could hardly become the norm until organized religion lost its power to execute as heretics those entrepreneurs who would upset the status quo. After the Reformation, Europeans soon recognized another fundamental tenet of capitalism: the role of interest as a return for the use of capital. Capitalism required that political leaders allow private hands to hold power as well as wealth; in turn, power flowed from the rural nobility to merchants in cities. European states also permitted banks, insurance firms, and stock markets to develop. The "yeast" in this recipe lay in the notion that private as well as state organizations could mobilize and reallocate society's resources—an idea with profound social, political, and economic implications today.

Most of Europe's leading powers did not rely on private initiative alone but adopted mercantilism to promote their development. This strategy used state power to create a trading system that would raise national income, permitting the government to enhance its own power through additional taxes. Even though corruption was sometimes a side effect, the system generally worked well. Venice was the early leader, from about 1000 to 1500; the Dutch followed in the sixteenth and seventeenth centuries; Britain became dominant in the eighteenth century. In Britain, as in the other cases, mercantilist export promotion was associated with a dramatic rise in state spending and employment (especially in the navy), as well as "crony capitalism." After World War II, export-promotion regimes were adopted by Japan, South Korea, Singapore, and Taiwan with similar success. Today, of course, such strategies are condemned as violations of global trade rules, even for poor countries.

Finally, geography played a pivotal role in Europe's rise, providing a temperate climate, navigable rivers, accessible coastline, and defensible boundaries for future states. In addition, Europe lacked the conditions for the production of labor-intensive commodities such as coffee, cotton, sugar, or tobacco—production that might have induced the establishment of slavery. Like in the American North, European agriculture was largely rain-fed, diversified, and small-scale.

Europe's rise, then, was partly due to the creation and diffusion of technological innovations and the gradual accumulation of capital. But the underlying causes were political and social. The creation of the nation-state and institutionalized state rivalry fostered government accountability. Scientific enlightenment and upward social mobility, spurred by healthy competition, also helped Europe achieve such transformations. But many of today's developing countries still lack these factors crucial for economic transformation.

PLAYING CATCH-UP

GLOBALIZATION offers opportunities for all nations, but most developing countries are very poorly positioned to capitalize on them. Malarial climates, limited access to navigable water, long distances to major markets, and unchecked population growth are only part of the problem. Such countries also have very unequal income structures inherited from colonial regimes, and these patterns of income distribution are hard to change unless prompted by a major upheaval such as a war or a revolution. But as serious as these disadvantages are, the greatest disadvantage has been the poor quality of government.

If today's global opportunities are far greater and potentially more accessible than at any other time in world history, developing countries are also further behind than ever before. Realistic political logic suggests that weak governments need to

show that they can manage their affairs much better before they pretend to have strategic ambitions. So what kind of catch-up models could they adopt?

Substituting domestic goods for imports was the most popular route to economic development prior to the 1980s. But its inward orientation made those who adopted it unable to take advantage of the new global opportunities and ultimately it led to a dead end. Although the United States enjoyed success with such a strategy from 1790 until 1940, no developing country has a home market large enough to support a modern economy today. The other successful early growth model was European mercantilism, namely export promotion, as pioneered by Venice, the Dutch republic, Britain, and Germany. Almost all of the East Asian success stories, China included, are modern versions of the export-oriented form of mercantilism.

For its part, free trade remains the right model for rich countries because it provides decentralized initiatives to search for tomorrow's market opportunities. But it does not necessarily promote development. Britain did not adopt free trade until the 1840s, long after it had become the world's leading industrial power. The prescription of lower trade barriers may help avoid even worse strategies at the hands of bad governments, but the Washington-consensus model remains best suited for those who are ahead rather than behind.

Today's shareholder capitalism brings additional threats to poor countries, first by elevating compensation for successful executives, and second by subordinating all activities to those that maximize shareholder value. Since 1970, the estimated earnings of an American chief executive have gone from 30 times to 450 times that of the average worker. In the leading developing countries, this ratio is still less than 50. Applying a similar "market-friendly" rise in executive compensation within the developing world would therefore only aggravate the income gap, providing new ammunition for populist politicians. In addition, shareholder capitalism calls for narrowing the managerial focus to the interests of shareholders, even if this means dropping activities that offset local market imperfections. A leading South African bank has shed almost a million small accounts—mostly held by blacks—to raise its earnings per share. Should this bank, like its American counterparts, have an obligation to serve its community, including its black members, in return for its banking license?

Poor nations must improve the effectiveness of their institutions and bureaucracies in spite of entrenched opposition and poorly paid civil servants. As the journalist Thomas Friedman has pointed out, it is true that foreign-exchange traders can dump the currencies of poorly managed countries, thereby helping discipline governments to restrain their fiscal deficits and lax monetary policies. But currency pressures will not influence the feudal systems in Pakistan and Saudi Arabia, the theocracies in Afghanistan and Iran, or the kleptocracies in Kenya or southern Mexico. The forces of capital markets will not restrain Brazilian squatters as they take possession of "public lands" or the slums of Rio de Janeiro or São Paulo, nor will they help discipline landlords and vigilantes in India's Bihar as they fight for control of their state. Only strong, accountable government can do that.

LOOKING AHEAD

INCREASED TRADE and investment have indeed brought great improvements in some countries, but the global economy is hardly a win-win situation. Roughly one billion people earn less than $1 per day, and their numbers are growing. Economic resources to ameliorate such problems exist, but the political and administrative will to realize the potential of these resources in poor areas is lacking. Developing-nation governments need both the pressure to reform their administrations and institutions, and the access to help in doing so. But sovereignty removes much of the external pressure, while immigration barriers reduce key internal motivation. And the Washington consensus on the universality of the rich-country model is both simplistic and self-serving.

The world needs a more pragmatic, country-by-country approach, with room for neomercantilist regimes until such countries are firmly on the convergence track. Poor nations should be allowed to do what today's rich countries did to get ahead, not be forced to adopt the laissez-faire approach. Insisting on the merits of comparative advantage in low-wage, low-growth industries is a sure way to stay poor. And continued poverty will lead to rising levels of illegal immigration and low-level violence, such as kidnappings and vigilante justice, as the poor take the only options that remain. Over time, the rich countries will be forced to pay more attention to the fortunes of the poor—if only to enjoy their own prosperity and safety.

Still, the key initiatives must come from the poor countries, not the rich. In the last 50 years, China, India, and Indonesia have led the world in reducing poverty. In China, it took civil war and revolution, with tens of millions of deaths, to create a strong state and economic stability; a de facto coup d'état in 1978 brought about a very fortunate change of management. The basic forces behind Chinese reform were political and domestic, and their success depended as much on better using resources as opening up markets. Meanwhile, the former Soviet Union and Africa lie at the other extreme. Their economic decline stems from their failure to maintain effective states and ensure the rule of law.

It will not be surprising if some of today's states experience failure and economic decline in the new century. Argentina, Colombia, Indonesia, and Pakistan will be obvious cases to watch, but other nations could also suffer from internal regional failures—for example, the Indian state of Bihar. Income growth depends heavily on the legal, administrative, and political capabilities of public actors in sovereign states. That is why, in the end, external economic advice and aid must go beyond formal models and conform to each country's unique political and social context.

BRUCE R. SCOTT is Paul W. Cherington Professor of Business Administration at Harvard Business School.

Prisoners of Geography

Economic-development experts promise that with the correct mix of pro-market policies, poor countries will eventually prosper. But policy isn't the problem—geography is. Tropical, landlocked nations may never enjoy access to the markets and new technologies they need to flourish in the global economy.

By Ricardo Hausmann

So you are a Scorpio. Then you must be passionate. So the barometer says that the atmospheric pressure is declining. Then it is going to rain. So your latitude is less than 20 degrees. Then your country must be poor.

There may be some debate about which of these statements is true, but only one is truly offensive—the last one. Indeed, the notion that a country's geography determines its level of economic development is fraught with controversy. People take offense at such a connection because it smacks of racism and undermines the notion of equal opportunity among nations and individuals. It is also paralyzing and defeatist: What can policymakers and politicians do or promise if nothing can overcome geography? From World War II through the mid-1980s, these sentiments prompted a backlash against the study of economic geography in much of the academic world. Today, however, new theories of economic growth coupled with empirical research have brought economic geography back to the forefront of the development debate. Speaking at the United Nations Conference on Women and Development in June 2000, U.S. Treasury Secretary Lawrence Summers decried "the tyranny of geography," particularly in African countries, and warned against concluding that "the economic failures of isolated, tropical nations

with poor soil, an erratic climate and vulnerability to infectious disease can be traced simply to the failure of governments to put in place the right enabling environment." The prevailing development paradigm—according to which market-oriented economic policies and the rule of law alone suffice to make all countries rich—appears to be losing credibility. What if geography gets in the way of the Promised Land?

LOCATION, LOCATION, LOCATION

Closing the income gap between rich and poor countries has been a stated objective of the international community for the last 50 years. This commitment spawned the creation or redesign of institutions such as the World Bank, specialized United Nations agencies such as the United Nations Development Programme and the United Nations Conference on Trade and Development, regional development banks such as the Inter-American Development Bank (IDB), bilateral aid agencies in the governments of the most advanced economies, and innumerable foundations, research centers, and other non-governmental organizations.

But the global gap between rich and poor countries has not closed. Instead, it has widened. Economist Angus Maddison estimates that, in 1820, Western Europe was 2.9 times richer than Africa. By 1992, this gap had risen to 13.2 times. The trend continues—albeit less dramatically—in South Asia, the Middle East, Eastern Europe, and Latin America. In 1997, the richest 20 percent of the world's population enjoyed 74 times the income of the lowest 20 percent, compared to 30 times in 1960.

Tropical diseases do not "merit" the sort of R&D investments that a cure for baldness attracts in Western markets.

The countries left behind have distinguishing geographical characteristics: They tend to be located in tropical regions or, because of their location, face large transportation costs in accessing world markets—or both.

In 1995, tropical countries had an average income equivalent to roughly one third of the income of temperate-zone countries. Of the 24 countries classified as "industrial," not one lies between the Tropics of Cancer and Capricorn, except for the northern part of Australia and most of the Hawaiian Islands. Among the richest 30 economies in the world, only Brunei, Hong Kong, and Singapore are in tropical zones, and their geographical locations leave them ideally suited for growth through trade. Tropical nations tend to have annual rates of economic growth that are between one half and a full percentage point lower than temperate countries. A recent IDB study found that after considering the quality of institutions and economic policies, geography explained about a quarter of the income difference between industrialized and Latin American countries in 1995. Tropical countries also have poorer health conditions than their nontropical counterparts. After considering income levels and female education, life expectancy in tropical regions is seven years lower than in temperate zones. Nations in tropical areas often display especially skewed income distributions. In Africa and Latin America, the richest 5 percent of the population earn nearly 25 percent of the national income, while in industrial countries they earn only 13 percent. Latitude alone can explain half of this difference. Even within regions of the same country, living standards are strongly linked to geography. For example, in Mexico, the southern states of Chiapas, Oaxaca, and Guerrero have twice the infant mortality rate and half the educational attainment of the country's northern states.

Nations with populations far from a coastline also tend to be poorer and show lower rates of economic growth than coastal countries. A country whose population is farther than 100 kilometers from the sea grows 0.6 percent slower per year than nations in which the entire population is within 100 kilometers of the coast. That means, for example, that the post-Soviet republics will experience as

much difficulty battling their geographical disadvantages as they will overcoming the aftereffects of communism. Countries that are tropical, far from the coast, and landlocked have three geographical strikes against them. Many countries in Africa are handicapped by one or all of these factors.

There is still much we do not understand about the links between geography and economic growth. But what we do know suggests that the challenges of economic development must be examined from a very new perspective. Denying the impact of geography will only lead to misguided policies and wasted effort. Geography may pose severe constraints on economic growth, but it need not be destiny.

LATITUDE PROBLEMS

To understand why geography can matter so much for economic development, consider what economists regard as the main engines of growth: Access to markets (based on the work of Scottish economist Adam Smith) and technological progress (drawn from the writings of U.S. economist Joseph Schumpeter).

For Adam Smith, productivity gains achieved through specialization are the secret to the wealth of nations. But for these gains to materialize, producers must have access to markets where they can sell their specialized output and buy other goods. The larger the market, the greater the scope for specialization. In today's global marketplace, most industrial products require inputs from various locations around the world. Therefore, if transportation costs are high, local companies will be at a disadvantage in accessing the imported inputs they need and in getting their own goods to foreign markets.

Unfortunately, transportation costs are often determined by a country's geography. A recent study found that shipping goods over 1 additional kilometer of land costs as much as shipping them over 7 extra kilometers of sea. Maritime shipping is particularly suited to the bulky, low-value-added goods that developing nations tend to produce; therefore, countries lacking cheap access to the sea will be shut out of many potential markets. Moreover, if countries far from the sea do not enjoy the physical infrastructure (the system of roads, railways, and ports) needed for access to navigable rivers or the sea, they will not develop the very industries that could help maintain such an infrastructure.

Land transportation is especially costly for landlocked countries whose products need to cross borders, which are a much more costly hurdle than previously thought. Studies on trade between U.S. states and Canadian provinces find that simply crossing the U.S.-Canadian border is equivalent to adding from 4,000 to 16,000 kilometers worth of transportation costs. Little wonder, then, that the median landlocked country pays up to 50 percent more in transportation costs than the median coastal na-

Living Between the Poverty Lines

Gross Domestic Product Per Capita (1995, U.S.$)

- 400–4,000
- 4,000–7,500
- 7,500–15,000
- 15,000–32,000

SOURCE: John Luke Gallup, Jeffery D. Sachs, and Andrew D. Mellinger "Geography and Economic Development"
(National Bureau of Economic Research Working Paper No. W6849, December 1998)

tion. In practical terms, these differences can be enormous: Shipping a standard container from Baltimore to the Ivory Coast costs about $3,000, while sending that same container to the landlocked Central African Republic costs $13,000.

Governments in landlocked countries face the additional challenge of coordinating infrastructure expenditures with neighboring countries. Sometimes, political or commercial problems inhibit passage to the sea. For example, the agricultural potential of the upper Parana River basin in landlocked Paraguay remained dormant until a Mercosur agreement in the mid-1990s facilitated barge transportation through Brazil and Argentina. Jordan's access to the Mediterranean requires crossing the Israeli border or those of Syria and Lebanon. These instances illustrate why landlocked nations suffer from sluggish economic growth. Countries and territories like Hong Kong, Taiwan, and Singapore have an advantageous geographical position, but much of inland Africa, China, and India remains far from markets and maritime trade.

Geography harms developing countries in other ways. Joseph Schumpeter showed that technological innovations, through research and development (R&D), are powerful engines of economic growth. (This notion is what Schumpeter had in mind when he coined his famous term, "creative destruction.") R&D displays increasing returns: The more people who use and pay for a new idea, the greater its market value. (For example, a new computer program or novel may cost a lot to pro-

duce, but subsequent copies are extremely cheap.) In order to recoup their initial costs, R&D investors will tend to focus on innovations for which potential customers abound. Unsurprisingly, rich countries with large, middleclass populations are more lucrative markets than poor nations with little purchasing power.

Even though innovations such as computers or cellular phones work in many geographical conditions and are therefore easily adopted by developing countries, technologies in other sectors often require research that is very location-specific. Many technologies are not universally applicable; their effectiveness depends on the geographical or climatic conditions in which they are used.

Consider agriculture. The divergence in agricultural productivity between the developed and developing world is grounded in dramatically different R&D capabilities. Governments in advanced economies spend up to five times more (as a percentage of total agricultural production) on agriculture-related R&D than their counterparts in developing countries. Rich nations also benefit from the expenditures of private agricultural producers—a source of funding that is virtually nonexistent in developing nations. Geography aggravates this disparity. Plant varieties need to be adapted to the local climate, meaning that R&D geared toward rich, temperate-zone agriculture is of little use in tropical areas. Countries like Argentina, Chile, Australia, New Zealand, and South Africa can enjoy thriving export sectors in fruit, wine, cereals, oilseeds, and salmon thanks to the technologies developed for these products in temperate zones in the

Northern Hemisphere. But the tropical countries—with their production of coffee, cocoa, sugar cane, and cassava—are left out of the modern-technology club. The result is that the agricultural sector is much less dynamic in tropical areas than in temperate zones. Since unproductive agricultural workers can produce little more than what they require for personal subsistence (and therefore cannot support large urban populations), rural areas remain sparsely populated, have small, poor markets, and suffer from high transportation costs—all of which hamper economic growth.

Climate differences and economies of scale have long played a powerful role in the development of agriculture in different geographical zones. In his Pulitzer Prize-winning book *Guns, Germs, and Steel*, physiologist Jared Diamond explains how Eurasia's east-west geographical layout and the north-south layout in Africa and the Americas determined these regions' historical patterns of economic growth. Since climate changes little with longitude but quite rapidly with latitude, the Eurasian landmass enjoyed fairly uniform climatic conditions. Hence, agricultural innovations developed in one region could travel long distances and be shared by many people, resulting in a large set of plant and animal varieties available throughout the region. By contrast, new varieties developed in the Americas or in Africa could not migrate very far since climates change swiftly, limiting the technological opportunities available to these regions and stunting economic growth.

Of course, agricultural productivity and transportation cost advantages do not necessarily go together. As historian David S. Landes points out in *The Wealth and Poverty of Nations*, the ancient civilizations of Mesopotamia and Egypt had their most fertile lands along rivers. This location—far removed from the seashore—limited their ability to expand their economies through trade. Their power eventually waned and they were supplanted by the seafaring Phoenicians, Greeks, and Romans. More recently, in India and China, agricultural conditions encouraged large populations to cluster along riverbeds far away from the sea, hurting the countries' long-term prospects for economic growth and development through trade.

Investments in health research and technology are also very sensitive to geography. Diseases such as malaria, hookworm, schistosomiasis, river blindness, and yellow fever are hard to control in tropical regions because the lack of seasons makes the reproduction of mosquitoes and other disease transmitters rather constant throughout the year. Since the afflicted countries tend to be poor, tropical diseases do not "merit" the sort of R&D investments that a cure for baldness or erectile dysfunction can attract in Western markets. (Of the aforementioned tropical diseases, only yellow fever has been controlled through an effective vaccine.) Technological development is skewed away from the needs of geographically disadvantaged countries. Thus, children in tropical re-

gions often die of gastrointestinal and other infectious diseases, while many nations still suffer from endemic tropical ailments. Economists John Luke Gallup and Jeffrey Sachs estimate that per capita economic growth in countries with severe malaria is more than a full percentage point lower than in nations where this illness is not prevalent, and that a 10 percent reduction in the incidence of malaria is associated with 0.3 percent higher growth.

The enormous divide in agricultural productivity ensures that standards of living in tropical areas are likely to remain stagnant.

The costs of not dealing with disease in tropical countries go far beyond higher healthcare expenses and reduced worker productivity. Disease can no longer be considered a mere public health problem, but a socioeconomic development issue that affects everything from trade flows to migration patterns. The 1991 cholera outbreak in Peru cost the country's fishing sector nearly $800 million in lost revenues because of a temporary ban on seafood exports. The 1994 plague outbreak in Surat, India, prompted 500,000 people to move from the region and led to work stoppages across several industries, as well as new restrictions on international trade. Estimates of the cost India bore for this plague reach $2 billion.

BORDERING ON POVERTY

The dominant development paradigm these days holds that market-oriented economic policies and the rule of law are all that matter for economic progress. In other words, Mozambique could become Singapore if it would only get its institutions and policies in order; in the meantime, we could alleviate poverty through targeted social spending for the poor, such as the financing of education for girls. But this mantra vastly oversimplifies the challenges of development. If a region is poor because its geography undermines agricultural productivity, impedes market access, and facilitates endemic disease, then good domestic policies will hardly suffice to foster growth. Poverty will not disappear because of expanded nutrition programs or improvements in the teaching materials available in schools. (At best, better trained students simply will migrate to more prosperous regions.)

From this perspective, it may be more important to devote time and resources to transportation infrastructure, which lowers the costs of trading, new technologies for agriculture and public health, and economic integration projects than to focus solely on areas like health, education, and the rule of law.

Infrastructure Development If small, rural communities in developing countries are to experience economic

Locational Correctness

Economic geography offends people because it seems to imply an immutable destiny—if you live in one area, you are poor; if you live in another, you are rich. When the Inter-American Development Bank dared highlight economic geography in its *Economic and Social Progress in Latin America* report in 2000, the Brazilian media attacked the institution for reviving racist and determinist theses. "Ideas from Another Century" screamed the headline in *Gazeta Mercantil*, Brazil's leading business newspaper.

This virulent reaction was not lacking in irony, particularly since income differences in Brazil are closely related to latitude, with the tropical northeast being very poor while the more temperate south is much richer. But these attacks should not be surprising. Since the Enlightenment, economic geography has been a matter of great debate and controversy among scholars and political leaders throughout the world. Their interpretations of the issue have ranged from sensible to silly to outright dangerous: Adam Smith regarded ports, navigable rivers, and canals as essential for industrialization—assets that Great Britain possesses but that

places like Africa and Siberia lack. Montesquieu saw a close relationship between geography and politics, concluding that democracy was fine for Switzerland because of its low agricultural productivity, but that wealthier nations such as France needed a monarchy. During the European imperialist expansion of the 19th century, and under the impact of social Darwinism, geography became a way to justify notions of white racial supremacy. The "fittest" race had become so because, among other reasons, the temperate climate where it developed helped forge populations more prone to thoughtfulness and responsibility than to ebullient pleasure seeking.

Such racially charged views became increasingly unacceptable after the rise of the Nazi regime and the horror of the Holocaust. The reputations of 20th-century geographers such as the famed Ellsworth Huntington of Yale University, author of the landmark 1915 work *Civilization and Climate*, suffered greatly (and unfairly) by association. Historian David S. Landes attributes this reaction not so much to weaknesses in geographers' analyses, of

which there were plenty, but to their pessimistic message that nature, like life, is unfair. Victimized by this backlash, the geography departments at Harvard, Michigan, Northwestern, Chicago, and Columbia universities were shut down in short order following World War II. As a result, several generations of academics disregarded geography as a key factor in socioeconomic development.

In recent years, however, geography has slowly made its way back into mainstream economic thinking; new theories and techniques for studying trade, growth, and the environment have contributed to this resurgence. And interest in geography as a discipline is also rising: In the United States alone, the number of bachelor's degrees awarded in geography rose from about 3,000 in 1985–86 to nearly 4,300 in 1994–95. In the academic arena, economic geography is no longer taboo. It is only a matter of time before the discipline becomes acceptable in broader circles—maybe even among Brazilian editorialists.

—R.H

growth, it is crucial to connect them with the rest of their country and the world through investments in roads and other transportation infrastructure. Many of these investments must be made outside of the particular countries in question. For example, for Rwandan and Ugandan goods to reach new markets, the Kenyan rail system must be improved. This complication poses severe coordination and political challenges; it is not clear, for instance, that such an improvement should be a priority for Kenyan authorities. Unfortunately, the major regional development banks operate with this same narrow focus, granting loans to national governments on the basis of perceived national priorities. Important region-oriented projects remain chronically underfunded. To overcome this problem, bilateral or multilateral organizations should

provide financial incentives to national governments to encourage them to cofinance investment projects that benefit themselves as well as neighboring countries.

Technological Development Although it is fashionable (and accurate) to decry the "digital divide" between advanced and developing economies, this information-technology gap need not be a major concern for poor countries since they benefit from global innovations in these arenas. For instance, Latin American countries soon will have more cellular phones than regular telephone lines, allowing for a major expansion in the region's telecommunications system by skipping the need to install underground cables. By contrast, the dramatic difference between rich and poor countries in agricultural and phar-

13

maceutical R&D ensures that standards of living in tropical areas are likely to remain low and stagnant. Governments in developing nations lack sufficient resources to address this problem by themselves, and the world's private sector allocates very little financing to agricultural R&D for developing nations. Although the well-known difficulties in enforcing intellectual property rights create a significant disincentive for this sort of investment, there may be ways to enlist the knowledge and research capabilities of corporations such as Pfizer and Archur Daniels Midland. Economists Michael Kremer and Jeffrey Sachs have proposed contests so companies can compete to develop effective vaccines. The Clinton administration included in its 2001 budget proposal a tax credit to U.S. pharmaceutical companies that developed vaccines for diseases prevalent in the developing world. However, the vast needs in this area suggest that multilateral financing will be needed to compensate private firms for such initiatives.

Integration National borders, as they are currently conceived, make nations artificially more distant and only accentuate the costs already imposed by geographical conditions. Borders limit the movement of goods, capital, and labor and thus limit access to markets. Some regions—most notably Western Europe—have already begun eliminating internal borders. But for the last 50 years we have witnessed the creation of more and more nations in the developing world, with their own new borders, making these countries effectively more distant than their physical geography implies. Can poor nations afford this additional source of remoteness?

If shipping goods across the U.S.-Canadian border adds the equivalent of thousands of miles in transportation costs, then the commercial logistics of trading between countries with weak political institutions and a history of cross-border animosity will prove to be infinitely more expensive problems for importers and exporters. And borders do not merely complicate the movement of goods and the coordination of cross-country infrastructure; capital also has trouble crossing borders. Since investment contracts are often enforced at the national level, sovereignty can shelter borrowers who are able but unwilling to repay. This situation introduces "sovereign risk" into financial markets, limiting capital movements and rendering them increasingly fickle.

Borders also prevent people in poorer areas from moving to more prosperous regions. For example, the decline in agricultural employment in the United States prompted significant regional migration, and when Europe went through a similar process at the end of the 19th and beginning of the 20th century, it had an escape valve in the form of a wide-open immigration policy in the United States. Today's geographically trapped peoples seldom enjoy such opportunities. Not that they don't search for them: About one third of the landlocked Burkinabes and one fifth of Bolivians work in neighboring nations. Not only does immigration offer poor people a chance to have a better life but it also allows them to send money to their families at home. For nations such as El Salvador, the Dominican Republic, and Egypt, worker remittances from abroad often exceed the value of those countries' annual manufacturing exports.

Finally, borders limit the possibilities for risk-sharing in the face of natural disasters. In the United States, the Federal Emergency Management Agency is funded by federal taxes; therefore, when disaster strikes a particular state or region, the rest of the nation can help mitigate the damage. Small countries have a smaller geographical space than large countries in which to share risks. When earthquakes destroyed Managua, Nicaragua, in 1972, and when a hurricane devastated Honduras in 1998, the national tax base was destroyed, making it impossible to marshal national resources to deal with the lost infrastructure. Countries that are small and vulnerable to hurricanes, floods, and earthquakes may become nonviable after a major disaster wipes out their productive capacity. Poor nations usually bear the brunt of such emergencies: Ninety-six percent of all deaths from natural disasters occur in developing countries.

The current conceptions of borders compound the problems attributable to geography. The world has been quite willing to create new nation-states under the banner of self-determination. But unless borders can be made less problematic for economic integration, they may condemn geographically distant countries to an independent oblivion.

GEO-GLOBALIZATION

If distance and geography did not matter for economic development, then we would witness much greater convergence of income levels and standards of living across regions and countries. Instead, we are witnessing divergence, because geography prevents poor nations from fully participating in the global division of labor. If current trends persist, countries that face high transportation costs and a high dependence on tropical agriculture will be left far behind, mired in poverty and income inequality. Will the rest of the world find this outcome morally acceptable? Will it find it efficient? Or will the fallout from these destitute regions be seen as endangering the quality of life for the rest of us? In a sense, we have already asked and answered these questions; the existence of myriad development institutions around the globe attests to the world's desire to meet the challenges of economic development. But all our answers have fallen short. The gap between rich and poor has only widened.

Many people blame economic globalization for poverty and injustice in the developing world. Yet it is the absence of globalization—or an insufficient dose of it—that is truly to blame for these inequities. The solution to geography's poverty trap is for developing countries to become more globalized. We need transnational arrangements to make borders less of an impediment to moving people, goods, and capital. We need agreements that can facilitate the development of international transportation infrastructure. And we need global mechanisms to harness the R&D capabilities of the world in health and agricultural technology. In short, we need more globalized governance.

Want to Know More?

Recent major works have emphasized the crucial role of geography in human history. In particular, see David S. Landes's *The Wealth and Poverty of Nations: Why Some Are So Rich and Some So Poor* (New York: W.W. Norton & Company, 1999); William H. McNeill's *Plagues and Peoples* (New York: Anchor Books, 1998); Jared Diamond's *Guns, Germs, and Steel: The Fates of Human Societies* (New York: W.W. Norton & Company, 1997); and Lawrence E. Harrison and Samuel Huntington, eds. *Culture Matters: How Values Shape Human Progress* (New York: Basic Books, 2000), particularly Jeffrey Sachs's chapter titled **"Notes on a New Sociology of Economic Development."**

John Luke Gallup, Jeffrey D. Sachs, and Andrew D. Mellinger explore the link between geographical factors and socioeconomic progress in **"Geography and Economic Development"** (Cambridge: National Bureau of Economic Research Working Paper W6849, December 1998). The Inter-American Development Bank's *Development Beyond Economics: Economic and Social Progress in Latin America 2000 Report* (Washington: Inter-American Development Bank, 2000) also examines this relationship, particularly in Chapter 3, titled **"Geography and Development."** Raymond Arsenault describes the impact of the air conditioner on socioeconomic conditions in the southern United States in **"The Cooling of the South"** (*Wilson Quarterly*, Summer 1984).

For an assessment of the connection between geography and income distribution, see **"Nature, Development, and Distribution in Latin America: Evidence on the Role of Geography, Climate, and Natural Resources"** (Washington: Inter-American Development Bank Working Paper No. 378, August 1998) by Michael Gavin and Ricardo Hausmann. Stanley Engerman and Kenneth Sokoloff assess the impact of natural resources on institutional development in **"Factor Endowments, Institutions, and Differential Paths of Growth Among New World Economies: A View from Economic Historians of the United States"** in Stephen Haber, ed. *How Latin America Fell Behind: Essays on the Economic Histories of Brazil and Mexico, 1800–1914* (Stanford: Stanford University Press, 1997). Paul Krugman surveys the interplay between geography and contemporary economic thought in *Development, Geography, and Economic Theory* (Cambridge: MIT Press, 1995).

* For links to relevant Web sites, as well as a comprehensive index of related FOREIGN POLICY articles, access **www.foreignpolicy.com.**

Ricardo Hausmann is professor of the practice of economic development at the John F. Kennedy School of Government at Harvard University and former chief economist of the Inter-American Development Bank.

The Poor Speak Up

Leaders of the developing world are rising up with a strength not seen since Tito, Nasser and Nehru, challenging the rules of globalization as defined by both Western governments and Western activists

Rana Foroohar
With Ian MacKinnon in Delhi, Mac Margolis in Rio, Paul Mooney in Beijing and Mark Ashurst in London

They drew blanket press coverage and a watchful audience of New York police, but mainly on the strength of past protest performances. From Seattle to Davos, the riotous anti-globalization road show had pressed its case that rising trade and capital flows are bringing nothing but oppression and instability to the developing world. Their ranks boomed with the global economy, but faded with it as well. The thousands who descended on the World Economic Forum last year in Davos dwindled to a few hundred by the time the fete reconvened last week in New York.

Inside the party at the Waldorf-Astoria, the protesters still inspired a degree of soul-searching among the gathered stars of the global age—CEOs, prime ministers, princes and celebs—as if they had noticed that the era and the opposition have changed. But leadership of the effort to alter the course of globalization has shifted away from Western street rebels—and toward those they have long claimed to represent. The main actors now are leaders of the poor nations, who have united against the global trade elite and are demanding radical changes in the rules of international commerce.

This is now a far more powerful movement, particularly in the grim wake of an age of plenty. Unlike anarchists, greens and reds, the presidents and ministers of the developing world cannot be humored and dismissed as a bag of mixed nuts. They have a seat at the table, real influence. At the November trade summit in Doha, Qatar, they stepped forward in a show of unity and strength that, arguably, had not been seen from poor nations since the days of Tito, Nehru and Nasser. Led by India, Brazil and South Africa, this emerging front is opposed to globalization as defined by Western governments and Western activists. In short, they are for free trade (unlike the activists) but on their own terms (not those of the West). Born of long frustration with the postwar commercial order, the mystery is why this movement is erupting so publicly only now. "There was a clear improvement in the awareness of developing countries at Doha,

and in their capacity to deal," says EU Trade Commissioner Pascal Lamy, who now expects this front to "go on pushing us."

In Doha, a bloc of developing nations won unprecedented victories. They helped force Europe to consider phasing out subsidies for its farms, which elbow out produce from poor, farm-based economies. They compelled the United States to consider limits on anti-dumping laws, which are often used against cheap exports from emerging manufacturing powers. They won the right to ignore Western drug patents when necessary to fight developing world scourges like AIDS, and then kept right on fighting. They recently pushed for new rules that would give their own ambassadors the lead role in the new trade round launched at Doha, and lost. But they'll be back. Many want no less than major reform of global financial institutions that date to the Bretton Woods agreement of 1944, which they believe has failed crisis-torn emerging nations like Argentina. "Financial capital is globalized, but not policy decisions," Brazilian President Fernando Henrique Cardoso said recently. "The Bretton Woods system is obsolete."

The new alliance has distant roots in the Third World solidarity of the early postwar years. The Group of 77 was founded in 1955 as a united front against the injustices of the colonial era. Many of its members were still colonies, barred by their rulers from working as full partners in the emerging international trading regime. Once they won independence, the emerging nations set out to develop in solidarity with one another, as a union of the poor against the empires they had just thrown off. They shut out imports in the hope of developing local industry, sealing themselves off in a socialist Third World.

Many of the poor were still living in this state of partial isolation at the start of the 1980s, when Reaganism and Thatcherism set off a boom in trade and in the complexity of trade agreements. In 1986 the Uruguay Round expanded the talks beyond simple tariffs to subjects like financial services and intellectual property. By this time, many poor nations had begun to drop state capitalism and to prosper as export powers in the global markets—but few had the sophistication to analyze issues like the trade-related aspects of intellectual property rights, or TRIPS. "Many countries didn't understand what they were signing up for," says Leif Pagrotsky, Sweden's trade minister.

The poor would soon suffer the consequences of obliviousness. By 1995 the United States was demanding that poor nations live up to the detailed letter of Uruguay. Even those prospering as export powers couldn't always enforce new rules around issues like intellectual property. The United States wound up suing India and Pakistan for patent-law violations. "It became clear that no matter what we had thought, the other side would take all this quite seriously," says Rashid S. Kaukab, a former Pakistani diplomat now at the South Centre in Geneva. "We couldn't afford not to look at these issues very carefully."

The more they studied, the angrier they became. Even booming nations like India and Brazil came to believe that, while they benefit from global trade, the West was benefiting more. In fact, a growing body of research showed that the global tariff system favored the rich. According to the World Bank, the poor countries pay average Uruguay Round tariffs of more than 14 percent, a rate more than twice as high as everyone else. Worse, developing-country exports tend to be farm goods or simple manufactures, two of the markets most heavily protected in the West by nontariff barriers. For example, African sugar and Brazilian steel both face stiff quotas and heavy subsidies in the developed world. "Developing countries are encouraged to diversify their exports," says Brazilian Trade Minister Sergio Amaral. "But precisely when they do so, and become competitive, they hit barriers in the developed world."

By the mid- to late 1990s there was a growing perception among emerging countries that they had been bamboozled in Uruguay and that the West was still pressing its advantage in trade expertise and clout. "Throughout the 1980s and 1990s, it was easier for the U.S. and the EU to intimidate the developing countries," says Alan Winters, a fellow at the Centre for Economic Policy Research in London. "They'd pick a few at a time and work on them individually, getting them to agree to enforcement of trade issues, like patent laws." To a growing number of developing-world trade ministers, it looked like a Western strategy of divide and conquer.

The obvious response was a more united front, even if the interests of poor nations didn't always mesh. As early as a 1996 WTO meeting in Singapore, developing countries were beginning to unite behind a rough agenda: just say no to whatever the rich nations ask. By the time of the 1999 WTO summit in Seattle, trade alliances were emerging among the states of Africa, Asia and Latin America, with increasingly sophisticated demands for greater access to rich-country markets, and a greater say in trade talks. When several small nations were locked out of the "green room" negotiations of the big trade powers in Seattle, several coalitions of lesser powers threatened to walk out. Their backroom revolt had at least as much to do with the collapse of the Seattle talks as protests in the streets, and many were left more alienated than ever. After Seattle, says trade ex-

pert Kent Hughes, many developing nations were left wondering, "What were these meetings doing for them?"

In mounting frustration, various alliances of poor nations began plotting for the next big trade summit. Jolted by South African President Thabo Mbeki's critique of its economic failures, Africa began uniting. In July 2001 the ineffectual Organization for African Unity was replaced by the African Union, boasting a clear agenda for trade liberalization. Led by Tanzania, the least-developed countries met in Zanzibar to hash out a plan for Doha. Geneva ambassadors of the Like Minded Group, a coalition of 13 countries from Asia, Latin America and Africa, began dropping by each other's offices more frequently for tea, and targeted issues to press at the upcoming summit. China's Zhu Rongji led a pre-Doha relationship-building delegation to India. Leaders of five developing countries—India, Brazil, South Africa, Malaysia and Egypt—held informal talks to set common goals. And a bloc of 50 nations began lobbying on an old grievance: TRIPS. "There was constant, coordinated pressure by the various developing countries and missions in Geneva," says Egyptian trade negotiator Magdi Farahat. "Everyone kept everyone else in the loop."

The sweep of their victory changed the rules of the game. They pushed the envelope not only on intellectual property, farms and dumping, but also on their complaint that Uruguay left them with trade commitments they could not afford to enforce. The northern countries made concessions on 50 of these "implementation" issues and agreed to help poor nations build up the capacity to carry out their trade-related responsibilities. That could include everything from helping nations pay for technology to carry out stricter customs checks, to boosting their ability to carry out policy research and analysis. At the center of it was Murasoli Maran, who had worked as a screenwriter before becoming trade minister of India, and had scripted an eleventh-hour drama in Doha. Pressured by European and American negotiators to either sign a deal or bump the final agreement up to the presidential level for approval, Maran held out until he got his way. Asked whether he'd been intimidated by the West, Maran shot back, "No. I intimidated them."

It was a public declaration of victory for the new southern front, heralding a new era in the battle over globalization. The elite trade powers know it. Europe's lead negotiator, Lamy, says that trade has not only "risen on the agenda of these governments," but trade ministers are increasingly powerful figures in the developing world. Trade ministers like Maran and South Africa's Alec Erwin are "punching above their weight," says Lamy. This new strength will force Europe and America to spend more time courting opinion in the developing world. At the World Economic Forum in New York last week, U.S. Trade Representative Robert Zoellick stressed the importance of the new developing-nation alliances in the post-Doha world. "It is critical to look at the role that developing nations play and the networks they are creating," he said. In order to make the next round successful, Zoellick noted, "We need consensus-building."

But this is not the '50s, when one leader like Tito could claim to speak for the Third World. The movement was simpler then, united by animus against colonial rule, and asking only for a broad redistribution of wealth. It is no longer a unit bound by ideology, but a front of shifting alliances that change from issue to issue. "As long as you're opposed to something, you can hold together," says Columbia University economist Jagdish Bhagwati. "Once you move forward, interests are bound to diverge."

The splits are already appearing. While developing nations agree broadly on opening up richer markets, they differ on the details. India, Malaysia and Egypt want Europeans to cut agricultural subsidies while they keep on protecting their own farmers. Brazil and South Africa are opposed to protection for anyone. The recent European decision to grant preferred trade status to its former colonies pleased the former colonies, but no one else. "It's all a matter of what's convenient," says Tattamangalam Vishwanath, international trade adviser to the Confederation of Indian Industry. "In trade talks, there are no permanent friends and no permanent enemies."

The new battlelines don't necessarily pit rich versus poor anymore, either. In Doha, the United States backed the developing nations that protested against European farm subsidies. And the EU backed the stand against U.S. anti-dumping rules.

WTO director-general Michael Moore welcomes this new movement of pragmatic, nonideological developing nations as "healthy." It's no longer so much a revolt against colonial masters as a spat about stuff like steel quotas and banana tariffs. "There is no Third World anymore," says Brazil's chief trade negotiator, Sergio Amaral. "But there is a unanimity among the developing countries that protectionism is the common enemy."

This is a sharp departure from anti-globalization as championed by activists, whether from rich or poor nations. The activists aim to temper the destabilizing effects of rising trade and capital flows by protection, if necessary, and by requiring multinationals to raise wages and improve labor conditions in the developing world. This vision of managed trade would slow the pace of globalization, and was ridiculed at Davos 2000 by former Mexican president Ernesto Zedillo as a plan "to save the people of developing countries from developing." So the game is on. It pits the emerging front of poor nations against rich governments, who once made the rules, but also against Western activists, who once claimed to speak for them. The victors will shape the changing global order.

The Rich Should Not Forget the ROW (Rest of the World)

JOSE RAMOS-HORTA WON THE NOBEL PEACE PRIZE IN 1996.

DILI, EAST TIMOR—The world's economy is many times larger than it was only 50 years ago. Particularly in the Northern countries, which have the greatest concentration of personal wealth, the quality of life has improved dramatically. Mind-boggling advances have occurred in vast fields of human endeavor, from genetics to computers. Human beings have walked on the moon, Mars is being studied at ever closer range.

Yet the same human intelligence that has produced such advances seems so far unable to eliminate extreme poverty or tropical diseases such as malaria and cannot provide clean water to hundreds of millions in Africa, Asia and Latin America.

> ## Child labor, prostitution and sex slavery are rampant in the impoverished but aspiring societies.

And the gap between rich and poor has grown, not diminished. Hundreds of millions survive on less than $1 a day, children walk miles to go to school, if at all, or to fetch firewood and water for the household. Child labor, prostitution and sex slavery are rampant in the impoverished but aspiring societies.

In their pursuit of even greater prosperity, weapons-producing countries aggressively push arms in developing countries that cannot even afford to provide clean water to most of their people, fueling local, often ethnic, conflicts.

Do we have some answers to these challenges from the dark side of the human condition?

There is no dispute that abject poverty, child labor and prostitution are indictments of all humanity. However, poverty should not only touch our conscience: It is also a matter of peace and security because it destabilizes entire countries and regions. That in turn threatens the integration of the global economy that is vital if the rich are to stay rich or if the poor are to move up, if only an inch.

One need not be original to propose some key elements of a solution to these problems, as those more enlightened than I have.

Drawing on the ideas already in circulation, here is the agenda I propose:

DEBT CANCELLATION

The G-8, European Union and World Bank should lead the initiative in writing off the entire public-sector debt of all non-oil-producing countries whose per capita income is less than $1,000.

In addition, a special fund should be developed by the World Bank and the UN Development Agency (UNDP) to assist these countries in improving governance and generating employment for the poorest.

Other highly indebted countries (for instance, Indonesia, Nigeria, Brazil and Mexico) should also benefit from a special debt-relief package of up to half of their public-sector debt if the proceeds are aimed at poverty reduction and education.

> ## The United States, Canada and Japan should open up their markets for goods from the HIC (highly indebted countries) and ease some of the stringent quality-control and quarantine rules that make it impossible for the poorest countries to export their goods and commodities.

For all cases there must be strict conditions involving reduction in defense expenditures, democratic reforms (including of the security forces), good governance and accountability, and allocation of saved resources to eradicating poverty.

Debt cancellation or relief can be phased in tandem with the reform policies being adopted and implemented by the targeted country.

INCREASE OVERSEAS DEVELOPMENT ASSISTANCE

All rich countries should increase the percentage of their overseas development assistance within the next 10 years to the UN-recommended 0.8 percent of GDP. Perhaps on a dollar-for-dollar basis where applicable, such aid could match reduction in military expenditures associated with debt relief.

IMPROVE MARKET ACCESS

Following the example set by the European Union, the United States, Canada and Japan should open up their markets for goods from the HIC (highly indebted countries) and ease some of the stringent quality-control and quarantine rules that make it impossible for the poorest countries to export their goods and commodities.

BUILD AN ANTI-POVERTY COALITION

The violent demonstrations that greet every gathering of world leaders from Seattle to Prague and Davos to Gothenberg and Genoa reflect the justified frustration of those who are genuinely concerned about the effects of globalization on the poorest of the world.

However, one can also see the opportunistic manipulation of these people by Communist-era hard liners who, seeing their world revolution agenda discredited with the collapse of the Soviet Union, now try to hijack what is otherwise a genuine anti-poverty movement.

This past year I attended the World Economic Forum in Davos, Switzerland, during the last week of January. Looking around me, I saw the richest and the most powerful people in the world and realized that I was the poorest among that lot. Yet I did not see, hear or read any complicated plot by the rich to rule the world.

From that modern-day Robin Hood, George Soros, to Bill Gates and James Wolfensohn, the World Bank president, I heard genuine concern and motivation to help the poor.

An ocean away in Porto Alegre, Brazil, thousands were meeting in defiance of the rich running the world from Davos. From that snowy perch in the Alps, as one of the poorest among the richest, I concluded there was enough good will on both sides of the divide to meet halfway.

> **The poor will not see their lot improved if we opt for the arrogant and discredited Marxist dogma still trotted out by far too many as a solution to the world's ills. The rich will not be able to continue to reap the profits of their investment in globalization if they do not seriously address the issues of poverty on a world scale.**

The poor will not see their lot improved if we opt for the arrogant and discredited Marxist dogma still trotted out by far too many as a solution to the world's ills. The rich will not be able to continue to reap the profits of their investment in globalization if they do not seriously address the issues of poverty on a world scale.

To the end of establishing this middle position, I propose a world summit bringing together representatives from the G-8, World Bank, Group of 77 (developing nations' group in the United Nations), development and human rights NGOs (nongovernmental organizations) as well as the corporate world to debate and fashion a global strategy. The ultimate aim should be to boost the poorest with a sort of global Marshall Plan that involves debt cancellation or relief as well as proactive programs to reduce poverty.

Globalization has tied the G-8 to the ROW (rest of the world). To sink or to swim is the choice they now have to make together.

From *New Perspectives Quarterly*, Fall 2001, pp. 16-18. © 2001 by New Perspectives Quarterly. Reprinted by permission.

Putting a human face on development

Rubens Ricupero

There are better ways of ending a century (indeed, a millennium) than with a war superimposed on a major economic crisis amidst recurring bouts of food panic.

Wars, crises, food panics, share a common effect: they produce fear, anxiety, insecurity. They do it not only by inflicting pain but also by threatening to take away the possibility of having a future. As we live almost as much for the future as the present, without that perspective it is hard to contemplate life itself.

If we want to avoid Eliot's terrible conclusion; that "The cycles of Heaven in twenty centuries // Bring us farther from God and nearer to the Dust",[1] we have to go back to where we started. At a minimum, the security of human beings has to be assured in all of life's dimensions.

But security is not enough: no one can live exclusively from security or stability. These only provide human beings with the possibility of having a future. We need the dream that tomorrow will be better than today and yesterday; that our children, and their children, shall be free from fear and want, that they will not only be secure but that they will be able to fulfil their lives through productive, creative work and through love, affection, solidarity, and cooperation.

However, the century is ending with failure to solve two major threats to that future: mass unemployment and growing inequality. No system of organising production has ever been able to provide a productive job for every man and woman who wanted to work. Moreover, disparities in the distribution of wealth and income are on the rise, both within and between nations.

In the poor parts of the world, covering much of the planet, the very possibility of sustainable development has been called into question by the economic crisis that began in Asia in 1997. The fifth serious monetary and financial crisis of the last 20 years, it truly deserves to be called a "crisis of development", for three main reasons. First, it hit almost exclusively developing countries, at the same time sparing and even benefiting the industrial economies through falling prices for commodities, capital flight, and cheap manufactured imports, because of currency devaluations. Second, and paradoxically, it was much more destructive of the most advanced of the developing na-

tions, raising serious doubts about whether, as had long been assumed, development is a process that reduces the vulnerability of economies to external shocks. Third, it created uncertainty whether it would be possible, once the crisis was over, to regain the levels of economic performance that constituted the only convincing demonstration so far of the possibility of development over several decades; that is, the experience of that group of countries once dubbed "the Asian tigers".

In the course of the crisis, millions of people lost their jobs in the affected countries; 30 years of progress in fighting poverty wiped out in a matter of weeks; and anguish, desperation, insecurity and, in some cases, political disintegration and violence returned in force. For the first time in many years, in 1998 and 1999 economic growth in the rich countries has been significantly higher than in poor nations, widening rather than narrowing the gap between the two.

Contrasts in the impact of crises resulting from differences in power and knowledge have reappeared in the economy, as they have in the areas of political or environmental security. Currency devaluations in the UK or Italy in the early 1990s did not st off a financial meltdown or an investors' stampede, as they did in Thailand or the Republic of Korea in 1997. Was it because the two European industrial economies had more economic power and better "fundamentals"? Or was it because they had more knowledge and skills about how to regulate and supervise financial markets?

This situation forces us to reflect on the whole concept and experience of development in the last few years, and to examine closely the recipes or formulas that have been advanced for economic development.

Of course, it is not only a problem of economic development or of poverty; the impact of this crisis has been much broader. In reality, it affected the world economy in general. It also raised questions regarding financial liberalisation even in industrialised countries. However, this article will concentrate mostly on questions related to economic development and its social implications, while keeping in mind that there is an underlying link connecting economic development, poverty, and globalisation.

We are witnessing a search for an alternative to the paradigm of development that has been hegemonic in the last 12 years, the so-called "Washington Consensus". This took its name from a famous article by John Williamson, an economist, who tried to codify the paradigm in a series of 10 principles. His approach to development was based on three major areas. First, sound macroeconomic policies, that is, low inflation, minimal budget deficits, and balanced external accounts. Second, the advice that countries should open up and follow the path of trade and financial liberalisation. (Although the Washington Consensus did not make a major distinction between the concepts of trade and financial liberalisation, financial liberalisation is much more difficult to deal with than trade liberalisation, as the Asian experience has shown. The Asian countries were very successful in trade liberalisation, but not in financial liberalisation.) Finally, the third element was to promote the role of the market much more than that of the state, through privatisation and reducing the action of the state to essential tasks, deregulation, and related matters.

These three categories of measure have been enforced by the IMF and the World Bank over the last 12 to 13 years in a top-down approach imposed through the conditionalities of the loans of the two institutions; and the underlying principles were the inspiration for the so-called structural adjustment programmers, applied over many years in many different countries. Those assumptions are now coming under increasing scrutiny.

We at UNCTAD are trying to look beyond the Washington Consensus in order to reach one that can be widely shared, a consensus built on the need for balance, equilibrium, and a sense of proportion, that will not reopen the old ideological battles of the 1970s or the 1980s, but will integrate more fully the complexity and diversity of conditions influencing development. There is still a great deal to be done in trying to reconcile apparently contradictory approaches, such as the role of the state and the market, price stability and economic growth, flexibility of the labour market and job security, or integration into the world economy and the building up of a national industrial base. These have often been presented as antagonistic, mutually exclusive positions, but the search now under way for alternatives is inspired precisely by the need to take a more multidisciplinary approach, to see to what extent we can make these goals mutually reinforcing and complementary.

We need, then, a thorough and comprehensive study of the experience of development over the last few decades, with three basic objectives. The first is to take stock of what went right and what went wrong in development. The second is to identify what was missing in the original approaches and concepts. In the 1950s and 1960s, the approach was much too macroeconomic; it emphasised aspects such as economic growth, capital accumulation, and productivity increases, but it did not give sufficient attention to the quality of development, the quality of life and social aspects such as the distribution of income or wealth. Other aspects that were totally ignored in the 1960s were the environmental dimension, the so-called sustainable quality of development, the role of women in the economy and the role of minorities and indigenous communities. The third objective of the study should be to identify the challenges

ahead: what is the challenge facing development in the next century, in the next millennium? To realise these objectives is of course a tall order, and I would refrain from giving the false impression that anybody in the world has a ready recipe.

In a world of globalised finance, one of the biggest challenges is to reach a harmonious complementarity between social and economic development. One should beware of thinking that as soon as the developing Asian economies recover, their societies will immediately and automatically reach the social level they enjoyed in pre-crisis times. That has not been the experience in Latin America, where, even today, 17 years after the beginning of the Mexican foreign debt crisis, followed by crises in Argentina, Brazil, Peru and others, the continent has yet to recapture the pre-crisis level in social indicators.

An excellent recent report by the Economic Commission for Latin America and the Caribbean (ECLAC) entitled "The social panorama of Latin America" shows that even now the level of poverty in Latin America stands at 39% of the population—in other words, 209 million people in Latin America live below the poverty line—4 percentage points above the 1982 precrisis level of 35%. For some countries, such as Chile, the results are better, but ECLAC refers to the average of the continent as a whole.

One of the aspects of the emerging paradigm for development, which will only materialise if we work together to produce it, is the central importance of knowledge and information in the economy of the future. Professor Stiglitz, Chief Economist of the World Bank, and a strong dissenter from some of the basic premises of the Washington Consensus, has made an important contribution to a new branch of economics called "information economics". This does not refer, as some people think, to electronics or the way we transmit information through telecommunications. It refers to information in economic terms. The classical economists tended to consider that information had a zero or negligible cost; that is, every actor in the market had equal access to information about the market, and so the cost of acquiring it could be considered to be zero or negligible. The main contribution of the information economists has been to show that this is not true; information does have a cost, the so-called "transaction cost", which sometimes makes the difference between success and failure. Those who have had a good education or training tend to be better at dealing with information and to succeed where others fail. The problem is, what to do with the failures—the legion of unskilled workers in industrial countries, or the poor countries that are not able to compete in the market place because they do not have appropriate access to information.

This is a particularly acute and important problem today, for the simple reason that we are moving towards a new kind of economy and a new form of development, where the decisive factors are no longer capital, cheap labour, or an abundance of natural resources. More and more, the central, crucial factor is knowledge, information, patents; how to deal with the knowledge that is being constantly generated. As we move towards a knowledge-intensive economy, access to information becomes the difference between prosperity and poverty, and between domination and liberation. This is why information and knowledge will have to be increasingly considered in the rule-making negotiations on trade and on investment and in relation to economic life in general.

Competition of the intensified kind we are seeing as a result of globalisation has many analogies to a game; and it is not by accident that game theory, with its mathematical formulation, is nowadays frequently applied to competition. As in every game, competition certainly needs fair rules, such as the norms of the World Trade Organization, and it also needs an impartial arbiter, as in the WTO's dispute-settlement mechanism. Governments and trade negotiators tend to think that once we have fair rules and an impartial arbiter, the ideal conditions for competition will be in place. They forget a third and fundamental element of competition. In order to play a game, it is not enough to know the rules and to obey the arbiter, you have to learn how to play it; you need to be educated and trained. Nobody can run the 100 metres in the Olympic Games just because there are rules and umpires! So how can we include this element of learning and training as an integral part of competition in order to have a truly level playing field?

At the very least, it is necessary to provided each and every beginner with an equal opportunity to learn to play the game, with training time, during which the newcomer would not be crushed by veterans. Even a relatively level playing field may not be enough when inequality and poverty are such that countries and individuals are starting from such widely disparate levels.

As R. H. Tawney, the British historian, wrote: "… opportunities to rise are not a substitute for a large measure of practical equality of income and social condition. The existence of such opportunities… depends not only upon an open road but upon an equal start". In the *New Statesman* essay by the former Labour Minister Roy Hattersley,[2] in which I found the above quotation, the author comments that, 130 years after William Gladstone removed the institutional barriers to civil service appointments and military commissions in Great Britain, the same people are still getting most of the jobs!

Recognising that in an unequal society, families below the poverty line are doomed to remain poor in absolute as well as relative terms, some countries have resorted to US-style affirmative action, equal opportunity laws and other pro-active measures to correct gross disparities at the start. The same rationale, I would suggest, holds true in terms of the continuous need for special and differential treatment for developing countries, redefined from its previous form in a more concrete and up-to-date way.

Federico Mayor, former Director-General of UNESCO, has said that "trade not aid" should be the instrument of development. Everybody agrees with this. Thus, you would think that trade-related technical cooperation would be a very significant part of what happens in the field of technical cooperation. But that is not the case. OECD figures show that in fact only 2% of technical cooperation is trade-related. Nobody is really trying

hard to teach countries how to produce, how to trade, or how to compete. This is why information economics should be an important element in the revision of the rules concerning development, conceived and recast as a continuous learning process.

At the root of all movement towards globalisation, there has always been a revolution in ideas, science, and technology. It was so at the beginning of the expansion of the West with the Galilean revolution in the sixteenth century; again with the Newtonian revolution in the eighteenth century leading up to the Industrial Revolution; and once more today. The difference this time is that the revolution is about time and space: the previous revolutions were about matter and energy. This time it is the very concept of distance and time that is being changed by telecommunications and by informatics, and this is why the problem of access to information becomes central.

The fact that information, technology, and science are fundamental components of human development does not in itself, however, guarantee that these elements will not again be used for oppression or domination. This we must guard against. In the past, scientific knowledge has too often been used for oppression and domination. We should not be so naive as to think that access to information is just a matter of pedagogy, of learning, of education. There is an element of power—both market and political power—in controlling information. But access to information will remain the crucial condition for development.

Norbert Wiener, the founder of cybernetics, used to say that to be informed is to be free. He meant freedom to make choices, to choose among options. But in order to be able to choose, one needs to have knowledge about what the options are, if indeed there are options, and about the relative costs and benefits of each of them, because in political life, as in culture and in the economy, you always have a trade-off—you win some and lose some. In order to choose an option, you need information. Let us hope that, this time, information will serve not to oppress and to exploit humankind, but to liberate it and promote true human development.

Notes

1. T. S. Eliot, "The Rock", in *The Complete Poems and Plays 1909–1950*. Harcourt Brace World Inc, New York, 1971, p. 96.
2. R. Hattersley, "Up and down the social ladder". *New Statesman* (London), 22 January 1999.

Rubens Ricupero is Secretary-General of the United Nations Conference on Trade and Development (UNCTAD). Previously, he was Brazil's Minister of Environment and Amazonian Affairs, and Minister of Finance. He has also represented his country at the UN, World Bank, IMF, and GATT. He has been Professor of International Relations at the University of Brasilia, and Professor at the Rio Branco Institute. He is author of several books and essays on international relations.

From *International Social Science Journal*, UNESCO 2000, pp. 441-446. © 2000 by Blackwell Publishers Ltd.

UNIT 2
Political Economy and the Developing World

Unit Selections

6. **The Free-Trade Fix**, Tina Rosenberg
7. **Trading for Development: The Poor's Best Hope**, Jagdish Bhaqwati
8. **Learning From Doha: A Civil Society Perspective From the South**, Walden Bello
9. **Do as We Say, Not as We Do**, Jack Beatty
10. **Spreading the Wealth**, David Dollar and Aart Kraay
11. **Surmounting the Challenges of Globalization**, Eduardo Aninat
12. **The Sacking of Argentina**, Tim Frasca
13. **Safe Deposit: The Case for Foreign Aid**, Gregg Easterbrook
14. **The Cartel of Good Intentions**, William Easterly
15. **Nongovernmental Organizations (NGOs) and Third World Development: An Alternative Approach to Development**, J. Wagona Makoba
16. **Fishermen on the Net**, *The Economist*

Key Points to Consider

- In what ways can globalization be regulated to help the poor?

- What steps can developing countries themselves take to improve their benefits from trade? How did the 2001 WTO ministerial conference affect developing countries?

- Are rich countries' demands that the poor open their markets hypocritical?

- How does globalization help the poor?

- To what can Argentina's economic collapse be attributed?

- How has foreign aid helped the developing world?

- In what ways can foreign aid be made more effective?

 **Links: www.dushkin.com/online/**
These sites are annotated in the World Wide Web pages.

Center for Third World Organizing
http://www.ctwo.org/

ENTERWeb
http://www.enterweb.org

Grameen Bank (GB)
http://www.grameen-info.org

International Monetary Fund (IMF)
http://www.imf.org

TWN   Third World Network
http://www.twnside.org.sg/

U.S. Agency for International Development (USAID)
http://www.info.usaid.gov

The World Bank
http://www.worldbank.org

Economic issues are among the developing world's most pressing concerns. Economic growth and stability are essential to progress on the variety of problems confronting developing countries. The developing world's position in the international economic system contributes to the difficulty of achieving consistent economic growth. From their incorporation into the international economic system during colonialism to the present, the majority of developing countries have been primarily suppliers of raw materials, agricultural products, and inexpensive labor. Dependence on commodity exports has meant that developing countries have had to deal with fluctuating, and frequently declining, prices for their exports. At the same time, prices for imports have remained constant or have increased. At best, this decline in the terms of trade has made development planning difficult; at worst, it has led to economic stagnation and decline.

With some exceptions, developing nations have had limited success in breaking out of this dilemma by diversifying their economies. Efforts at industrialization and export of light manufactures have led to competition with less efficient industries in the industrialized world. The response of industrialized countries has often been protectionism and demands for trade reciprocity, which can overwhelm markets in developing countries. Although the World Trade Organization (WTO) was established to standardize trade regulations and increase international trade, critics charge that the WTO continues to disadvantage developing countries through a selective dismantling of trade barriers.

The economic situation in the developing world, however, is not entirely attributable to colonial legacy and protectionism on the part of industrialized countries. Developing countries have sometimes constructed their own trade barriers or relied on preferential trade relationships. Industrialization schemes involving heavy government direction were often ill-conceived or resulted in corruption and mismanagement. Industrialized countries frequently point to these inefficiencies in calling for market-oriented reforms, but the emphasis on privatization does not adequately recognize the role of the state in developing countries and may result in foreign control of important sectors of the economy as well as a loss of jobs.

Debt has further compounded economic problems for many developing countries. During the 1970s, developing countries' prior economic performance and the availability of petrodollars encouraged extensive commercial lending. Developing countries sought these loans to fill the gap between revenues from exports and foreign aid, and development expenditures. The second oil price hike in the late 1970s, declining export earnings, and worldwide recession in the early 1980s left many developing countries unable to meet their debt obligations. The commercial banks weathered the crisis, and some actually showed a profit. Commercial lending declined and international financial institutions became the lenders of last resort for many developing countries. Access to World Bank and International Monetary Fund financing became conditional on the adoption of structural adjustment programs that involved steps such as reduced public expenditures, devaluation of currencies, and export promotion, all geared to debt reduction. The consequences of these programs have been painful for developing countries. Fewer public

services, higher prices, and greater exploitation of resources have resulted.

The poorest countries continue to struggle with heavy debt burdens, and the IMF and World Bank have come under increasing criticism for their programs in the developing world. Though they have made some efforts to shift the emphasis to poverty reduction, some critics charge that the reforms are superficial.

The Heavily Indebted Poor Countries (HIPC) initiative, organized by industrialized countries to provide debt relief to poor countries that follow sound economic practices, has provided limited relief to a few countries, but the conditions attached to this relief are stringent. In contrast, a campaign called Jubilee 2000, based on the scriptural prescription to forgive debts every 50 years, called for full debt forgiveness for the poorest countries. This campaign has picked up momentum, and antiglobalization protestors who regularly show up at WTO and IMF/World Bank meetings have made debt forgiveness a central theme of their protests.

Globalization has emerged as a controversial issue with differing views regarding benefits and costs taking center stage. Advocates claim that closer economic integration, especially trade and financial liberalization, increases economic prosperity in developing countries and encourages good governance, transparency, and accountability. Critics respond that globalization's requirements, as determined by the powerful nations and reinforced through the international financial institutions, impose difficult and perhaps counterproductive policies on struggling economies. They also charge that globalization undermines workers' rights and encourages environmental degradation. Moreover, most of the benefits of globalization have gone to those countries that were already growing.

In part due to the realization that poverty in the developing world contributes to the despair and resentment that leads some to terrorism, there has been increased attention focused on foreign aid. As a result, the United States has recently committed to increasing foreign aid. Although aid has often been criticized, it does produce benefits; those benefits could reach even more needy recipients if aid is channeled through nongovernmental organizations.

The Free-Trade Fix

So far, **globalization** has failed the world's poor.
But it's not trade that has hurt them. It's a rigged system.

By Tina Rosenberg

Globalization is a phenomenon that has remade the economy of virtually every nation, reshaped almost every industry and touched billions of lives, often in surprising and ambiguous ways. The stories filling the front pages in recent weeks—about economic crisis and contagion in Argentina, Uruguay and Brazil, about President Bush getting the trade bill he wanted—are all part of the same story, the largest story of our times: what globalization has done, or has failed to do.

Globalization is meant to signify integration and unity—yet it has proved, in its way, to be no less polarizing than the cold-war divisions it has supplanted. The lines between globalization's supporters and its critics run not only between countries but also through them, as people struggle to come to terms with the defining economic force shaping the planet today. The two sides in the discussion—a shouting match, really—describe what seem to be two completely different forces. Is the globe being knit together by the Nikes and Microsofts and Citigroups in a dynamic new system that will eventually lift the have-nots of the world up from medieval misery? Or are ordinary people now victims of ruthless corporate domination, as the Nikes and Microsofts and Citigroups roll over the poor in nation after nation in search of new profits?

The debate over globalization's true nature has divided people in third-world countries since the phenomenon arose. It is now an issue in the United States as

well, and many Americans—those who neither make the deals inside World Trade Organization meetings nor man the barricades outside—are perplexed.

When I first set out to see for myself whether globalization has been for better or for worse, I was perplexed, too. I had sympathy for some of the issues raised by the protesters, especially their outrage over sweatshops. But I have also spent many years in Latin America, and I have seen firsthand how protected economies became corrupt systems that helped only those with clout. In general, I thought the protesters were simply being sentimental; after all, the masters of the universe must know what they are doing. But that was before I studied the agreements that regulate global trade—including this month's new law granting President Bush a free hand to negotiate trade agreements, a document redolent of corporate lobbying. And it was before looking at globalization up close in Chile and Mexico, two nations that have embraced globalization especially ardently in the region of the third world that has done the most to follow the accepted rules. I no longer think the masters of the universe know what they are doing.

The architects of globalization are right that international economic integration is not only good for the poor; it is essential. To embrace self-sufficiency or to deride growth, as some protesters do, is to glamorize poverty. No nation has ever developed over the long term without trade. East Asia is the most recent exam-

ple. Since the mid-1970's, Japan, Korea, Taiwan, China and their neighbors have lifted 300 million people out of poverty, chiefly through trade.

But the protesters are also right—no nation has ever developed over the long term under the rules being imposed today on third-world countries by the institutions controlling globalization. The United States, Germany, France and Japan all became wealthy and powerful nations behind the barriers of protectionism. East Asia built its export industry by protecting its markets and banks from foreign competition and requiring investors to buy local products and build local know-how. These are all practices discouraged or made illegal by the rules of trade today.

The World Trade Organization was designed as a meeting place where willing nations could sit in equality and negotiate rules of trade for their mutual advantage, in the service of sustainable international development. Instead, it has become an unbalanced institution largely controlled by the United States and the nations of Europe, and especially the agribusiness, pharmaceutical and financial-services industries in these countries. At W.T.O. meetings, important deals are hammered out in negotiations attended by the trade ministers of a couple dozen powerful nations, while those of poor countries wait in the bar outside for news.

The International Monetary Fund was created to prevent future Great Depressions in part by lending countries in recession money and pressing them to adopt expansionary policies, like deficit spending and low interest rates, so they would continue to buy their neighbors' products. Over time, its mission has evolved into the reverse: it has become a long-term manager of the economies of developing countries, blindly committed to the bitter medicine of contraction no matter what the illness. Its formation was an acknowledgment that markets sometimes work imperfectly, but it has become a champion of market supremacy in all situations, echoing the voice of Wall Street and the United States Treasury Department, more interested in getting wealthy creditors repaid than in serving the poor.

It is often said that globalization is a force of nature, as unstoppable and difficult to contain as a storm. This is untrue and misleading. Globalization is a powerful phenomenon—but it is not irreversible, and indeed the previous wave of globalization, at the turn of the last century, was stopped dead by World War I. Today it would be more likely for globalization to be sabotaged by its own inequities, as disillusioned nations withdraw from a system they see as indifferent or harmful to the poor.

Globalization's supporters portray it as the peeling away of distortions to reveal a clean and elegant system of international commerce, the one nature intended. It is anything but. The accord creating the W.T.O. is 22,500 pages long—not exactly a *free* trade agreement. All globalization, it seems, is local, the rules drawn up by, and written to benefit, powerful nations and powerful interests within those nations. Globalization has been good for the United States, but even in this country, the gains go disproportionately to the wealthy and to big business.

It's not too late for globalization to work. But the system is in need of serious reform. More equitable rules would spread its benefits to the ordinary citizens of wealthy countries. They would also help to preserve globalization by giving the poor of the world a stake in the system—and, not incidentally, improve the lives of hundreds of millions of people. Here, then, are nine new rules for the

global economy—a prescription to save globalization from itself.

1. Make the State a Partner

If there is any place in Latin America where the poor have thrived because of globalization, it is Chile. Between 1987 and 1998, Chile cut poverty by more than half. Its success shows that poor nations can take advantage of globalization—if they have governments that actively make it happen.

> The previous wave of globalization was stopped dead by World War I. Today it would be more likely for globalization to be sabotaged by its own inequities, as disillusioned nations withdraw from a system they see as indifferent or harmful to the poor.

Chile reduced poverty by growing its economy—6.6 percent a year from 1985 to 2000. One of the few points economists can agree on is that growth is the most important thing a nation can do for its poor. They can't agree on basics like whether poverty in the world is up or down in the last 15 years—the number of people who live on less than $1 a day is slightly down, but the number who live on less than $2 is slightly up. Inequality has soared during the last 15 years, but economists cannot agree on whether globalization is mainly at fault or whether other forces, like the uneven spread of technology, are responsible. They can't agree on how to reduce inequality—growth tends not to change it. They can't agree on whether the poor who have not been helped are victims of globalization or have simply not yet enjoyed access to its benefits—in other words, whether the solution is more globalization or less. But economists agree on one thing: to help the poor, you'd better grow.

For the rest of Latin America, and most of the developing world except China (and to a lesser extent India), globalization as practiced today is failing,

and it is failing because it has not produced growth. Excluding China, the growth rate of poor countries was 2 percent a year lower in the 1990's than in the 1970's, when closed economies were the norm and the world was in a recession brought on in part by oil-price shocks. Latin American economies in the 1990's grew at an average annual rate of 2.9 percent—about half the rate of the 1960's. By the end of the 1990's, 11 million more Latin Americans lived in poverty than at the beginning of the decade. And in country after country, Latin America's poor are suffering—either from economic crises and market panics or from the day-to-day deprivations that globalization was supposed to relieve. The surprise is not that Latin Americans are once again voting for populist candidates but that the revolt against globalization took so long.

When I visited Eastern Europe after the end of Communism, a time when democracy was mainly bringing poverty, I heard over and over again that the reason for Chile's success was Augusto Pinochet. Only a dictator with a strong hand can put his country through the pain of economic reform, went the popular wisdom. In truth, we now know that inflicting pain is the easy part; governments democratic and dictatorial are all instituting free-market austerity. The point is not to inflict pain but to lessen it. In this Pinochet failed, and the democratic governments that followed him beginning in 1990 have succeeded.

What Pinochet did was to shut down sectors of Chile's economy that produced goods for the domestic market, like subsistence farming and appliance manufacturing, and point the economy toward exports. Here he was following the standard advice that economists give developing countries—but there are different ways to do it, and Pinochet's were disastrous. Instead of helping the losers, he dismantled the social safety net and much of the regulatory apparatus that might have kept privatization honest. When the world economy went into recession in 1982, Chile's integration into the global marketplace and its dependence on foreign capital magnified the crash. Poverty soared, and unemployment reached 20 percent.

Pinochet's second wave of globalization, in the late 1980's, worked better, because the state did not stand on the side. It regulated the changes effectively and aggressively promoted exports. But Pinochet created a time bomb in Chile: the country's exports were, and still are, nonrenewable natural resources. Chile began subsidizing companies that cut down native forests for wood chips, for example, and the industry is rapidly deforesting the nation.

Chile began to grow, but inequality soared—the other problem with Pinochet's globalization was that it left out the poor. While the democratic governments that succeeded Pinochet have not yet been able to reduce inequality, at least it is no longer increasing, and they have been able to use the fruits of Chile's growth to help the poor.

Chile's democratic governments have spread the benefits of economic integration by designing effective social programs and aiming them at the poor. Chile has sunk money into revitalizing the 900 worst primary schools. It now leads Latin America in computers in schools, along with Costa Rica. It provides the very low-income with housing subsidies, child care and income support. Open economy or closed, these are good things. But Chile's government is also taking action to mitigate one of the most dangerous aspects of global integration: the violent ups and downs that come from linking your economy to the rest of the world. This year it created unemployment insurance. And it was the first nation to institute what is essentially a tax on short-term capital, to discourage the kind of investment that can flood out during a market panic.

The conventional wisdom among economists today is that successful globalizers must be like Chile. This was not always the thinking. In the 1980's, the Washington Consensus—the master-of-the-universe ideology at the time, highly influenced by the Reagan and Thatcher administrations—held that government was in the way. Globalizers' tasks included privatization, deregulation, fiscal austerity and financial liberalization. "In the 1980's and up to 1996 or 1997, the state was considered the devil," says Juan Martin, an Argentine economist at the

United Nations' Economic Commission for Latin America and the Caribbean. "Now we know you need infrastructure, institutions, education. In fact, when the economy opens, you need *more* control mechanisms from the state, not fewer."

And what if you don't have these things? Bolivia carried out extensive reforms beginning in 1985—a year in which it had inflation of 23,000 percent—to make the economy more stable and efficient. But in the words of the World Bank, "It is a good example of a country that has achieved successful stabilization and implemented innovative market reforms, yet made only limited progress in the fight against poverty." Latin America is full of nations that cannot make globalization work. The saddest example is Haiti, an excellent student of the rules of globalization, ranked at the top of the I.M.F.'s index of trade openness. Yet over the 1990's, Haiti's economy *contracted*; annual per capita income is now $250. No surprise—if you are a corrupt and misgoverned nation with a closed economy, becoming a corrupt and misgoverned nation with an open economy is not going to solve your problems.

Import Know-How Along With the Assembly Line

2. If there is a showcase for globalization in Latin America, it lies on the outskirts of Puebla, Mexico, at Volkswagen Mexico. Every New Beetle in the world is made here, 440 a day, in a factory so sparkling and clean that you could have a baby on the floor, so hightech that in some halls it is not evident that human beings work here. Volkswagen Mexico also makes Jettas and, in a special hall, 80 classic Beetles a day to sell in Mexico, one of the last places in the world where the old Bug still chugs.

The Volkswagen factory is the biggest single industrial plant in Mexico. Humans do work here—11,000 people in assembly-line jobs, 4,000 more in the rest of the factory—with 11,000 more jobs in the industrial park of VW suppliers across the street making parts, seats, dashboards and other components. Perhaps 50,000 more people work in other companies around Mexico that supply VW. The average monthly wage in the

plant is $760, among the highest in the country's industrial sector. The factory is the equal of any in Germany, the product of a billion-dollar investment in 1995, when VW chose Puebla as the exclusive site for the New Beetle.

Ahhh, globalization.

Except… this plant is not here because Mexico has an open economy, but because it had a closed one. In 1962, Mexico decreed that any automaker that wanted to sell cars here had to produce them here. Five years later, VW opened the factory. Mexico's local content requirement is now illegal, except for very limited exceptions, under W.T.O. rules; in Mexico the local content requirement for automobiles is being phased out and will disappear entirely in January 2004.

The Puebla factory, for all the jobs and foreign exchange it brings Mexico, also refutes the argument that foreign technology automatically rubs off on the local host. Despite 40 years here, the auto industry has not created much local business or know-how. VW makes the point that it buys 60 percent of its parts in Mexico, but the "local" suppliers are virtually all foreign-owned and import most of the materials they use. The value Mexico adds to the Beetles it exports is mainly labor. Technology transfer—the transmission of know-how from foreign companies to local ones—is limited in part because most foreign trade today is intracompany; Ford Hermosillo, for example, is a stamping and assembly plant shipping exclusively to Ford plants in the United States. Trade like this is particularly impenetrable to outsiders. "In spite of the fact that Mexico has been host to many car plants, we don't know how to build a car," says Huberto Juárez, an economist at the Autonomous University of Puebla.

Volkswagen Mexico is the epitome of the strategy Mexico has chosen for globalization—assembly of imported parts. It is a strategy that makes perfect sense given Mexico's proximity to the world's largest market, and it has given rise to the maquila industry, which uses Mexican labor to assemble foreign parts and then re-export the finished products. Although the economic slowdown in the United States is hurting the *maquila* industry, it still employs a million people and brings the country $10 billion a year in foreign exchange. The factories have turned Mexico into one of the developing world's biggest exporters of medium- and high-technology products. But the maquila sector remains an island and has failed to stimulate Mexican industries—one reason Mexico's globalization has brought disappointing growth, averaging only 3 percent a year during the 1990's.

In countries as varied as South Korea, China and Mauritius, however, assembly work has been the crucible of wider development. Jeffrey Sachs, the development economist who now directs Columbia University's Earth Institute, says that the maquila industry is "magnificent." "I could cite 10 success stories," he says, "and every one started with a maquila sector." When Korea opened its export-processing zone in Masan in the early 1970's, local inputs were 3 percent of the export value, according to the British development group Oxfam. Ten years later they were almost 50 percent. General Motors took a Korean textile company called Daewoo and helped shape it into a conglomerate making cars, electronic goods, ships and dozens of other products. Daewoo calls itself "a locomotive for national economic development since its founding in 1967." And despite the company's recent troubles, it's true—because Korea made it true. G.M. did not tutor Daewoo because it welcomed competition but because Korea demanded it. Korea wanted to build high-tech industry, and it did so by requiring technology transfer and by closing markets to imports.

Maquilas first appeared in Mexico in 1966. Although the country has gone from assembling clothing to assembling high-tech goods, nearly 40 years later 97 percent of the components used in Mexican maquilas are still imported, and the value that Mexico adds to its exports has actually declined sharply since the mid-1970's.

Mexico has never required companies to transfer technology to locals, and indeed, under the rules of the North American Free Trade Agreement, it cannot. "We should have included a technical component in Nafta," says Luis de la Calle, one of the treaty's negotiators and later Mexico's under secretary of economy for foreign trade. "We should be getting a significant transfer of technology from the United States, and we didn't really try."

Without technology transfer, maquila work is marked for extinction. As transport costs become less important, Mexico is increasingly competing with China and Bangladesh—where labor goes for as little as 9 cents an hour. This is one reason that real wages for the lowest-paid workers in Mexico dropped by 50 percent from 1985 to 2000. Businesses, in fact, are already leaving to go to China.

Sweat the Sweatshops—But Sweat Other Problems More

3. When Americans think about globalization, they often think about sweatshops—one aspect of globalization that ordinary people believe they can influence through their buying choices. In many of the factories in Mexico, Central America and Asia producing American-brand toys, clothes, sneakers and other goods, exploitation is the norm. The young women who work in them—almost all sweatshop workers are young women—endure starvation wages, forced overtime and dangerous working conditions.

In Chile, I met a man who works at a chicken-processing plant in a small town. The plant is owned by Chileans and processes chicken for the domestic market and for export to Europe, Asia and other countries in Latin America. His job is to stand in a freezing room and crack open chickens as they come down an assembly line at the rate of 41 per minute. When visitors arrive at the factory (the owners did not return my phone calls requesting a visit or an interview), the workers get a respite, as the line slows down to half-speed for show. His work uniform does not protect him from the cold, the man said, and after a few minutes of work he loses feeling in his hands. Some of his colleagues, he said, are no longer able to raise their arms. If he misses a day he is docked $30. He earns less than $200 a month.

Is this man a victim of globalization? The protesters say that he is, and at one point I would have said so, too. He—and all workers—should have dignified conditions and the right to organize. All companies should follow local labor laws, and activists should pressure companies to pay their workers decent wages.

But today if I were to picket globalization, I would protest other inequities. In a way, the chicken worker, who came to the factory when driving a taxi ceased to be profitable, is a beneficiary of globalization. So are the millions of young women who have left rural villages to be exploited gluing tennis shoes or assembling computer keyboards. The losers are those who get laid off when companies move to low-wage countries, or those forced off their land when imports undercut their crop prices, or those who can no longer afford life-saving medicine—people whose choices in life *diminish* because of global trade. Globalization has offered this man a hellish job, but it is a choice he did not have before, and he took it; I don't name him because he is afraid of being fired. When this chicken company is hiring, the lines go around the block.

Get Rid of the Lobbyists

4. The argument that open economies help the poor rests to a large extent on the evidence that closed economies do not. While South Korea and other East Asian countries successfully used trade barriers to create export industries, this is rare; most protected economies are disasters. "The main tendency in a sheltered market is to goof off," says Jagdish Bhagwati, a prominent free-trader who is the Arthur Lehman professor of economics at Columbia University. "A crutch becomes a permanent crutch. Infant-industry protection should be for infant industries."

Anyone who has lived or traveled in the third world can attest that while controlled economies theoretically allow governments to help the poor, in practice it's usually a different story. In Latin America, spending on social programs largely goes to the urban middle class. Attention goes to people who can organize, strike, lobby and contribute money. And in a closed economy, the "state" car factory is often owned by the dictator's son and the country's forests can be chopped down by his golf partner.

Free trade, its proponents argue, takes these decisions away from the government and leaves them to the market, which punishes corruption. And it's true that a system that took corruption and undue political influence out of economic decision-making could indeed benefit the poor. But humans have not yet invented such a system—and if they did, it would certainly not be the current system of globalization, which is soiled with the footprints of special interests. In every country that negotiates at the W.T.O. or cuts a free-trade deal, trade ministers fall under heavy pressure from powerful business groups. Lobbyists have learned that they can often quietly slip provisions that pay big dividends into complex trade deals. None have been more successful at getting what they want than those from America.

The most egregious example of a special-interest provision is the W.T.O.'s rules on intellectual property. The ability of poor nations to make or import cheap copies of drugs still under patent in rich countries has been a boon to world public health. But the W.T.O. will require most of its poor members to accept patents on medicine by 2005, with the very poorest nations following in 2016. This regime does nothing for the poor. Medicine prices will probably double, but poor countries will never offer enough of a market to persuade the pharmaceutical industry to invent cures for their diseases.

The intellectual-property rules have won worldwide notoriety for the obstacles they pose to cheap AIDS medicine. They are also the provision of the W.T.O. that economists respect the least. They were rammed into the W.T.O. by Washington in response to the industry groups who control United States trade policy on the subject. "This is not a trade issue," Bhagwati says. "It's a royalty-collection issue. It's pharmaceuticals and software throwing their weight around." The World Bank calculated that the intellectual-property rules will result in a transfer of $40 billion a year from poor countries to corporations in the developed world.

No Dumping

5. Manuel de Jesús Gómez is a corn farmer in the hills of Puebla State, 72 years old and less than five feet tall. I met him in his field of six acres, where he was trudging behind a plow pulled by a burro. He farms the same way *campesinos* in these hills have been farming for thousands of years. In Puebla, and in the poverty belt of Mexico's southern states—Chiapas, Oaxaca, Guerrero—corn growers plow with animals and irrigate by praying for rain.

Before Nafta, corn covered 60 percent of Mexico's cultivated land. This is where corn was born, and it remains a symbol of the nation and daily bread for most Mexicans. But in the Nafta negotiations, Mexico agreed to open itself to subsidized American corn, a policy that has crushed small corn farmers. "Before, we could make a living, but now sometimes what we sell our corn for doesn't even cover our costs," Gómez says. With Nafta, he suddenly had to compete with American corn—raised with the most modern methods, but more important, subsidized to sell overseas at 20 percent less than the cost of production. Subsidized American corn now makes up almost half of the world's stock, effectively setting the world price so low that local small farmers can no longer survive. This competition helped cut the price paid to Gómez for his corn by half.

Because of corn's importance to Mexico, when it negotiated Nafta it was promised 15 years to gradually raise the amount of corn that could enter the country without tariffs. But Mexico voluntarily lifted the quotas in less than three years—to help the chicken and pork industry, Mexican negotiators told me unabashedly. (Eduardo Bours, a member of the family that owns Mexico's largest chicken processor, was one of Mexico's Nafta negotiators.) The state lost some $2 billion in tariffs it could have charged, and farmers were instantly exposed to competition from the north. According to ANEC, a national association of campesino cooperatives, half a million corn farmers have left their land and moved to Mexican cities or to America. If it were not for a weak peso, which keeps the price of imports relatively

high, far more farmers would be forced off their land.

The toll on small farmers is particularly bitter because cheaper corn has not translated into cheaper food for Mexicans. As part of its economic reforms, Mexico has gradually removed price controls on tortillas and tortilla flour. Tortilla prices have nearly tripled in real terms even as the price of corn has dropped.

Is this how it was supposed to be? I asked Andrés Rosenzweig, a longtime Mexican agriculture official who helped negotiate the agricultural sections of Nafta. He was silent for a minute. "The problems of rural poverty in Mexico did not start with Nafta," he said. "The size of our farms is not viable, and they get smaller each generation because farmers have many children, who divide the land. A family in Puebla with five hectares could raise 10, maybe 15, tons of corn each year. That was an annual income of 16,000 pesos," the equivalent of $1,600 today. "Double it and you still die of hunger. This has nothing to do with Nafta.

"The solution for small corn farmers," he went on, "is to educate their children and find them jobs outside agriculture. But Mexico was not growing, not generating jobs. Who's going to employ them? Nafta."

One prominent antiglobalization report keeps referring to farms like Gómez's as "small-scale, diversified, self-reliant, community-based agriculture systems." You could call them that, I guess; you could also use words like "malnourished," "undereducated" and "miserable" to describe their inhabitants. Rosenzweig is right—this is not a life to be romanticized.

But to turn the farm families' malnutrition into starvation makes no sense. Mexico spends foreign exchange to buy corn. Instead, it could be spending money to bring farmers irrigation, technical help and credit. A system in which the government purchased farmers' corn at a guaranteed price—done away with in states like Puebla during the free-market reforms of the mid-1990's—has now been replaced by direct payments to farmers. The program is focused on the poor, but the payments are symbolic—

$36 an acre. In addition, rural credit has disappeared, as the government has effectively shut down the rural bank, which was badly run, and other banks won't lend to small farmers. There is a program—understaffed and poorly publicized—to help small producers, but the farmers I met didn't know about it.

FREE TRADE IS a religion, and with religion comes hypocrisy. Rich nations press other countries to open their agricultural markets. At the urging of the I.M.F. and Washington, Haiti slashed its tariffs on rice in 1995. Prices paid to rice farmers fell by 25 percent, which has devastated Haiti's rural poor. In China, the tariff demands of W.T.O. membership will cost tens of millions of peasants their livelihoods. But European farmers get 35 percent of their income from government subsidies, and American farmers get 20 percent. Farm subsidies in the United States, moreover, are a huge corporate-welfare program, with nearly 70 percent of payments going to the largest 10 percent of producers. Subsidies also depress crop prices abroad by encouraging overproduction. The farm bill President Bush signed in May—with substantial Democratic support—provides about $57 billion in subsidies for American corn and other commodities over the next 10 years.

Wealthy nations justify pressure on small countries to open markets by arguing that these countries cannot grow rice and corn efficiently—that American crops are cheap food for the world's hungry. But with subsidies this large, it takes chutzpah to question other nations' efficiency. And in fact, the poor suffer when America is the supermarket to the world, even at bargain prices. There is plenty of food in the world, and even many countries with severe malnutrition are food exporters. The problem is that poor people can't afford it. The poor *are* the small farmers. Three-quarters of the world's poor are rural. If they are forced off their land by subsidized grain imports, they starve.

Help Countries Break the Coffee Habit

6. Back in the 1950's, Latin American economists made a simple cal-

culation. The products their nations exported—copper, tin, coffee, rice and other commodities—were buying less and less of the high-value-added goods they wanted to import. In effect, they were getting poorer each day. Their solution was to close their markets and develop domestic industries to produce their own appliances and other goods for their citizens.

The strategy, which became known as import substitution, produced high growth—for a while. But these closed economies ultimately proved unsustainable. Latin American governments made their consumers buy inferior and expensive products—remember the Brazilian computer of the 1970's? Growth depended on heavy borrowing and high deficits. When they could no longer roll over their debts, Latin American economies crashed, and a decade of stagnation resulted.

At the time, the architects of import substitution could not imagine that it was possible to export anything but commodities. But East Asia—as poor or poorer than Latin America in the 1960's—showed in the 1980's and 1990's that it can be done. Unfortunately, the rules of global trade now prohibit countries from using the strategies successfully employed to develop export industries in East Asia.

American trade officials argue that they are not using tariffs to block poor countries from exporting, and they are right—the average tariff charged by the United States is a negligible 1.7 percent, much lower than other nations. But the rules rich nations have set—on technology transfer, local content and government aid to their infant industries, among other things—are destroying poor nations' abilities to move beyond commodities. "We are pulling up the ladder on policies the developed countries used to become rich," says Lori Wallach, the director of Public Citizen's Global Trade Watch.

The commodities that poor countries are left to export are even more of a dead end today than in the 1950's. Because of oversupply, prices for coffee, cocoa, rice, sugar and tin dropped by more than 60 percent between 1980 and 2000. Because of the price collapse of commodi-

ties and sub-Saharan Africa's failure to move beyond them, the region's share of world trade dropped by two-thirds during that time. If it had the same share of exports today that it had at the start of the 1980's, per capita income in sub-Saharan Africa would be almost twice as high.

Let the People Go

7. Probably the single most important change for the developing world would be to legalize the export of the one thing they have in abundance—people. Earlier waves of globalization were kinder to the poor because not only capital, but also labor, was free to move. Dani Rodrik, an economist at Harvard's Kennedy School of Government and a leading academic critic of the rules of globalization, argues for a scheme of legal short-term migration. If rich nations opened 3 percent of their work forces to temporary migrants, who then had to return home, Rodrik says, it would generate $200 billion annually in wages, and a lot of technology transfer for poor countries.

Free the I.M.F.

8. Globalization means risk. By opening its economy, a nation makes itself vulnerable to contagion from abroad. Countries that have liberalized their capital markets are especially susceptible, as short-term capital that has whooshed into a country on investor whim whooshes out just as fast when investors panic. This is how a real-estate crisis in Thailand in 1997 touched off one of the biggest global conflagrations since the Depression.

The desire to keep money from rushing out inspired Chile to install speed bumps discouraging short-term capital inflows. But Chile's policy runs counter to the standard advice of the I.M.F., which has required many countries to open their capital markets. "There were so many obstacles to capital-market integration that it was hard to err on the side of pushing countries to liberalize too much," says Ken Rogoff, the I.M.F.'s director of research.

Prudent nations are wary of capital liberalization, and rightly so. Joseph

Stiglitz, the Nobel Prize-winning economist who has become the most influential critic of globalization's rules, writes that in December 1997, when he was chief economist at the World Bank, he met with South Korean officials who were balking at the I.M.F.'s advice to open their capital markets. They were scared of the hot money, but they could not disagree with the I.M.F., lest they be seen as irresponsible. If the I.M.F. expressed disapproval, it would drive away other donors and private investors as well.

In the wake of the Asian collapse, Prime Minister Mahathir Mohamad imposed capital controls in Malaysia—to worldwide condemnation. But his policy is now widely considered to be the reason that Malaysia stayed stable while its neighbors did not. "It turned out to be a brilliant decision," Bhagwati says.

Post-crash, the I.M.F. prescribed its standard advice for nations—making loan arrangements contingent on spending cuts, interest-rate hikes and other contractionary measures. But balancing a budget in recession is, as Stiglitz puts it in his new book, "Globalization and Its Discontents," a recommendation last taken seriously in the days of Herbert Hoover. The I.M.F.'s recommendations deepened the crisis and forced governments to reduce much of the cushion that was left for the poor. Indonesia had to cut subsidies on food. "While the I.M.F. had provided some $23 billion to be used to support the exchange rate and bail out creditors," Stiglitz writes, "the far, far, smaller sums required to help the poor were not forthcoming."

IS YOUR INTERNATIONAL financial infrastructure breeding Bolsheviks? If it does create a backlash, one reason is the standard Bolshevik explanation—the I.M.F. really is controlled by the epicenter of international capital. Formal influence in the I.M.F. depends on a nation's financial contribution, and America is the only country with enough shares to have a veto. It is striking how many economists think the I.M.F. is part of the "Wall Street-Treasury complex," in the words of Bhagwati. The fund serves "the interests of global finance," Stiglitz says. It listens to the "voice of the markets," says

Nancy Birdsall, president of the Center for Global Development in Washington and a former executive vice president of the Inter-American Development Bank. "The I.M.F. is a front for the U.S. government—keep the masses away from our taxpayers," Sachs says.

I.M.F. officials argue that their advice is completely equitable—they tell even wealthy countries to open their markets and contract their economies. In fact, Stiglitz writes, the I.M.F. told the Clinton administration to hike interest rates to lower the danger of inflation—at a time when inflation was the lowest it had been in decades. But the White House fortunately had the luxury of ignoring the I.M.F.: Washington will only have to take the organization's advice the next time it turns to the I.M.F. for a loan. And that will be never.

Let the Poor Get Rich the Way the Rich Have

9. The idea that free trade maximizes benefits for all is one of the few tenets economists agree on. But the power of the idea has led to the overly credulous acceptance of much of what is put forward in its name. Stiglitz writes that there is simply no support for many I.M.F. policies, and in some cases the I.M.F. has ignored clear evidence that what it advocated was harmful. You can always argue—and American and I.M.F. officials do—that countries that follow the I.M.F.'s line but still fail to grow either didn't follow the openness recipe precisely enough or didn't check off other items on the to-do list, like expanding education.

Policy makers also seem to be skipping the fine print on supposedly congenial studies. An influential recent paper by the World Bank economists David Dollar and Aart Kraay is a case in point. It finds a strong correlation between globalization and growth and is widely cited to support the standard rules of openness. But in fact, on close reading, it does not support them. Among successful "globalizers," Dollar and Kraay count countries like China, India and Malaysia, all of whom are trading and growing but still have protected economies and could not be doing more to misbehave by the received wisdom of globalization.

Dani Rodrik of Harvard used Dollar and Kraay's data to look at whether the single-best measure of openness—a country's tariff levels—correlates with growth. They do, he found—but not the way they are supposed to. High-tariff countries grew *faster*. Rodrik argues that the countries in the study may have begun to trade more because they had grown and gotten richer, not the other way around. China and India, he points out, began trade reforms about 10 years after they began high growth.

When economists talk about many of the policies associated with free trade today, they are talking about national averages and ignoring questions of distribution and inequality. They are talking about equations, not what works in messy third-world economies. What economic model taught in school takes into account a government ministry that stops work because it has run out of pens? The I.M.F. and the World Bank—which recommends many of the same austerity measures as the I.M.F. and frequently conditions its loans on I.M.F.-advocated reforms—often tell countries to cut subsidies, including many that do help the poor, and impose user fees on services like water. The argument is that subsidies are an inefficient way to help poor people—because they help rich people too—and instead, countries should aid the poor directly with vouchers or social programs. As an equation, it adds up. But in the real world, the subsidies disappear, and the vouchers never materialize.

The I.M.F. argues that it often saves countries from even more budget cuts. "Countries come to us when they are in severe distress and no one will lend to them," Rogoff says. "They may even have to run surpluses because their loans are being called in. Being in an I.M.F. program means less austerity." But a third of the developing world is under I.M.F. tutelage, some countries for decades, during which they must remodel their economies according to the standard I.M.F. blueprint. In March 2000, a panel appointed to advise Congress on international financial institutions, named for its head, Allan Meltzer of Carnegie Mellon University, recommended unanimously that the I.M.F. should undertake only short-term crisis assistance and get out of the business of long-term economic micromanagement altogether.

The standard reforms deprive countries of flexibility, the power to get rich the way we know can work. "Most Latin American countries have had deep reforms, have gone much further than India or China and haven't gotten much return for their effort," Birdsall says. "Many of the reforms were about creating an efficient economy, but the economic technicalities are not addressing the fundamental question of why countries are not growing, or the constraint that all these people are being left out. Economists are way too allergic to the wishy-washy concept of fairness."

THE PROTESTERS IN the street, the Asian financial crisis, criticism from respected economists like Stiglitz and Rodrik and those on the Meltzer Commission and particularly the growing realization in the circles of power that globalization is sustainable for wealthy nations only if it is acceptable to the poor ones are all combining to change the rules—slightly. The debt-forgiveness initiative for the poorest nations, for all its limitations, is one example. The Asian crisis has modified the I.M.F.'s view on capital markets, and it is beginning to apply less pressure on countries in crisis to cut government spending. It is also debating whether it should be encouraging countries to adopt Chile's speed bumps. The incoming director of the W.T.O. is from Thailand, and third-world countries are beginning to assert themselves more and more.

But the changes do not alter the underlying idea of globalization, that openness is the universal prescription for all ills. "Belt-tightening is not a development strategy," Sachs says. "The I.M.F. has no sense that its job is to help countries climb a ladder."

Sachs says that for many developing nations, even climbing the ladder is unrealistic. "It can't work in an AIDS pandemic or an endemic malaria zone. I don't have a strategy for a significant number of countries, other than we ought to help them stay alive and control disease and have clean water. You can't do this purely on market forces. The prospects for the Central African Republic are not the same as for Shanghai, and it doesn't do any good to give pep talks."

China, Chile and other nations show that under the right conditions, globalization can lift the poor out of misery. Hundreds of millions of poor people will never be helped by globalization, but hundreds of millions more could be benefiting now, if the rules had not been rigged to help the rich and follow abstract orthodoxies. Globalization can begin to work for the vast majority of the world's population only if it ceases to be viewed as an end in itself, and instead is treated as a tool in service of development: a way to provide food, health, housing and education to the wretched of the earth.

Tina Rosenberg writes editorials for The Times. Her last article for the magazine was about human rights in China.

Trading for development: The poor's best hope

Removing trade barriers is not just a job for the rich. The poor must do the same in order to prosper, says Jagdish Bhagwati

WASHINGTON, DC

THE launch of a new round of multilateral trade negotiations (MTN) at Doha dealt a needed blow to the anti-globalisers who triumphed at Seattle just two years ago. But it was also important for a different reason. The word "development" now graces the name of the new round. This is unconventional, but it underlines the fact that development of the poor countries will be the round's central objective.

Pleasing rhetoric aside, however, we must ask: What does this mean? The question is not idle. For if the current thinking among policymakers and NGOs is any guide, the answer they would give is not the right one. And that is cause for alarm.

Of course, proponents of trade have always considered that trade is the policy and development the objective. The experience of the post-war years only proves them right. The objections advanced by a handful of dissenting economists, claiming that free-traders exaggerate the gains from trade or forget that good trade policy is best embedded within a package of reforms, are mostly setting up and knocking down straw men.

But if trade is indeed good for the poor countries, what can be done to enhance its value for them? A great deal. But not until we confront and discard several misconceptions. Among them:

- the world trading system is "unfair": the poor countries face protectionism that is more acute than their own;
- the rich countries have wickedly held on to their trade barriers against poor countries, while using the Bretton Woods institutions to force down the poor countries' own trade barriers; and

- it is hypocritical to ask poor countries to reduce their trade barriers when the rich countries have their own.

In fact, asymmetry of trade barriers goes the other way. Take industrial tariffs. As of today, rich-country tariffs average 3%; poor countries' tariffs average 13%. Nor do peaks in tariffs—concentrated in textiles and clothing, fisheries and footwear, and clearly directed at the poor countries—change the picture much: the United Nations Council for Trade, Aid and Development (UNCTAD) has estimated that they apply to only a third of poor-country exports. Moreover, the trade barriers of the poor countries against one another are more significant restraints on their own development than those imposed by the rich countries.

The situation is little different when it comes to the use of anti-dumping actions, the classic "fair trade" instrument that has ironically been used "unfairly" to undermine free trade. The "new" users, among them Argentina, Brazil, India, South Korea, South Africa and Mexico, are now filing more anti-dumping complaints than the rich countries (see chart 1). Between July and December 2001 alone, India carried out more anti-dumping investigations than anywhere else.

The wicked rich?

These facts fly in the face of the populist myth that the rich countries, often acting through the conditionality imposed by the World Bank and the International Monetary Fund (IMF), have demolished the trade barriers of the poor countries while holding on to their own. Indeed, both the omnipotence of the Bretton Woods institutions, and the wickedness of the rich countries, have been grossly exaggerated.

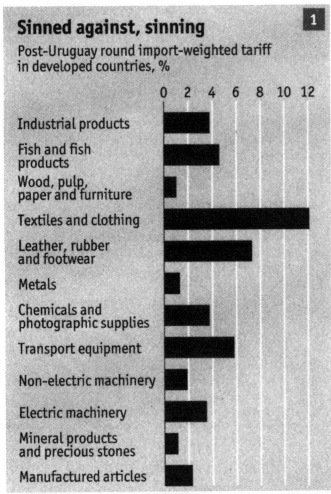

Sinned against, sinning

1

Post-Uruguay round import-weighted tariff in developed countries, %

Industrial products
Fish and fish products
Wood, pulp, paper and furniture
Textiles and clothing
Leather, rubber and footwear
Metals
Chemicals and photographic supplies
Transport equipment
Non-electric machinery
Electric machinery
Mineral products and precious stones
Manufactured articles

Source: UNCTAD

The World Bank's conditionality is so extensive and diffused, and its need to lend so compelling, that it can in fact be bypassed. Many client states typically satisfy some conditions while ignoring others. Besides, countries go to the IMF when there is a stabilisation crisis. Since stabilisation requires that the excess of expenditures over income be brought into line, the IMF has often been reluctant to suggest tariff reductions. These could reduce revenues, exacerbating the crisis.

Then again, since countries are free to return to their bad ways once the crisis is past and the loans repaid, tariff reforms can be reversed. Countries do not "bind" their tariff reductions under the IMF programmes, as they do at the World Trade Organisation (WTO). Equally, tariff reductions may be reversed when a stabilisation crisis recurs and the tariffs are reimposed to increase revenues. My student Ravi Yatawara, who has studied what he calls "commercial policy switches", documents several instances of such tariff-reduction reversals by countries borrowing from the IMF. For instance, Uruguay in 1971 increased trade protection during an IMF programme that began the year before, and even managed to get another credit tranche the year after. Kenya's 1977 liberalisation

was reversed in 1979, the year in which another arrangement was negotiated with the IMF.

Moreover, the comparatively higher trade barriers against labour-intensive products are not usually the result of wickedness, but of simple political economy. Unilateral reductions of trade barriers are in fact not uncommon, and I document them for many countries and several sectors in the post-war period in my new book, "Going Alone: The Case for Relaxed Reciprocity in Freeing Trade" (MIT Press, July). But the fact remains that the developing countries were exempted by the economic ideology of the time, which embraced "Special and Differential" treatment for them, from having to make trade concessions of their own at the successive multilateral trade negotiations that reduced trade barriers after the war. The rich countries, denied reciprocal concessions from the poor countries, wound up concentrating on liberalising trade in products of interest largely to themselves, such as machinery, chemicals and manufactures, rather than textiles and clothing.

The situation changed when the poor countries became full participants. In 1995 in Marrakesh, where the Uruguay round was concluded, action was taken at last to dismantle the infamous Multi-fibre Arrangement (MFA), which—from its birth in 1961 as the Short-term Cotton Textile Arrangement—had grown by 1974 into a Frankenstein monster incorporating several separate agreements restricting world trade in all textiles. At Marrakesh the MFA was put on the block, and was scheduled to end in ten years.

But even if rich-country protectionism were asymmetrically higher, it would be dangerous to argue that it is therefore hypocritical to suggest that poor countries should reduce their own trade barriers. Except in the few cases of oligopolistic competition, such as that between Fuji and Eastman Kodak (hardly applicable to poor countries) where strategic tit-for-tat is credible, the net effect of matching other people's protection with one's own is to hurt oneself twice over. But there is ample evidence that many leaders of the poor countries have predictably made the wrong inference: that rich-country protectionism excuses, and justifies, going easy on relaxing their own barriers to trade.

In fact, the protectionism of the poor and the rich countries must be viewed together symbiotically to ensure effective exports by the poor countries. Thus, even if the doors to the markets of the rich countries were fully open to imports, exports from the poor countries would have to get past their own doors.

We know from numerous case studies dating back to the 1970s (which only corroborated elementary economic logic) that protection is often the cause of dismal export, and hence economic, performance. It creates a "bias against exports" by sheltering domestic markets that then become more lucrative. Just ask yourself why, though India and the far-eastern countries faced virtually the same external trade barriers in the quarter-century after the

1960s, inward-looking India registered a miserable export performance while outward-looking South Korea, Taiwan, Singapore and Hong Kong chalked up spectacular exports. Just as charity begins at home, so exports begin with a good domestic policy. In the near-exclusive focus on rich-country protectionism, this dramatic lesson has been lost from view.

A strategy for change

Rich-country protectionism matters too, of course. And it must be assaulted effectively. But here, too, we witness folly. The current fashion is to shame the rich countries by arguing that their protection hurts the poor countries, whose poverty is the focus of renewed international efforts. And where action is actually undertaken, the preference is for granting preferences to the poorer countries, with yet deeper preferences for the poorest among them (the least-developed countries, or LDCs, as they are now called). But the former solution is woefully inadequate, and the latter is downright wrong.

If shame were sufficient, there would be no rich-country protectionism left. Trade economists and international institutions such as UNCTAD and the General Agreement on Tariffs and Trade (the GATT) have denounced the rich countries on this count over three decades. Added support, from charities such as Oxfam, could help in principle. But these charities need both expertise and a talent for strategy, not simply a conscience and a voice. They fall short. By subscribing to the counterproductive language of "hypocrisy" and the rhetoric of "unfair trade" to attack protection by the rich, a charity such as Oxfam, splendid at fighting plagues and famines, does more harm than good.

The argument to rich countries should be made in quite a different way: If you hold on to your own protection, no matter how much smaller, and in fact even raise it as the United States did recently with steel tariffs and the farm bill, you are going to undermine seriously the efforts of those poor-country leaders who have turned to freer trade in recent decades. It is difficult for such countries to reduce protection if others, more prosperous and fiercer supporters of free trade, are breaking ranks.

Beyond this, an effective tariff-reduction strategy requires that we handle labour-intensive goods such as textiles separately from agriculture. The differences between them dwarf the commonalities. Labour-intensive manufactures in the rich countries typically employ their own poor, the unskilled. To argue that we should eliminate protection, harming them simply because it helps yet poorer folk abroad, runs into evident ethical (and hence political) difficulties. The answer must be a gradual, but certain, phase-out of protection coupled with a simultaneous and substantial adjustment and retraining programme. That way, we address the problems of the poor both at home and abroad.

Once this is done, church groups and charities can be asked to endorse a programme that is balanced and just. Such a strategy is morally more compelling than either marching against free trade to protect workers in the labour-intensive industries of the rich nations—while forgetting the needs of poor workers in poor countries—or asking for trade restrictions to be abolished without providing for workers in such industries in the rich countries.

The removal of agricultural protection does not raise the same ethical problems; production and export subsidies in the United States and the European Union go mainly to large farmers. That should make it easier to dismantle farm protection on the grounds of helping the poor. At the same time, however, agricultural protectionism is energetically defended as necessary for preserving greenery and the environment. With the greens in play, protectionism becomes more difficult to remove. But, just as income support can be de-linked from increasing production and exports, so measures to support greenery can be de-linked too. Such new measures, and other environmental protections added as sweeteners, must be part of the strategic assault on agricultural protection.

The target date of Jubilee 2000 helped greatly to focus efforts on the objective of debt relief. Following that example, I and Arvind Panagariya of the University of Maryland suggested well over a year ago—with a nod from Kofi Annan, the UN's secretary general—a Jubilee 2010 movement to eliminate protection on labour-intensive products by 2010. Since agricultural protection is politically a harder nut to crack, 2020, rather than 2010, is probably a more realistic date for its demise. Leaders of rich and poor countries could endorse both targets at the mammoth UN Conference on Sustainable Development in Johannesburg in August.

The perils of preferences

A final word is necessary on the efforts to open rich-country doors. This is often done not by dismantling barriers on a most-favoured-nation (MFN) basis, which reduces them in a non-discriminatory manner, but through grants of preferences to the poor countries. This approach goes back to the Generalised System of Preferences (GSP), introduced in 1971 through a waiver and then granted legal status in 1979 with an enabling clause at the GATT. Under this, the eligible poor countries were granted entry at preferentially lower tariff rates.

GSP did little for the poor countries. The eligible products often excluded those on which poor countries had pinned their hopes of increasing exports. Thus the United States' GSP scheme excluded textiles, clothing and footwear. Upper caps were also introduced. The United States imposed a $100m limit on exports per tariff line, per year, per country; beyond this limit, the preferential rate vanished. Even the benefits granted were not "bound", and could be varied at a rich country's displea-

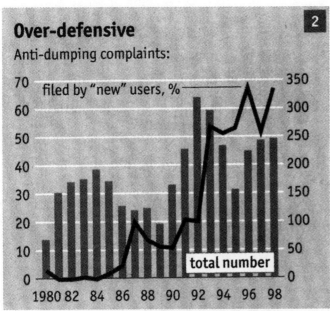

Over-defensive

Anti-dumping complaints:

filed by "new" users, %

total number

1980 82 84 86 88 90 92 94 96 98

Source: "Worldwide Use of Anti-Dumping, 1980–1998", by Thomas Prusa and Susan Skeath. NBER Working Paper no. W8424, Aug. 2001

local-content specifications (for example, shoes had to have uppers, soles and laces made locally) to qualify for GSP benefits.

The rich countries are still going down this preferential route today. The United States has introduced the Africa Growth and Opportunity Act (AGOA), while the EU has brought in the "everything but arms" initiative, properly known as the EBE, to eliminate trade barriers for the 49 LDCs. Yet virtually every drawback of GSP applies to these schemes as well. If anything, they are worse. Under the AGOA, for example, preferences for African garments are tightly linked to reverse preferences for American fabrics.

Since preferences typically divert trade away from non-preferred countries, they tend to pitch poor nations against each other. They are also a wasting asset, since they are relative to an MFN tariff that will probably decline with further multilateral liberalisation. And since they are non-binding and can be readily withdrawn for political reasons, investors are not likely to be impressed by them.

Preferences sound attractive and generous, and the poor countries have accepted them as such. But this has been a mistake. There is no good substitute for the MFN reduction of trade barriers in the rich countries. It should go hand in hand with enhanced technical and financial assistance. By focusing this help preferentially on the poor nations, the poor should be able to exploit the trade opportunities that are opened up for them by non-preferential treatment. This is the only way ahead.

sure. Thus, when India was put on the Special 301 list in 1991 and the United States trade representative determined unilaterally that India's intellectual-property protection was "unreasonable", President George Bush senior suspended duty-free privileges under GSP for $60m in trade from India in April 1992.

Preferences were also often dropped for commodities when they began to be successfully exported, a fact documented in a forthcoming study by Caglar Ozden and Eric Reinhardt of Emory University. Rules of origin served to curb exports, too. Exported items had to satisfy stringent

Jagdish Bhagwati is University Professor at Columbia University and Andre Meyer Senior Fellow in International Economics at the Council on Foreign Relations. His most recently published book is "Free Trade Today" (Princeton).

Learning from Doha: a civil society perspective from the South

Walden Bello

The fate of the Fourth Ministerial of the World Trade Organization (WTO), held in Doha, Qatar, in November 2001, hung in the balance until literally the last minute. Shortly after the adoption of the ministerial declaration,[1] WTO director general Mike Moore thanked the delegates for "saving the WTO," while U.S. trade representative Robert Zoellick congratulated them for wiping away "the stain of Seattle."

The relief expressed by Moore and Zoellick reflected the severe crisis of legitimacy the WTO was experiencing prior to Doha. A deadlock between North and South could very well have spelled the end of the WTO as an effective engine of trade liberalization. This crisis, however, was part of the larger crisis of the paradigm of corporate and market-driven globalization, which became particularly acute in the period between the earlier summit in Seattle and the tragic events of September 11, 2001.

The Doha declaration may have curiously saved the WTO, but it has not surmounted its crisis of legitimacy. Indeed, the very methods by which the trading powers extracted a "consensus" at Doha may leave a legacy of bitterness and resentment that will be felt in the critical run-up to the Fifth Ministerial, to be held in Mexico in 2003.

The Key Outcomes

Doha put the WTO back on its feet after the disaster in Seattle. C. Fred Bergsten, a prominent partisan, once said that the WTO is like a bicycle: It collapses if it does not move forward. By agreeing on a declaration giving momentum to new negotiations for liberalization, the Doha meeting got the bicycle upright and moving again.

What resulted may not be a new round in the sense of immediate negotiations on a wide range of issues. But it was a major step toward further liberalization. The Doha declaration affirmed ongoing negotiations on certain existing agreements, such as agriculture and the General Agreement on Trade in Services (GATS), and opened negotiations to review other existing agreements, like the antidumping agreement. It also initiated negotiations for new agreements in such new areas as industrial tariffs. But perhaps most ominously, by putting them as the centerpiece of the declaration, Doha gave momentum to what may be the eventual launching of negotiations to bring new, nontrade areas within WTO jurisdiction. These are the so-called new or Singapore issues of investment, competition policy, government procurement, and trade facilitation.

Doha was a clear setback for developing countries, most of which had wanted to focus the ministerial and its aftermath on resolving outstanding issues of implementation from the Uruguay Round, of which there are at least 104 according to the Group of 77.[2] The declaration simply acknowledged these concerns in perfunctory fashion and outlined a vague process for their resolution. Even in key areas of implementation specified in the text, such as agriculture and textiles and garments, developing countries came out as losers. The European Union (EU) managed to water down the Cairns Group's demand that there be a quick phaseout of agricultural export subsidies, and the United States and other developed countries did not commit to an early removal of quotas on textile and garment imports of critical importance to the developing countries.

It is important to stress that the South lost out because much of the influential Northern press has been saying that Doha proved that developing countries can "win" in WTO negotiations. The Doha resolution of the trade-related intellectual property rights (TRIPs) and public health issue is often cited as proof of this "win." However, it is important not to exaggerate its significance.

That "the TRIPS Agreement does not and should not prevent Members from taking measures to protect public health" is a political statement.[3] There is nothing in the Doha compromise that commits members to change the text of TRIPs—in effect, a victory for Washington, which held that the current text is sufficiently flexible to accommodate the public health concerns of developing countries. Even The Economist, an untiring promoter of the idea that the WTO is good for the developing countries, admitted that "the declaration is political and not legally binding."[4] This is important since in the long run it is the legal text that counts, and that will serve as the basis on which the pharmaceuticals may sue developing countries when they deem the political climate more favorable.

While Doha was a defeat for the South, developing countries were much better prepared for it than for previous ministerials. The united front of developing countries against the new, nontrade issues did hold until the end, when it crumbled under the onslaught of terrific pressure from the Quad (the United States, the EU, Japan, and Canada) and other developed countries. India stood alone in active opposition, with the tacit backing—it must be noted—of Cuba, Jamaica, Zimbabwe, and the Dominican Republic. Nonetheless, the consciousness of common interests and the need for a common strategy vis-a-vis the dominant trading powers is something that is greater now. It provides a good basis for common action in the hard months ahead.

Context and Tactics

Doha took place amid conditions that were already unfavorable from the point of view of developing country interests. September 11 provided an unanticipated opportunity for U.S. trade representative Robert Zoellick and EU trade commissioner Pascal Lamy to step up the pressure on developing countries to agree to launch a new trade round. They invoked the rationale that it was necessary to counter a global downturn that had been worsened by the terrorist actions. The location was also unfavorable, Qatar being a monarchy where dissent could be easily controlled. The WTO's secretariat had the authority to grant entry visas to Qatar to whomever it pleased, thus allowing it to radically limit the number of legitimate nongovernmental organizations (NGOs). It therefore prevented the kind of explosive interactions of resentment and massive street protest that had taken place in Seattle and Genoa.

Still, these factors would not have been sufficient to bring about an unfavorable outcome. Tactics mattered, and the major trading powers succeeded in breaking the fragile unity of developing countries, largely because they were cloaked by a nontransparent process.

Doha represented a low point in the WTO's—and GATT's (General Agreement on Tariffs and Trade) before it—history of backroom intimidation, threats, bribery, and nontransparency. Pascal Lamy described the WTO decisionmaking process in Se-

attle as "medieval." The same could be said of Doha. There are no records of the actual decisionmaking process because the formal sessions of the ministerial—which is where decisionmaking is made in a democratic system—were reserved for speeches. The real decisions took place in informal groupings whose meeting places kept shifting and were not known to all. There are no records. Moreover, there is precious little accountability in the process, as the principals in any deals can deny questionable behavior.

Among the tactics employed by the United States, the EU, and other major trading powers in Doha were the following:

- Pushing a highly unbalanced draft declaration and presenting it to the ministerial as a "clean text" on which there allegedly was consensus, thus restricting the arena of substantive discussion and making it difficult for developing countries to register fundamental objections without seeming "obstructionist."
- Pitting developing country representatives from capitals against their counterparts from Geneva, with the latter being characterized as "recalcitrant" or "narrow," thus isolating the latter.
- Employing direct threats, as the United States did when it warned Haiti and the Dominican Republic to cease opposition to its position on government procurement or risk cancellation of their preferential trade arrangements.
- Buying off countries, as the European Union did when, in return for their agreeing to the final declaration, it assured countries in the Africa, Caribbean, and Pacific (ACP) Group that the WTO would respect the so-called ACP waiver that allows them to export their agricultural commodities to Europe at preferential terms relative to other developing countries. Pakistan, a stalwart among developing countries in Geneva, was notably quiet at Doha. Apparently, this had something to do with the United States granting Pakistan a massive aid package of grants, loans, and debt reduction owing to its special status in Washington's war against terrorism. Nigeria, which was coordinating the African Group in Doha, had taken the step of issuing an official communique denouncing the draft declaration before Doha but came out loudly supporting it on 14 November—a flip-flop that is difficult to separate from Washington's promise of a big economic and military aid package in the interim.
- Reinstituting the infamous "Green Room" on 13 and 14 November, when some twenty handpicked countries were isolated from the rest and "delegated" by the WTO secretariat, the United States, the EU, and the other big trading powers, to come up with the final declaration. These countries were selected undemocratically, and efforts by some developing country representatives to insert themselves into this select group were rebuffed, some gently, others quite explicitly, as was the case with a delegate from Uganda.
- Finally, pressuring developing countries by telling them that they would bear the onus for causing the collapse of another ministerial, the collapse of the WTO, and the deepening of the global recession that would allegedly be the consequence of these two events.

These negotiating tactics wore down the South's resistance, split their initial fragile unity, and pushed them to confer what has been described as a "negative consensus" to the final declaration.

Counterstrategy

The bicycle is upright but wobbly, partly because of lingering resentment among developing countries about the lack of transparency in the process of being bamboozled into accepting a declaration that contradicts their interests. Nonetheless, unlike the aftermath of Seattle, developing countries and international civil society are now at a strategic disadvantage. How can they reverse the momentum arising from Doha toward further trade liberalization?

A potentially effective weapon for the South lies in a statement made by Youssef Hussain Kamal, conference chairman and Qatar's trade minister, about negotiations concerning new, nontrade issues, which will be the main area of contention. Appended to the final declaration, which was issued to end India's opposition to the document, the statement reads:

Let me say that with respect to the reference to an "explicit consensus" being needed, in these paragraphs, for a decision to be taken at the Fifth Session of the Ministerial Conference, my understanding is that, at that session, a decision would indeed need to be taken by explicit consensus, before negotiations on trade and investment and trade and competition policy, transparency in government procurement, and trade facilitation could proceed.

In my view, this would also give each member the right to take a position on the modalities that would prevent negotiations from proceeding after the Fifth Session of the Ministerial Conference until that member is prepared to join in an explicit consensus.[5]

In the future, much weight will be attached to the chairman's statement and whose interpretation of it will prevail. The position of the EU, for instance, is that Doha launched negotiations in investment and other new issues. Given this not-too-subtle offensive to define the process, it is important for developing countries and civil society to loudly insist at every possible forum that negotiations cannot start until after the Fifth Ministerial; that negotiations can take place only if there is an explicit consensus drawn from each and every country to undertake them; that the absence of an explicit consensus to start negotiations from even one member country will be sufficient to keep negotiations from proceeding; and that the chairman's statement carries the same legal weight as the Doha declaration.

The Missing Element

Governments are not the only participants in WTO proceedings. Although civil society organizations are not recognized actors in the WTO's formal process, their role since Seattle has been visible and significant. Good empirical research and analysis countering the alleged benefits of trade liberalization and globalization have been one form of effective civil society pressure. Massive street protests at the site of strategic meetings of the WTO and other key international organizations or associations have been another. Highly professional efforts in Geneva to assist developing countries in upgrading their capacity to understand and counter developed country pressures is still another.

Doha, however, revealed a missing element: consistent domestic pressure from social movements on governments. Doha underlined that there is simply no substitute for movements on the ground. In country after country, government's feet have to be put and kept to the fire. A strong, concerted civil society movement against further global liberalization, for example, accounted for the Indian government's refusal to sign a ministerial declaration that was unqualified by the chairman's statement.

Civil society certainly has its work cut out. Only fifteen to eighteen months remain for the Fifth Ministerial to get its act together. The agenda is to formulate and set into motion a strategy that innovatively brings together continuing pressure against the trade superpowers, street protests, intense delegation-by-delegation lobbying in Geneva, and massive domestic grassroots pressure on Southern governments to hold the line in negotiations.

Notes

1. DOHA 4th WTO Ministerial 2001: Ministerial Declaration, Fourth Session, Nov. 9-14, 2001 can be found online at http://www.wto.org/english/thewto.e/minist_e/min01_e/mindecl_e.htm

2. The Final Acts of the 1986-1994 Uruguay Round of trade negotiations can be found online at http://www.wto.org/english/docs_e/legal_e/final_e.htm.

3. DOHA 4th WTO Ministerial 2001: Declaration on the TRIPS Agreement and Public Health, Fourth Session, Nov. 9-44 2001. The full text can be found online at http://www.wto.org/english/thewto_e/min01_e/mindecl_trips_e.htm.

4. "Seeds Sown for Future Growth," The Economist, 17-23 November 2001, P. 77.

5. Chairman's Statement, Nov 14, 2001. Text provided by WTO Secretarial at Conference Media Center.

Walden Bello is professor of sociology and public information at the University of the Philippines and executive director of the Bangkok-based Focus on the Global South, a program of the Chulalongkorn University Social Research Institute.

From *Global Governance: A Review of Multilateralism and International Organizations*, Vol. 8, No. 3, July-September 2002, pp. 273-279. © 2002 by Lynne Reinner Publishers. Used with permission.

COMMENT

DO AS WE SAY, NOT AS WE DO

*Globalization might actually be good for poor countries,
if only rich countries played by the rules*

JACK BEATTY

America's leaders have to think beyond guns if the war on terror is to succeed in freeing us from the fear of terrorism. That Osama bin Laden was born into his millions and that the mass murderers of September 11 appear to have come from comfortable circumstances should not blind us to the danger posed by Third World poverty. Writing in these pages some years ago, Robert D. Kaplan said that we in the West were like the occupants of a limousine being driven through a slum, with the desperate masses, their patience thinned by global communications, peering through the windows. In order to strip the aura of Armageddon from what Paul Kennedy and Matthew Connelly, also in these pages, described as a contest of "the West against the Rest," the Rest must share in the prosperity of the West. As long as the only way to help them was by sending tax dollars abroad in the form of development aid, the Rest had little hope. Taxpayers would never support the amount necessary to make a difference. They barely support the current level of U.S. aid, which amounts to one tenth of one percent of GNP. Now there is another way: trade plus aid. But globalization is no painless elixir. Unlike the cost of development aid, which is borne by all taxpayers, the costs of trade will be paid by a relative few Americans. Textile workers, sugar-production workers, and others employed in labor-intensive industries will lose their

jobs so that the Rest can narrow the life-chances gap with the West.

That, at any rate, is the unacknowledged message of two new studies of the effects of globalization on poor countries. "Eight Broken Promises: Why the WTO Isn't Working for the World's Poor," produced by the British development charity Oxfam, and "Global Economic Prospects and the Developing Countries 2002," produced by the World Bank, detail the "blatant hypocrisy and double standards that govern the behaviour of rich countries toward poor countries" (to quote from the Oxfam study). Basically, the West has required the Rest to open their markets without reciprocating commensurately. Developing countries lose about $100 billion a year owing to Western export subsidies and trade barriers. For agriculture alone these amounted to $245 billion in 2000—about five times what the West spent on development assistance that year. Tariff barriers against manufactured goods from the developing world are, on average, four times as high as those against products from the industrialized world. What the World Bank calls "deeply poor" nations, where people live on less than a dollar a day, face tariffs more than twice as high as do "non-poor" nations, where people live on more than two dollars a day. Leather, footwear, textiles, and other products coming into the United States from the forty-nine least-developed countries—where some 300 million people live

below a UN-defined poverty line—face tariffs so high that they cost those countries more than the countries get from the United States in aid. Trade is far more important than aid for such countries. Every 0.7 percent increase in their exports generates as much income as they now receive in aid. "International trade has the potential to act as a powerful source of economic growth and poverty reduction," the Oxfam report states. "Yet that potential is being lost." Overall, in recent years there has been a 10 percent drop in demand for imports from the developing world.

In the Uruguay Round of tariff reductions, concluded in 1994, the West pledged to reduce agricultural subsidies by 36 percent; in return, the developing countries would lower their tariffs on agricultural imports. The developing countries met their part of the bargain by halving their average tariffs. The developed countries reneged: subsidies have in recent years made up almost 40 percent of the value of Western farm output—about the same as when Uruguay started. Of the $90 billion spent on crop supports over the past five years, some $60 billion went to the top 10 percent of recipients—Fortune 500 companies, city-dwelling farm owners, and big agribusiness. The average subsidy to the bottom 80 percent, meanwhile, was $5,830. These subsidies price the developing countries' foodstuffs not only out of possible export markets but also out of their home markets.

On the import side, Western governments levy tariffs on developing countries' agricultural products that are five times as high as those on all other goods. In the European Union, for example, meat and dairy imports face a tariff of more than 100 percent.

If farm exports are about the only escape from rural poverty in the Third World, textile and other labor-intensive exports are the only escape from urban poverty. Yet here, too, the West has failed to live up to its promises. At Uruguay the industrialized countries agreed to phase out the protectionist Multi-Fibre Agreement by 2005, yet 80 percent of textile and clothing exports from the developing countries will still face restrictions in 2004. The average tariff on these goods is 11 percent—three times as high as the average industrial tariff. South Asia loses approximately $2 billion a year to these tariffs. The United States promised General Pervez Musharaff, the President of Pakistan, that it would accept more textile imports as one of the conditions for Pakistan's support in the war on terror, but Senator Ernest Hollings, of South Carolina, was not consulted, and the textile lobby must have whole congressional delegations in its pocket.

If the United States lives up to the promises it made to the developing world, U.S. textile workers will eventually have to find new jobs. Their sacrifice may be a matter of national security, to preserve the safety of the rest of us in Robert Kaplan's limo. Although Dick Armey, the House majority leader, says that helping workers displaced by terrorism is not "commensurate with the American spirit," the nation should prove him wrong in the treatment it accords to workers displaced by the economic war on terrorism—that is, by the urgent necessity to import more goods from the Third World.

Spreading the Wealth

David Dollar and Aart Kraay

A RISING TIDE

ONE OF THE MAIN CLAIMS of the antiglobalization movement is that globalization is widening the gap between the haves and the have-nots. It benefits the rich and does little for the poor, perhaps even making their lot harder. As union leader Jay Mazur put it in these pages, "globalization has dramatically increased inequality between and within nations" ("Labor's New Internationalism," January/February 2000). The problem with this new conventional wisdom is that the best evidence available shows the exact opposite to be true. So far, the current wave of globalization, which started around 1980, has actually promoted economic equality and reduced poverty.

Global economic integration has complex effects on income, culture, society, and the environment. But in the debate over globalization's merits, its impact on poverty is particularly important. If international trade and investment primarily benefit the rich, many people will feel that restricting trade to protect jobs, culture, or the environment is worth the costs. But if restricting trade imposes further hardship on poor people in the developing world, many of the same people will think otherwise.

Three facts bear on this question. First, a long-term global trend toward greater inequality prevailed for at least 200 years; it peaked around 1975. But since then, it has stabilized and possibly even reversed. The chief reason for the change has been the accelerated growth of two large and initially poor countries: China and India.

Second, a strong correlation links increased participation in international trade and investment on the one hand and faster growth on the other. The developing world can be divided into a "globalizing" group of countries that have seen rapid increases in trade and foreign investment over the last two decades—well above the rates for rich countries—and a "nonglobalizing" group that trades even less of its income today than it did 20 years ago. The aggregate annual per capita growth rate of the globalizing group accelerated steadily from one percent in the 1960s to five percent in the 1990s. During that latter decade, in contrast, rich countries grew at two percent and nonglobalizers at only one percent. Economists are cautious about drawing conclusions concerning causality, but they largely agree that openness to foreign trade and investment (along with complementary reforms) explains the faster growth of the globalizers.

Third, and contrary to popular perception, globalization has not resulted in higher inequality within economies. Inequality has indeed gone up in some countries (such as China) and down in others (such as the Philippines). But those changes are not systematically linked to globalization measures such as trade and investment flows, tariff rates, and the presence of capital controls. Instead, shifts in inequality stem more from domestic education, taxes, and social policies. In general, higher growth rates in globalizing developing countries have translated into higher incomes for the poor. Even with its increased inequality, for example, China has seen the most spectacular reduction of poverty in world history—which was supported by opening its economy to foreign trade and investment.

Although globalization can be a powerful force for poverty reduction, its beneficial results are not inevitable. If policymakers hope to tap the full potential of economic integration and sustain its benefits, they must address three critical challenges. A growing protectionist movement in rich countries that aims to limit integration with poor ones must be stopped in its tracks. Developing countries need to acquire the kinds of insti-

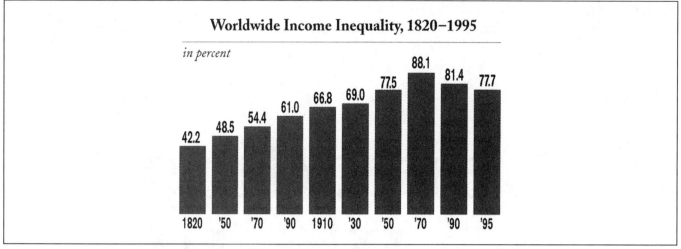

Worldwide Income Inequality, 1820–1995

in percent

1820	42.2
'50	48.5
'70	54.4
'90	61.0
1910	66.8
'30	69.0
'50	77.5
'70	88.1
'90	81.4
'95	77.7

Note: Figure represent the mean log deviation between a typical individual income and the average per capita income.

Sources: F. Bourguignon and C. Morrisson, "Inequality Among World Citizens, 1820–1992," working paper 2001-25 (Paris: Department and Laboratory of Applied and Theoretical Economics, 2001); and David Dollar, "Globalization, Inequality, and Poverty Since 1980," World Bank background paper, available at http://www.worldbank.org/research/global.

tutions and policies that will allow them to prosper under globalization, both of which may be different from place to place. And more migration, both domestic and international, must be permitted when geography limits the potential for development.

THE GREAT DIVIDE

OVER the past 200 years, different local economies around the world have become more integrated while the growth rate of the global economy has accelerated dramatically. Although it is impossible to prove causal linkage between the two developments—since there are no other world economies to be tested against—evidence suggests the arrows run in both directions. As Adam Smith argued, a larger market permits a finer division of labor, which in turn facilitates innovation and learning by doing. Some of that innovation involves transportation and communications technologies that lower costs and increase integration. So it is easy to see how integration and innovation can be mutually supportive.

Different locations have become more integrated because of increased flows of goods, capital, and knowledge. From 1820 to 1914, international trade increased faster than the global economy. Trade rose from about 2 percent of world income in 1820 to 18 percent in 1914. The globalization of trade took a step backward during the protectionist period of the Great Depression and World War II, and by 1950 trade (in relation to income) was lower than it had been in 1914. But thanks to a series of multilateral trade liberalizations under the General Agreement on Tariffs and Trade (GATT), trade dramatically expanded among industrialized countries between 1960 and 1980. Most developing countries remained largely isolated from this trade because of their own inward-focused policies, but the success of such notable exceptions as Taiwan and South Korea eventually helped encourage other developing economies to open themselves up to foreign trade and investment.

International capital flows, measured as foreign ownership of assets relative to world income, also grew during the first wave of globalization and declined during the Great Depression and World War II; they did not return to 1914 levels until 1980. But since then, such flows have increased markedly and changed their nature as well. One hundred years ago, foreign capital typically financed public infrastructure projects (such as canals and railroads) or direct investment related to natural resources. Today, in contrast, the bulk of capital flows to developing countries is direct investments tied to manufacturing and services.

The change in the nature of capital flows is clearly related to concurrent advances in economic integration, such as cheaper and faster transportation and revolutionary changes in telecommunications. Since 1920, seagoing freight charges have declined by about two-thirds and air travel costs by 84 percent; the cost of a three-minute call from New York City to London has dropped by 99 percent. Today, production in widely differing locations can be integrated in ways that simply were not possible before.

Another aspect of integration has been the movement of people. Yet here the trend is reversed: there is much more international travel than in the past but much less permanent migration. Between 1870 and 1910, about ten percent of the world's population relocated permanently from one country to another; over the past 25 years, only one to two percent have done so.

As economic integration has progressed, the annual growth rate of the world economy has accelerated, from 1 percent in the mid-nineteenth century to 3.5 percent in 1960–2000. Sustained over many years, such a jump in growth makes a huge difference in real living standards. It now takes only two to three years, for example, for the world economy to produce the same amount of goods and services that it did during the entire nineteenth century. Such a comparison is arguably a serious understatement of the true difference, since most of what is consumed today—airline travel, cars, televisions, synthetic fibers, life-extending drugs—did not exist 200 years ago. For any of these goods or services, therefore, the growth rate of output since 1820 is infinite. Human productivity has increased almost unimaginably.

All this tremendous growth in wealth was distributed very unequally up to about 1975, but since then growing equality has taken hold. One good measure of inequality among individuals worldwide is the mean log deviation—a measure of the gap between the income of any randomly selected person and a general average. It takes into account the fact that income distributions everywhere are skewed in favor of the rich, so that the typical person is poorer than the group average; the more skewed the distribution, the larger the gap. Per capita income in the world today, for example, is around $5,000, whereas a randomly selected person would most likely be living on close to $1,000—80 percent less. That gap translates into a mean log deviation of 0.8.

Taking this approach, an estimate of the world distribution of income among individuals shows rising inequality between 1820 and 1975. In that period, the gap between the typical person and world per capita income increased from about 40 percent to about 80 percent. Since changes in income inequality within countries were small, the increase in inequality was driven mostly by differences in growth rates across countries. Areas that were already relatively rich in 1820 (notably, Europe and the United States) grew faster than poor areas (notably, China and India). Global inequality peaked sometime in the 1970s, but it then stabilized and even began to decline, largely because growth in China and India began to accelerate.

Another way of looking at global inequality is to examine what is happening to the extreme poor—those people living on less than $1 per day. Although the percentage of the world's population living in poverty has declined over time, the absolute number rose fairly steadily until 1980. During the Great Depression and World War II, the number of poor increased particularly sharply, and it declined somewhat immediately thereafter. The world economy grew strongly between 1960 and 1980, but the number of poor rose because growth did not occur in the places where the worst-off live. But since then, the most rapid growth has occurred in poor locations. Consequently the number of poor has declined by 200 million since 1980. Again, this trend is explained primarily by the rapid income growth in China and India, which together in 1980 accounted for about one-third of the world's population and more than 60 percent of the world's extreme poor.

UPWARD BOUND

THE SHIFT in the trend in global inequality coincides with the shift in the economic strategies of several large developing countries. Following World War II, most developing regions chose strategies that focused inward and discouraged integration with the global economy. But these approaches were not particularly successful, and throughout the 1960s and 1970s developing countries on the whole grew less rapidly than industrialized ones. The oil shocks and U.S. inflation of the 1970s

created severe problems for them, contributing to negative growth, high inflation, and debt crises over the next several years. Faced with these disappointing results, several developing countries began to alter their strategies starting in the 1980s.

For example, China had an extremely closed economy until the mid-1970s. Although Beijing's initial economic reform focused on agriculture, a key part of its approach since the 1980s has involved opening up foreign trade and investment, including a drop in its tariff rates by two-thirds and its nontariff barriers by even more. These reforms have led to unprecedented economic growth in the country's coastal provinces and more moderate growth in the interior. From 1978 to 1994 the Chinese economy grew annually by 9 percent, while exports grew by 14 percent and imports by 13 percent. Of course, China and other globalizing developing countries have pursued a wide range of reforms, not just economic openness. Beijing has strengthened property rights through land reform and moved from a planned economy toward a market-oriented one, and these measures have contributed to its integration as well as to its growth.

Other developing countries have also opened up as a part of broader reform programs. During the 1990s, India liberalized foreign trade and investment with good results; its annual per capita income growth now tops four percent. It too has pursued a broad agenda of reform and has moved away from a highly regulated, planned system. Meanwhile, Uganda and Vietnam are the best examples of very low-income countries that have increased their participation in trade and investment and prospered as a result. And in the western hemisphere, Mexico is noteworthy both for signing its free-trade agreement with the United States and Canada in 1993 and for its rapid growth since then, especially in the northern regions near the U.S. border.

In general, countries that have become more open have grown faster.

These cases illustrate how openness to foreign trade and investment, coupled with complementary reforms, typically leads to faster growth. India, China, Vietnam, Uganda, and Mexico are not isolated examples; in general, countries that have become more open have grown faster. The best way to illustrate this trend is to rank developing countries in order of their increases in trade relative to national income over the past 20 years. The top third of this list can be thought of as the "globalizing" camp, and the bottom two-thirds as the "nonglobalizing" camp. The globalizers have increased their trade relative to income by 104 percent over the past two decades, compared to 71 percent for rich countries. The nonglobalizers, meanwhile, actually trade less today than they did 20 years ago. The globalizers have also cut their import tariffs by 22 percentage points on average, compared to only 11 percentage points for the nonglobalizers.

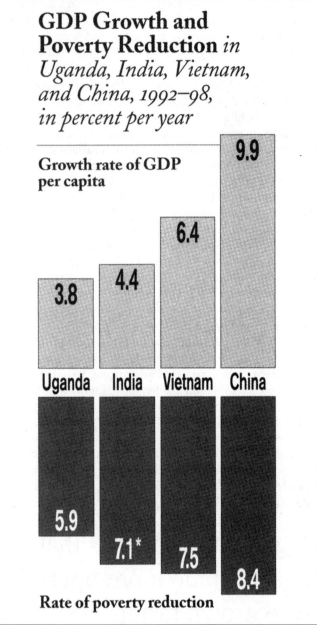

GDP Growth and Poverty Reduction *in Uganda, India, Vietnam, and China, 1992–98, in percent per year*

Growth rate of GDP per capita

Uganda	India	Vietnam	China
3.8	4.4	6.4	9.9

Rate of poverty reduction

Uganda	India	Vietnam	China
5.9	7.1*	7.5	8.4

*India poverty reduction figure is for 1993-99.

Source: David Dollar, "Globalization, Inequality, and Poverty Since 1980," World Bank background paper, available at http://www.worldbank.org/research/global.

How have the globalizers fared in terms of growth? Their average annual growth rates accelerated from 1 percent in the 1960s to 3 percent in the 1970s, 4 percent in the 1980s, and 5 percent in the 1990s. Rich countries' annual growth rates, by comparison, slowed to about 2 percent in the 1990s, and the nonglobalizers saw their growth rates decline from 3 percent in the 1970s to 1 percent in the 1980s and 1990s.

The same pattern can be observed on a local level. Within both China and India, the locations that are integrating with the global economy are growing much more rapidly than the disconnected regions. Indian states, for example, vary significantly in the quality of their investment climates as measured by gov-

ernment efficiency, corruption, and infrastructure. Those states with better investment climates have integrated themselves more closely with outside markets and have experienced more investment (domestic and foreign) than their less-integrated counterparts. Moreover, states that were initially poor and then created good investment climates had stronger poverty reduction in the 1990s than those not integrating with the global economy. Such internal comparisons are important because, by holding national trade and macroeconomic policies constant, they reveal how important it is to complement trade liberalization with institutional reform so that integration can actually occur.

The accelerated growth rates of globalizing countries such as China, India, and Vietnam are consistent with cross-country comparisons that find openness going hand in hand with faster growth. The most that these studies can establish is that more trade and investment is highly correlated with higher growth, so one needs to be careful about drawing conclusions about causality. Still, the overall evidence from individual cases and cross-country correlation is persuasive. As economists Peter Lindert and Jeffrey Williamson have written, "even though no one study can establish that openness to trade has unambiguously helped the representative Third World economy, the preponderance of evidence supports this conclusion." They go on to note that "there are no anti-global victories to report for the postwar Third World."

Contrary to the claims of the antiglobalization movement, therefore, greater openness to international trade and investment has in fact helped narrow the gap between rich and poor countries rather than widen it. During the 1990s, the economies of the globalizers, with a combined population of about 3 billion, grew more than twice as fast as the rich countries. The nonglobalizers, in contrast, grew only half as fast and nowadays lag further and further behind. Much of the discussion of global inequality assumes that there is growing divergence between the developing world and the rich world, but this is simply not true. The most important development in global inequality in recent decades is the growing divergence within the developing world, and it is directly related to whether countries take advantage of the economic benefits that globalization can offer.

THE PATH OUT OF POVERTY

THE ANTIGLOBALIZATION movement also claims that economic integration is worsening inequality within countries as well as between them. Until the mid-1980s, there was insufficient evidence to support strong conclusions on this important topic. But now more and more developing countries have begun to conduct household income and consumption surveys of reasonable quality. (In low-income countries, these surveys typically track what households actually consume because so much of their real income is self-produced and not part of the money economy.) Good surveys now exist for 137 countries, and many go back far enough to measure changes in inequality over time.

One way of looking at inequality within countries is to focus on what happens to the bottom 20 percent of households as globalization and growth proceed apace. Across all countries, incomes of the poor grow at around the same rate as GDP. Of course, there is a great deal of variation around that average relationship. In some countries, income distribution has shifted in favor of the poor; in others, against them. But these shifts cannot be explained by any globalization-related variable. So it simply cannot be said that inequality necessarily rises with more trade, more foreign investment, and lower tariffs. For many globalizers, the overall change in distribution was small, and in some cases (such as the Philippines and Malaysia) it was even in favor of the poor. What changes in inequality do reflect are country-specific policies on education, taxes, and social protection.

It is important not to misunderstand this finding. China is an important example of a country that has had a large increase in inequality in the past decade, when the income of the bottom 20 percent has risen much less rapidly than per capita income. This trend may be related to greater openness, although domestic liberalization is a more likely cause. China started out in the 1970s with a highly equal distribution of income, and part of its reform has deliberately aimed at increasing the returns on education, which financially reward the better schooled. But the Chinese case is not typical; inequality has not increased in most of the developing countries that have opened up to foreign trade and investment. Furthermore, income distribution in China may have become more unequal, but the income of the poor in China has still risen rapidly. In fact, the country's progress in reducing poverty has been one of the most dramatic successes in history.

Because increased trade usually accompanies more rapid growth and does not systematically change household-income distribution, it generally is associated with improved well-being of the poor. Vietnam nicely illustrates this finding. As the nation has opened up, it has experienced a large increase in per capita income and no significant change in inequality. Thus the income of the poor has risen dramatically, and the number of Vietnamese living in absolute poverty dropped sharply from 75 percent of the population in 1988 to 37 percent in 1998. Of the poorest 5 percent of households in 1992, 98 percent were better off six years later. And the improved well-being is not just a matter of income. Child labor has declined, and school enrollment has increased. It should be no surprise that the vast majority of poor households in Vietnam benefited immediately from a more liberalized trading system, since the country's opening has resulted in exports of rice (produced by most of the poor farmers) and labor-intensive products such as footwear. But the experience of China and Vietnam is not unique. India and Uganda also enjoyed rapid poverty reduction as they grew along with their integration into the global economy.

THE OPEN SOCIETIES

THESE FINDINGS have important implications for developing countries, for rich countries such as the United States, and for those who care about global poverty. All parties should recognize that the most recent wave of globalization has been a powerful force for equality and poverty reduction, and they should commit themselves to seeing that it continues despite the obstacles lying ahead.

It is not inevitable that globalization will proceed. In 1910, many believed globalization was unstoppable; they soon received a rude shock. History is not likely to repeat itself in the same way, but it is worth noting that antiglobalization sentiments are on the rise. A growing number of political leaders in the developing world realize that an open trading system is very much in their countries' interest. They would do well to heed Mexican President Vicente Fox, who said recently,

> We are convinced that globalization is good and it's good when you do your homework,... keep your fundamentals in line on the economy, build up high levels of education, respect the rule of law.... When you do your part, we are convinced that you get the benefit.

But today the narrow interests opposed to further integration—especially those in the rich countries—appear to be much more energetic than their opponents. In Québec City last spring and in Genoa last summer, a group of democratically elected leaders gathered to discuss how to pursue economic integration and improve the lives of their peoples. Antiglobalization demonstrators were quite effective in disrupting the meetings and drawing media attention to themselves. Leaders in developed and developing countries alike must make the proglobalization case more directly and effectively or risk having their opponents dominate the discussion and stall the process.

In addition, industrialized countries still raise protectionist measures against agricultural and labor-intensive products. Reducing those barriers would help developing countries significantly. The poorer areas of the world would benefit from further openings of their own markets as well, since 70 percent of the tariff barriers that developing countries face are from other developing countries.

If globalization proceeds, its potential to be an equalizing force will depend on whether poor countries manage to integrate themselves into the global economic system. True integration requires not just trade liberalization but wide-ranging institutional reform. Many of the nonglobalizing developing countries, such as Myanmar, Nigeria, Ukraine, and Pakistan, offer an unattractive investment climate. Even if they decide to open themselves up to trade, not much is likely to happen unless other reforms are also pursued. It is not easy to predict the reform paths of these countries; some of the relative successes in recent years, such as China, India, Uganda, and Vietnam, have come as quite a surprise. But as long as a location has weak institutions and policies, people living there are going to fall further behind the rest of the world.

Through their trade policies, rich countries can make it easier for those developing countries that do choose to open up and join the global trading club. But in recent years, the rich countries have been doing just the opposite. GATT was originally built around agreements concerning trade practices. Now, institutional harmonization, such as agreement on policies toward intellectual property rights, is a requirement for joining the WTO. Any sort of regulation of labor and environmental stan-

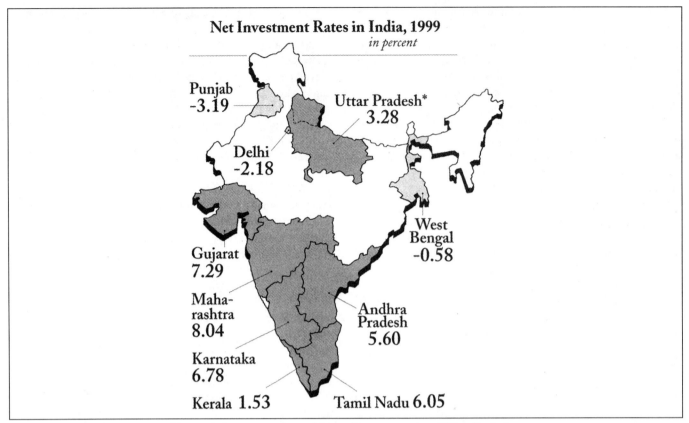

Net Investment Rates in India, 1999
in percent

Punjab
-3.19

Uttar Pradesh*
3.28

Delhi
-2.18

Gujarat
7.29

West
Bengal
-0.58

Maha-
rashtra
8.04

Andhra
Pradesh
5.60

Karnataka
6.78

Kerala 1.53

Tamil Nadu 6.05

*In September 2000, a new state, Uttaranchal, was created out of the northwestern section of Uttar Pradesh.

Note: Net investment rates represent the annual rate of growth of the capital stock of domestic and international firms. A negative rate implies that firms are pulling out.

Source: O. Goswami et al., "Competitiveness of Indian Manufacturing: Results of a Firm-Level Survey" (New Delhi: Confederation of Indian Industry, 2001).

dards made under the threat of WTO sanctions would take this requirement for harmonization much further. Such measures would be neoprotectionist in effect, because they would thwart the integration of developing countries into the world economy and discourage trade between poor countries and rich ones.

The WTO meeting in Doha was an important step forward on trade integration. More forcefully than in Seattle, leaders of industrial countries were willing to make the case for further integration and put on the table issues of central concern to developing nations: access to pharmaceutical patents, use of antidumping measures against developing countries, and agricultural subsidies. The new round of trade negotiations launched at Doha has the potential to reverse the current trend, which makes it more difficult for poor countries to integrate with the world economy.

A final potential obstacle to successful and equitable globalization relates to geography. There is no inherent reason why coastal China should be poor; the same goes for southern India, northern Mexico, and Vietnam. All of these locations are near important markets or trade routes but were long held back by misguided policies. Now, with appropriate reforms, they are starting to grow rapidly and take their natural place in the world. But the same cannot be said for Mali, Chad, or other countries or regions cursed with "poor geography"—i.e., distance from markets, in-

herently high transport costs, and challenging health and agricultural problems. It would be naive to think that trade and investment alone can alleviate poverty in all locations. In fact, for those locations with poor geography, trade liberalization is less important than developing proper health care systems or providing basic infrastructure—or letting people move elsewhere.

Migration from poor locations is the missing factor in the current wave of globalization that could make a large contribution to reducing poverty. Each year, 83 million people are added to the world's population, 82 million of them in the developing world. In Europe and Japan, moreover, the population is aging and the labor force is set to shrink. Migration of relatively unskilled workers from South to North would thus offer clear economic benefits to both. Most migration from South to North is economically motivated, and it raises the living standard of the migrant while benefiting the sending country in three ways. First, it reduces the South's labor force and thus raises wages for those who remain behind. Second, migrants send remittances of hard currency back home. Finally, migration bolsters transnational trade and investment networks. In the case of Mexico, for example, ten percent of its citizens live and work in the United States, taking pressure off its own labor market and raising wages there. India gets six times as much in remittances from its workers overseas as it gets in foreign aid.

Unlike trade, however, migration remains highly restricted and controversial. Some critics perceive a disruptive impact on society and culture and fear downward pressure on wages and rising unemployment in the richer countries. Yet anti-immigration lobbies ignore the fact that geographical economic disparities are so strong that illegal immigration is growing rapidly anyway, despite restrictive policies. In a perverse irony, some of the worst abuses of globalization occur because there is not enough of it in key economic areas such as labor flows. Human traffic, for example, has become a highly lucrative, unregulated business in which illegal migrants are easy prey for exploitation.

Realistically, none of the industrialized countries is going to adopt open migration. But they should reconsider their migration policies. Some, for example, have a strong bias in their immigration rules toward highly skilled workers, which in fact spurs a "brain drain" from the developing world. Such policies do little to stop the flow of unskilled workers and instead push many of these people into the illegal category. If rich countries would legally accept more unskilled workers, they could address their own looming labor shortages, improve living standards in developing countries, and reduce illegal human traffic and its abuses.

In sum, the integration of poor economies with richer ones over the past two decades has provided many opportunities for poor people to improve their lives. Examples of the beneficiaries of globalization can be found among Mexican migrants, Chinese factory workers, Vietnamese peasants, and Ugandan farmers. Many of the better-off in developing and rich countries alike also benefit. After all the rhetoric about globalization is stripped away, many of the policy questions come down to whether the rich world will make integrating with the world economy easy for those poor communities that want to do so. The world's poor have a large stake in how the rich countries answer.

DAVID DOLLAR and AART KRAAY are economists at the World Bank's Development Research Group. The views expressed here are their own.

Reprinted by permission of *Foreign Affairs*, January/February, 2002, pp. 120-133. © 2002 by the Council of Foreign Relations, Inc.

Surmounting the Challenges of Globalization

Eduardo Aninat

GLOBALIZATION—the process through which an increasingly free flow of ideas, people, goods, services, and capital leads to the integration of economies and societies—has brought rising prosperity to the countries that have participated. It has boosted incomes and helped raise living standards in many parts of the world, partly by making sophisticated technologies available to less advanced countries. Since 1960, for example, life expectancy in India has risen by more than 20 years, and illiteracy in Korea has gone from nearly 30 percent to almost zero. These improvements are due to a number of factors, but it is unlikely that they could have occurred without globalization. In addition, greater integration has promoted human freedom by spreading information and increasing choices.

But in recent years, concerns have grown about the negative aspects of globalization and especially about whether the world's poorest—the 1.2 billion people who still live on less than $1 a day—will share in its benefits. The beliefs that free trade favors only rich countries and that volatile capital markets hurt developing countries the most have led activists of many stripes to come together in an "antiglobalization" movement. The activists highlight the costs of rapid economic change, the loss of local control over economic policies and developments, the disappearance of old industries, and the related erosion of communities. They also criticize international organizations for moving too slowly in tackling these concerns.

The year 2001, however, saw the debate undergo a subtle but perhaps profound shift, with both sides seeming to step back from approaching it in terms of whether globalization was "good" or "bad"—an approach that seemed overly simplistic. This recognition gained momentum in the wake of the September 11 terrorist attacks in the United States, which exposed the vulnerability of globalization that stems in part—but only in part—from the sense of hopelessness present in some countries unwilling or unable to embrace it.

Both sides increasingly realized that the debate should center on how best to manage the process of globalization—at the national and international levels—so that the benefits are widely shared and the costs kept to a minimum. There is no question that greater integration into the world economy and more openness to other cultures offers all citizens of the global village a more hopeful future. Globalization, by offering a brighter future for all, provides perhaps the surest path to greater global security and world peace.

This understanding should attract support for the work needed to address the remaining challenges of globalization head-on. But there is an urgent need for a broad global debate on how these challenges can best be met and on who should play what role. This debate is under way through such initiatives as the United Nations Conference on Financing for Development being held March 18–22 in Monterrey, Mexico, and it will need to continue at many other venues. The IMF, along with the World Bank, has contributed significantly to the Monterrey conference by helping to focus the conference on global priorities, such as the Millennium Development Goals. The IMF, focusing on its mandate and areas of expertise, is also continuing to readapt itself to better help countries meet the challenges of globalization.

Globalization today

The world has experienced successive waves of what we now call globalization, going back as far as Marco Polo in the thirteenth century. These periods have all shared certain characteristics with our own: the expansion of trade, the diffusion of technology, extensive migration, and the cross-fertilization of diverse cultures—a mix that should give pause to those who perceive globalization narrowly, as a process nurtured strictly by economic forces.

By the end of the nineteenth century, the world was already highly globalized. Falling shipping costs had led to a rapid rise in trade, and in 1913 the ratio of world trade to world output reached a peak that would not be matched again until 1970. The growth of trade was accompanied by unprecedented flows of capital (as high as 10 percent of GDP, in net terms, in a number of both investor and recipient countries) and migration (for many countries, 1/2 of 1 percent of population a year), especially to the Americas.

Following the two world wars and the Great Depression, a new wave of globalization began, characterized by further declines in transport costs, which fell by half in real terms from 1940 to 1960; the expansion of modern multinational corporations, which are well suited to working around barriers to trade imposed by language, national commercial policies, and other factors; and unprecedented growth in output and living standards.

"Besides being a moral question, poverty reduction is now recognized as a necessity for peace and security."

More recently, globalization has been reinvigorated by the unprecedented ease with which information can be exchanged and processed thanks to breakthroughs in computer and telecommunications technologies, which since 1970 have reduced real computing and communications costs by 99 percent. This technological progress has steadily expanded the range and quality of services that can be traded, including those that support trade in goods, moving us toward a globally integrated economy.

Is this a development to be welcomed? Economic theory, as represented by the Heckscher-Ohlin-Samuelson model of trade, suggests that a fully integrated world economy provides the greatest scope for maximizing human welfare. This proposition is based on assumptions about the free international movement of goods and factors of production (capital and labor), the availability of information, and a high degree of competition. But benefits accrue even if capital and labor cannot move freely, so long as goods are freely traded.

In the real world, we know that there are still many barriers to the free movement of capital and labor. And, indeed, important barriers to trade remain. There has, however, been substantial progress in trade liberalization since the Second World War. The recently agreed Doha Development Round, for example, will be the tenth comprehensive trade round. Rising merchandise trade has been one of the hallmarks of the globalization process, and the gains from trade liberalization in recent decades have exceeded the costs by a very considerable margin. The Uruguay Round trade agreement reached in 1995 is estimated to have produced over $100 billion a year in net benefits, accruing mainly to those countries that have reduced trade barriers the most.

These trade gains have translated into faster economic growth and higher standards of living, as most clearly seen in East Asia: real incomes in Korea have doubled every 12 years since 1960. In the Spanish-speaking world, countries such as Spain, Mexico, and Chile have sharply boosted their shares of world trade and per capita incomes since 1980 by embracing globalization. A recent World Bank study also suggests that the countries that have opened themselves to trade in the last two decades have, on average, grown the fastest. These "new globalizers" among developing countries have reduced import tariffs, on average, by 34 percentage points since 1980, compared with only 11 percentage points for those developing countries that, on average, saw no growth in per capita incomes over the period.

Moreover, we know that faster growth goes hand in hand with bigger declines in poverty and larger increases in life expectancy. A recent World Bank study by David Dollar and Aart Kraay takes this full circle by deducing that since, in broad terms, trade is good for growth, and growth is generally good for the poor—they find that, on average, increased growth raises the incomes of the poor in proportion to those of the population as a whole—then trade is good for the poor.

Capital market integration has also advanced substantially in recent decades. But while the benefits of trade globalization are relatively clear, developing countries need to have a set of preconditions in place to benefit from financial globalization and not to succumb to an increased probability of a currency or banking crisis occurring. That is why capital account liberalization is being approached with much greater caution than during the bullish years of the mid-1990s. Capital inflows contribute to growth by stimulating investment and technical progress and promoting efficient financial development. Openness to capital flows, when combined with sound domestic policies, gives countries access to a much larger pool of capital with which to finance development. Foreign direct investment in particular—as opposed to potentially volatile portfolio flows—speeds up both capital accumulation and the absorption of foreign technologies and, like trade, has been shown to promote economic growth.

A new approach post–September 11

Clearly, globalization has the potential to make all people better off. The problem is that there is no assurance that all people will be better off or that all changes will be positive. The studies that show that, on average, poverty declines with economic growth are encouraging. But averages hide the negative impact on individual countries and on certain groups within them. In addition, there are important questions about the relationships between economic policies and outcomes, especially the impact of macroeconomic and structural reform policies on poverty. For example, when is growth especially beneficial to the poor? And when does growth not benefit the poor? How does trade generate growth? Does all foreign capital raise growth? How can we best ensure that capital flows do no harm?

These are all questions on which the IMF is seeking a better understanding, and as we gain further insights, we will, if necessary, adjust our policy recommendations accordingly. We are also committed to meeting four key challenges that fall in our areas of responsibility. The first is helping the poorest countries sustain the adjustment policies and structural reforms they need to reap the benefits of globalization. The second is increasing the stability of international financial markets—especially critical, given the importance of global financial stability as an international public good. The third is helping all of our members safely access these markets, including those countries that currently have no access. And the fourth is fostering a stable global macroeconomic environment. Only by addressing these challenges—in part through shared principles and rules—can we help our member countries accommodate the changes brought by globalization and cope with the dislocations such changes unavoidably bring.

But the atmosphere in which we are working has changed in some fundamental ways in the wake of the September 11 terrorist attacks—ways that provide an opportunity for a renewed dialogue. Even the antiglobalization movement that organized mass demonstrations in Seattle, Quebec, Genoa, and elsewhere has undergone profound change, as many of those who had been leading the protests against globalization—and against the IMF, the World Bank, and the World Trade Organization, in particular—are questioning whether such protests are an effective means to their ends. How have perceptions changed?

• It has become clear that the issues over which the debate has been conducted—issues central to the course of economic development—are governed by complex forces that defy glib generalization. It does not make sense to oppose globalization as such: the discussion must shift gears, aiming instead to identify effective ways to increase and spread the benefits of globalization while minimizing its costs.

• The importance of international cooperation has come into sharp focus across the broad spectrum of global issues. A byproduct seems to be a renewed appreciation of the role of the Bretton Woods institutions as forums for global economic cooperation and of the role of the IMF in particular.

• It has become even clearer that, in the words of IMF Managing Director Horst Köhler, "there will not be a good future for the rich if there is no prospect of a better future for the poor." Besides being a moral question, poverty reduction is now recognized as a necessity for peace and security. The decision to launch the Doha talks is the first evidence that this recognition will translate into greater attention to the requirements of economic development.

• The weakening of world economic growth, manifest in early 2001 but exacerbated by the September 11 attacks, has revealed the fragility of global economic prosperity. The need for the kind of high-quality analysis that the IMF provides, helping to keep the global economy on an even keel, has become more evident.

• Some of the protesters seem to have decided to channel their energies less against the international organizations themselves and more toward their member governments. They see that pressing national governments to increase foreign aid and market access for exports from poor countries can result in far greater benefits for the poor than changes in the policies of international lending institutions.

So how should all parties proceed? *First, besides finding solutions to problems, we need to find ways to implement them effectively.* This means keeping in mind that issues formerly seen as national—including financial markets, the environment, labor standards, and economic accountability—are now seen to have international aspects. The ripple effects of actions taken in one country tend to be far greater and to travel faster than ever before. A purely national approach to solving some problems risks merely pushing the problem across the frontier without providing a lasting solution even at the national level.

Second, we need to ensure that measures are taken to meet internationally agreed targets, such as the Millennium Development Goals, which include halving world poverty by 2015. Such measures would involve debt relief (especially for the heavily indebted poorest countries), social safety nets to cushion the short-term impact of economic reforms on the vulnerable, and higher social outlays, especially on health and education. In recent years, social outlays have been rising in countries with IMF-supported programs—significantly in countries benefiting from debt relief. Of course, this is only a modest beginning. For example, enormous additional resources are needed to improve health conditions in low-income and (for the poor) in middle-income countries, as pointed out in the World Health Organization's recent report of the Commission on Macroeconomics and Health.

Similarly, concerted action is needed to achieve the United Nations target that calls on rich countries to spend 0.7 percent of their GNP on development assistance. Action by the international community is also needed to open markets more broadly and effectively to exports from poor countries and to provide several of the poorest countries with lifesaving drugs at lower cost; the commitments made in Doha should serve as a minimum threshold for these goals.

> ## "Globalization holds the promise of enormous benefits for the peoples of the world. To make this promise a reality, however, we must find a way to carefully manage the process."

Third, we need to revisit the institutions of global governance, to establish mechanisms to implement global solutions to global problems and to ensure that governments become more accountable. On economic issues, the importance countries attach to the open and cooperative multilateral system is reflected in the now virtually universal membership of the IMF and the World Bank and the prospective accession of all major trading countries to the World Trade Organization. These three organizations address a very wide range of international economic issues, but they were not designed to be all-encompassing. Issues not central to any of their mandates are pressing and worthy of national and international attention. These include the environment, labor rights, international and local migration, and human rights, which must be addressed if globalization is to be sustained. As pointed out in the report by Michel Camdessus and others to the Bishops of the European Community, there are still many important institutional gaps in the present system.

Overall, this adds up to a weighty agenda for the international community, but perhaps never has so much been at stake, with so much potential within our reach. Globalization holds the promise of enormous benefits for the peoples of the world. To make this promise a reality, however, we must find a way to carefully manage the process. Better attention must be paid to reducing the negative effects and ensuring that the benefits are widely and fairly distributed. In this global village, we all need to work energetically toward that goal.

References

Eduardo Aninat, 2001, "Reflections on Globalization, Spain, and the IMF," speech given at the General Meeting of ELKARGI, San Sebastian, Spain, June 29.

Eduardo Aninat, Peter Heller, and Alfredo Cuevas, 2001, "Reflections on Globalization," Special Lecture for the XVIII Latin American Meetings of the Econometric Society.

Michel Camdessus and others, 2001, Global Governance, Bishops' Conferences of the European Community (COMECE).

David Dollar and Aart Kraay, 2001, "Trade, Growth, and Poverty," World Bank Policy Research Working Paper No. 199 *(Washington).*

Dolia Estévez, 2001, "Completar la globalización, clave para eliminar la pobreza: una entrevista con Sr. Eduardo Aninat," Mexico City: El Financiero, *December 10.*

Horst Köhler, 2001, "A Global Partnership for African Economic Development," address to the United Nations Economic and Social Council (ECOSOC), Geneva, July 16.

Michael Mussa, 2000, "Factors Driving Global Economic Integration," in Global Economic Integration: Opportunities and Challenges, *Federal Reserve Bank of Kansas City, pp. 9–55.*

World Bank, 2002, Globalization, Growth, and Poverty: Building an Inclusive World Economy, World Bank Policy Research Report (New York: Oxford University Press for the World Bank).

World Health Organization, 2001, Macroeconomics and Health: Investing in Health for Economic Development, Report of the Commission on Macroeconomics and Health (Geneva).

Ernesto Zedillo and others, 2001, Recommendations of the High-Level Panel on Financing for Development (New York: United Nations).

Eduardo Aninat is Deputy Managing Director of the IMF.

From *Finance & Development*, Vol. 39, No. 1, March 2002, pp. 4-7. © 2002 by International Monetary Funds. Reprinted by permission.

The Sacking of Argentina

THE IMF DESERVES TO BE BLAMED, BUT SO DOES THE COUNTRY'S WILLING POLITICAL CLASS.

TIM FRASCA

Buenos Aires

It is sobering to witness one of the greatest cities in the world slip, despite its deceptively placid surface, into a state of premodernity. Traffic lights in this metropolis of 12 million people still turn from red to green, newspapers in the kiosks report the latest bad news and Argentines occupy cafe tables, smoke a lot and shake their heads in disgust as they have for centuries. But money has ceased to exist. Oh, there are coins and bills, and cash still manages to facilitate exchange—except when the peso's value oscillates by 40 percent in three days, as it did in March. But money as the basis of a modern, capitalist economy, money that can be lent and borrowed, created or liquidated by central banks, money as the lubricating oil of investment and production—that has disappeared.

In normal times, María Esperanza Alvarez, 63, could be fairly considered a bit eccentric, if not a nut case. Standing outside the Spanish consulate where her niece is trying to get papers in order to abandon the country, she confides that for the past thirty years she has kept her savings in a box—thousands of dollars (which were always available) accumulated from her clothing business, which once employed twenty-three seamstresses. Having defied all common and expert sense for three decades, María Esperanza now deserves an Einstein award: She never for a moment believed the banks' basic pledge that they would give her back her money when they promised to do so. And she was right.

Lisandro Orlov, by contrast, was more trusting, perhaps in keeping with his professional outlook as a Lutheran pastor. Engaged for years in projects aimed at reintegrating social outcasts—street dwellers, drug users, people facing AIDS-related discrimination—the 59-year-old Orlov lost his pension fund in the December bank freeze-up. Like many Argentines, Orlov assumed that the austere entities in cavernous downtown palaces like HSBC, Citibank and BankBoston would honor their commitments. Now he's fighting in court to regain access to his dollars, already forcibly converted to 1.4 pesos each, or less than half the 3-to-1 rate the greenback now commands on the street.

In December of last year, Argentina's decade-long and highly celebrated experiment as the poster child of monetarist orthodoxy came to a crashing halt. While the International Monetary Fund is not the only responsible party, its spokespeople have now conveniently forgotten their laudatory worship of the main architect of the project, former President Carlos Saúl Menem, and the fund's gleeful funding of it throughout the 1990s. The IMF is now forcing Argentines to pay the price of its decadelong collusion with what it now says was a flawed performance all along.

So a once-wealthy country is suddenly seeing its sophisticated middle class driven into the streets, both to protest and to put food on the table. However, this extraordinary descent cannot simply be blamed on the opportunism of the free-trade globalizers. The sacking of Argentina could not have occurred had not a willing political class put out the FOR SALE signs long ago. Peronism, Argentina's peculiar form of nationalist populism from the 1940s and '50s, capitalized on the country's postwar largesse and lulled the populace into accepting political rot in exchange for fairly broad access to a share of the loot. After Menem came to power in 1989, he took a major detour from the Peronist vision. While preserving the rhetoric, the party structures and the patronage, he engineered huge privatizations, dismantled trade barriers and freed financial flows into and especially out of the country. The cornerstone of his monetarist policy, adopted at a time of hyperinflation, was the guarantee that a peso was a dollar was a peso, now and forever.

As a result of Menem's policies, what was once just old-fashioned corruption gave way to bargain-basement sales of the family silver. Now, with nothing left, citizens have awakened to the calamity. Although they may not be staging an active revolt for the moment, their unprecedented rejection of a whole generation of leaders has profoundly changed the political landscape.

Argentina's current plight is a lesson for those countries and their citizens that have toed the free-trade line and assume they will be rewarded accordingly. Now that things have fallen apart, the foreign beneficiaries of the fat years quickly wash their hands of responsibility and blame local elites, while turning their sights elsewhere for the next opportunities.

The Implosion

Last December the IMF, realizing that Argentina was a bottomless pit, turned off the cash spigots. Facing bankruptcy, President Fernando de la Rúa of the Radical Civic Union and his financial Rasputin, Economy Minister Domingo Cavallo, declared a *corralito* on bank deposits. This apt metaphor, suggesting cows liable to wander off and teams of neoliberal horsemen reining them in, meant, in practice, seizure. Confiscation. Argentines suddenly could withdraw only 1,000 pesos a month of their own money; otherwise, old-fashioned bank runs would have collapsed the system.

Implicit in the *corralito* was acknowledgment not only of the country's bankruptcy but also of the huge falsehood that had underpinned the entire economy for a decade: that the Argentine peso was worth one US dollar. Like the military dictatorship's assurances to the public in 1982 that the Falklands/Malvinas war would conclude with a glorious triumph, the political leadership simply couldn't give up its steady fix of convenient mendacity until it was far too late. The Argentine military still hasn't recovered from that debacle and the subsequent airing of its horrendous crimes during the local version of the war on terrorism, and in fact no one suggests the generals are in any way itching to get back into the political game, much less stage a coup.

But neither does the country's discredited political leadership, mostly Peronists and Radicals, seem to have much of a clue how to navigate the ship of state. "While the businessmen bankrupted the country," says Santiago Kovadloff, a former de la Rúa official, "the political leadership was directly complicit." Although not everyone would have subscribed to that radical indictment until recently, today Kovadloff's views are probably on the moderate side. The implications of this repudiation for Argentina's future are enormous. There is even a name for spontaneous outbursts of popular rage: the *escrache*, which in the local slang means, "in your face."

When the impact of the nationwide grand larceny first set in, the dispossessed victims surged into the streets. In December, police killed twenty-seven demonstrators before the teetering Radical government realized it had lost all legitimacy. De la Rúa had to be hauled from the presidential palace in a helicopter. His first replacement, a delusional provincial governor who immediately promised to create a million new jobs, lasted a week. Finally, on January 1, Eduardo Duhalde, a Peronist warhorse from Buenos Aires's suburban rust belt, took over. Duhalde, the loser in the 1999 presidential election, made immediate noises about breaking the deadly grip of the financiers and boosting the "productive" sector.

Virtually all progressive voices in Argentina today say a new strategy of this sort is urgently needed: pump-priming to generate internal demand, help for pulverized local industry and a break from the monetarist straitjacket. But even some of Duhalde's direct collaborators, who have since resigned, say it's all talk. Economist Héctor Valle, who worked for Duhalde for thirty-five days in the ministry of production, says his team's emergency recovery plan assumed that no foreign investors would touch Argentina with a ten-foot Brady bond and that the regime would have to cast its lot with local industry to reactivate the economy. "This meant confronting powerful interests and generating political support for more sacrifices," says Valle. "But Duhalde has no stomach for bold moves; he's a ribbon-cutter, and he's wasting crucial time."

Patacones and *Quebrachos*

What do governments do when they have no money? Under the theory on which convertibility of the dollar and peso was based, the governors of Argentina's twenty-three provinces should have slashed their payrolls and cut costs. But in some areas, over half the work force is state-employed, and salaries were often held up due to cash shortages even before the current disaster. With no jobs anywhere and no safety net, provincial bosses have been in no mood to commit political suicide for the IMF.

Instead, to avoid violent upheavals, the states have taken to issuing their own "bonds" or IOUs to pay their bills, including salaries. All over Buenos Aires, shop windows advise potential clients that they will accept one or more of the dozen quasi currencies that have sprung up to replace old-fashioned money: Chaco in the north issues *quebrachos*; Buenos Aires Province offers its employees *patacones*. At last count, there were fourteen of these funny moneys circulating. Desperate shopkeepers accept them, after knocking a percentage off their face value. The IMF wants them eliminated, as one of its many conditions for restarting the flow of emergency cash.

The retailers' desperation arises from the other predictable result of the anti-inflationary miracle: a devastating recession, now in its fourth year. It doesn't take an economics degree to see that the artificially expensive peso drove Argentine products off world markets and wrecked local industry through a flood of cheap imports. Argentina's economy has shrunk some 20 percent since 1998, and the free fall is just beginning.

Unemployment, now officially 22 percent, can only go higher. Just in the first two weeks of March, 20,000 businesses failed, each employing an average of ten people. Retail shops all over the capital are liquidating merchandise at a loss before closing their shutters for good. The only buyers in town are foreigners, including hordes of Chileans crossing the Andes to scoop up bargains. Poverty now affects 16 million Argentines, 43 percent of the population. Twenty percent are officially destitute. When a livestock truck overturned near the city of Rosario just before Easter, residents rushed to the highway and slaughtered the stunned cattle on the spot.

Another dramatic example of Argentina's accelerating creep toward prehistory are the barter clubs that have sprung up everywhere as a way around the fact that nobody can buy anything, while plenty of useful and needed products are in ample supply. While Spain sends charity food shipments—equivalent to sending donations of corn to Iowa—all over the capital long lines of people form with packages and shopping carts waiting patiently for one of the prized stalls in a church basement where the weekly session of frenzied commodity exchange is about to begin. In one club in upscale Palermo Viejo, at least 1,000 people clamored for a chance to crowd in and use their paper "credits"—another money substitute—to pick up food or used clothing or offer their skills as hairdressers, fumigators or aromatherapists. There are now upward of 4,000 of these clubs, with their deceptive air of a 1960s food co-op, generating the annual equivalent of $300 million in "commerce."

A recurrent slogan in the popular assemblies, pot-bangings and other demonstrations is "¡Que se vayan todos!"—the local equivalent of "Throw the bums out!" Some organizations actually promote dispensing with all leadership, a sign of the depth of revulsion for what has led them to this sorry pass. Others worry that the blanket rejection of "politicians" is a dangerously reactionary sentiment—readying Argentina for a Fujimori-type solution, a "nonpolitical" Bonaparte on a white horse.

There is certainly no shortage of motives for these sentiments. That the political class is corrupt to the bone, the civil service featherbedded beyond recognition and the union leadership complicit with every imaginable scam is no longer in question. Congressional deputies are so shameless that even during the current mass repugnance at their felonious antics, one Elsa Lofrano from Chubut Province could be appointed to a vacant Peronist seat, despite having received a pension for "physical and mental disability" for the past fifteen years. (The vote to seat her was 100 to 94.) But when a longtime Cassandra, congressional deputy Elisa Carrió from the center-left Movement for a Republic of Equals (ARI), and Otto Reich, George W. Bush's deputy for Latin America at the State Department, both put the blame for Argentina's problems on "corruption," they're not describing the same phenomenon.

'When the torturers go free, then anyone who steals or makes crooked deals can go free as well.' —Congressional deputy Elisa Carrió

Carrió, 43, was part of the alliance that brought de la Rúa into power in 1999. She broke with the ill-fated regime early on and is now one of the half-dozen recognizable politicians who can walk the streets without fear of an *escrache*. Carrió warned early on that corruption and incompetence were tearing the model apart. Wearing her signature mega-crucifix, Carrió says the ostentatious hand-washing by the IMF and Bush officials is just cynical amnesia. "Corruption in Argentina operates in complicity with foreigners. I had to go to Washington to denounce money-laundering because the US Embassy here was covering up for Citibank," she says. "All the biggest corruption scandals here involved American companies," she adds, noting that the laundering was enthusiastically carried out by top foreign

banks. If so, today's finger-pointers were happy to cash in while the corruption they denounce redounded to their allies' benefit.

Political Response

More than the current government, the Argentine state faces a vast crisis of legitimacy. Any plan with a chance of success will require even more pain and therefore patience from a severely battered populace—patience that can only be won by leaders they can believe in and trust. Those are so scarce that one economist suggested in dead seriousness importing a team of Finns to run the central bank.

The *escrache* is the most graphic symbol of this prostrate leadership. There are several varieties: Crowds may suddenly recognize a hated figure from the political or business elite at an airport and confront him with curses and even physical threats. An *escrache* can also be an old-fashioned demonstration, focusing on personal shaming rather than political demands. One took place on March 23, organized by HIJOS, or Children for Identity and Justice and Against Silence and Forgetting, composed of surprisingly youthful relatives of those disappeared and assassinated in the 1970s. One of their two targets was the former archbishop of Buenos Aires, known for his sympathies with the dictatorship.

Impunity is indeed the problem, says Carrio, but she worries that the *escrache* hints of private revenge. "We have to know what happened in the genocide of the 1970s, plus who robbed us blind and all the deals that were made." She blames the recent brigandage on the continuing impunity for human rights crimes. "Those who tortured were liberated! Those who disappeared people! That's where the system of truth and justice broke down. When the torturers go free, then anyone who steals or makes crooked deals can go free as well." Neither does it escape Argentines that the liberator of the generals and admirals convicted of some of the most vicious crimes in modern history was none other than Carlos Menem himself, who granted them an amnesty in 1990. The spin then was that Argentina needed "stability," because, as a Swiss banker elegantly told me at the time, "You can only stir the shit so far."

Given the total discredit of most political parties, people eager for action and participation have flocked to ad hoc neighborhood assemblies, now slowly institutionalizing themselves. These expressions of popular wrath have mushroomed, and their representatives appear with their increasingly worn banners at rallies and demonstrations of all sorts. It's not clear what sort of long-term impact the assemblies can have on broader policy decisions, and the weekly pot-banging sessions have already turned slightly routine. But their appearance is clearly breaking down the hegemony of the traditional party structures, as the experience of Gladys Quinteros illustrates.

Quinteros, a 40-year-old housewife in what was once the heavily Peronist suburb of Merlo, decided to join the demonstrations in December but found her own community "asleep."

After a few lonely vigils in the local plaza, she and some allies managed to drum up 300 people for a protest march against the situation. As they made their way through the streets, they were suddenly set upon by a well-organized band of thugs armed with homemade weapons. Twenty people were injured, and others were further abused in the local hospital where employees are beholden to the local bosses, clearly in no mood to tolerate spontaneous dissidence. "The idea was to intimidate us, and they succeeded," says Quinteros. "People sympathize, but now they are afraid to join in." Later, Quinteros's home was set afire and one room badly damaged.

Meanwhile, Peronism, if not its current representatives, still exercises a mystical lock on many Argentines. Portraits of Evita pop up unexpectedly: in union halls of the left-wing Argentine Workers Federation (CTA) or on the mantels of disgusted dissidents and lapsed Peronist Youth. Gabriel Guga, 31, was a Peronist captain in the La Matanza suburb until he became disaffected with Menem's neoliberal deviation. He recognizes that Peronism uses public coffers to buy political support and enrich the party elites, as well as its pronounced authoritarian streak, its vote-rigging, its links to the drug trade and organized crime. But Guga remains in his heart loyal to Peronist principles, "what our fathers and grandfathers stood for."

What Next

IMF teams were in Buenos Aires in March and April to dictate the terms for new loans. These include measures that worked so well for Herbert Hoover in the early 1930s, like more spending cuts, which will further reduce internal demand. In addition to seeking abolition of the provincial quasi currencies, the IMF wants civil service ranks reduced and bankers let off the hook. While the tiny US downturn last year was promptly treated with oceans of cash, the Argentines are supposed to swallow more neolib medicine. This is likely to occur, given the dearth of creative alternatives from Argentina's current leadership. With no popular support, they will probably turn to Washington to help prop themselves up, for now. The clear political quid pro quos involved in the next IMF loan, such as voting against Cuba at the UN or helping with Plan Colombia, are already done deals.

Some observers are convinced the United States wants to make an example of Argentina and undermine its MERCOSUR regional trade pact with Brazil, to clear the path for a US-dominated free-trade zone in the Americas. "We're the economic Guernica," says psychoanalyst Silvia Bleichman. "They have decided to punish us, to crush all resistance." But some of the IMF demands are in themselves reasonable. Any new government will have to deal with corruption, padded civil service rosters, the unsustainable provincial deficits. The question is who will take charge of the major surgery and to what ends.

The sheer scale of the rapine committed by Argentina's leaders over the past decades, sometimes in uniform, sometimes not, has left the nation ripe for a sharp break with the past. People who feel systematically tricked by their banks, their armies, their presidents and even their bishops will either despair or figure out how to put their trust in one another and construct something new. Carrió, one of the few who dare to predict anything, says radical change is now as unstoppable as a hurricane. "A whole class is disappearing, and after a brief period of anarchy, there will be new leaders, and not just political ones," she assured me.

Although not everyone shares her optimism and fighting spirit, Argentines are now alert to bullshit like never before. People show up at street-corner assemblies to patiently consider the activists' speeches or the comments of their neighbors, clearly inexperienced in such matters but of necessity eager to understand what's taking place. But for the most part, says Bleichman, Argentines simply refuse to look ahead—what has already happened is so unlikely, so implausible, that predictions or indeed logical faculties seem useless. Only one, usually unspoken, sentiment garners universal agreement in Buenos Aires today: The worst is yet to come.

Tim Frasca is a US journalist who has lived and worked in South America for twenty years.

The case for foreign aid.
Safe Deposit

By GREGG EASTERBROOK

PAUL O'NEILL AND BONO'S recent Africa trip—during which the odd couple wore fun clothes and inspected village wells—made headlines across the United States. But lost in the celebrity-induced excitement was a more significant development in the debate over foreign aid: George W. Bush and other prominent Republicans suddenly want to increase it. "I am here today," Bush told a United Nations conference last March, "to reaffirm the commitment of the United States to bring hope and opportunity to the world's poorest people and to call for a new compact for development defined by greater accountability for rich and poor nations alike." The president adroitly avoided the phrase "foreign aid," but everyone knew what he was talking about. Around the same time, Jesse Helms, long the Senate's determined roadblock to international assistance, announced that he too thought aid spending should rise. Bush pledged the United States to its first foreign aid increase in many a moon—a small one but significant for a nation where political discourse generally, and Republican discourse specifically, has been anti-foreign-aid for decades.

Bush's speech and Helms's conversion suggest that the politics of foreign aid may be changing. But they won't change enough as long as most Americans—in both parties—think foreign aid doesn't work. One reason many Americans think so is that they expect foreign aid to banish developing-world poverty and to build liberal democracies across the globe; by that standard, aid indeed has failed. But that's the wrong standard. The realistic benchmark is whether international assistance has made the world better than it would otherwise have been. And by that standard, foreign aid has not only been a success; it has been a triumph. In most developing nations, living standards are rising, and health care and education are improving, in part because of foreign aid. Billions of people are better off today thanks to Western help, however inconsistent and snafu-prone that help has often been. "On a world scale, the risk, intensity and severity of poverty has fallen more sharply in the past fifty years than in the preceding thousand," in part owing to aid from the West, Michael Lipton, a professor at the University of Sussex, has written. If the Bush administration spearheaded an ambitious new commitment to more accountable foreign aid, the cost to the United States would be small, but the benefits to humanity could be remarkable.

Some years ago I had a free day in Dar Es Salaam, Tanzania. I decided to walk around—walking being safer than boarding the perpetually overcrowded buses, off of which passengers often fell. Dotting the city were large structures—office buildings, high-rises—that appeared to be under construction but on closer inspection had been abandoned in the midst of assembly, their exposed beams rusting. Western money funded the building starts, but the aid ran out, or the loans were embezzled; and there the structures sat, worse than useless considering that the resources could have been spent on something simpler, like modest housing. Beat-up old cars, many 1950s-vintage models imported from the United States, bounced along cratered roads that lacked even the simplest maintenance. According to a 1998 World Bank study, Tanzania has received $2 billion from the West in road-maintenance aid since the late '70s, and during that period the condition of its roads has declined. Outside the city stood the hulk of a steel mill, built mainly with Western aid. It was modern, impressive-looking, and it didn't run.

TANZANIA IS NOT alone. World Bank officer William Easterly calculates in his 2001 book *The Elusive Quest for Growth* that based on the billions it has received in aid, the Zambian economy could by now have grown to a Western-like $20,000 per capita income. Instead Zambia's per capita income is $600, down in real terms since the '50s. Corruption explains some of this. Jeffrey Winters of Northwestern University estimates that of the $8 billion in World Bank funds channeled to Indonesia since about 1960, one-third has been stolen. In Congo (then

Zaire) the late dictator Mobutu Sese Seko siphoned off enough foreign aid to enjoy Nero-level opulence amid the desperation of his people. And though the World Bank and most Western donor nations have worked diligently in the last decade to reduce aid theft (Nobel Prize-winning economist Joseph Stiglitz has said one of the best arguments for international aid is that the corruption problem is finally being resolved), misappropriation is ongoing. A recent bribery scandal, for instance, marred the World Bank-funded Lesotho Highlands Water Project, a water-supply effort in South Africa. Estimates vary, but it is widely believed corruption has claimed at least 10 percent of global aid.

Combine that with the endless depressing statistics about life in the impoverished world, and you can understand why so many Americans consider foreign aid a failure. Some 1.2 billion people now live on $1 or less per day, and every year that quotient of misery increases by another million. Since World War II the West has allocated roughly $1 trillion to foreign aid, yet inequality is "accelerating everywhere on Earth," proclaims the leftist International Forum on Globalization. The gap between the developing and developed countries has grown so wide that the world's three richest individuals are now worth more than the 48 poorest nations.

But these depressing statistics obscure more than they reveal. In fact, much has improved in the developing world during the era of foreign aid. In 1975 the average income in developing nations was $1,300 (in 1985 dollars) per capita; today it is $2,500. In 1974 one-third of the world's nations regularly held genuine multiparty elections; today two-thirds do. In 1975 1.6 billion people lived at what the United Nations classifies as "medium development," meaning with reasonably decent living standards, education, and health care; today 3.5 billion people do— a stunning increase in the sheer number of human beings who are *not* destitute. Global adult literacy was 47 percent in 1970 and is 73 percent today, while school enrollment for girls has skyrocketed. Infant mortality has declined and life expectancy has risen in almost every developing nation. The world's population is growing not because of more births—fertility is down almost everywhere—but because of fewer deaths. Even in developing nations, it is becoming the norm for a person to survive to old age. At the beginning of the twentieth century, average life expectancy in the United States was 47 years; today it is 66 years—*for the world*.

Yes, each year brings more people living on $1 per day, but this is mainly because each year brings more people; the *percentage* of the global population living in such hardship is in steady decline. An estimated 800 million people in the world today are malnourished, but the *percentage* of people who are malnourished is dropping, and the world's average daily caloric intake continues to rise—so much so that for the first time in history, there are roughly as many people on Earth who are overweight as are underweight. A few decades ago the Stanford University population theorist Paul Ehrlich was predicting global famine, projecting that by now hundreds of millions would have died of hunger while Middle Ages-style plagues swept the globe. Respectable opinion echoed this view with a presidential report commissioned by Jimmy Carter, *Global*

2000, forecasting in 1980 that by the year 2000 vast numbers would starve or die of uncontrollable plagues while entire ecosystems collapsed. Not even the awfulness of the African AIDs epidemic comes close to what experts were expecting. Instead, for all the problems poor countries continue to experience— especially in Africa—for most people in most developing nations, life is mostly getting better.

In some ways this is the great story of the developing world over the last half-century: Things have not fallen apart. And one of the reasons is international assistance. Sure, many Third World countries have not enjoyed sustained economic growth. But asking whether foreign aid has brought growth is the wrong question, since Western governments often do poorly in picking and choosing winners in their own economies, too. The morally realistic standard is whether foreign aid brings humanitarian gains. And by that standard, it has been a success.

CONSIDER SOME OF foreign aid's accomplishments. First are the health care initiatives. Smallpox has been eradicated globally via a U.N.-run, Western-financed campaign that completed its work in 1977, the heyday of aid money. River blindness, a scourge of Africa, which once infected hundreds of thousands of people, has been nearly eradicated by Western-backed health initiatives. Fully 80 percent of the world's children are today vaccinated against polio, diphtheria, and other diseases, compared with almost no vaccinations in the developing world when foreign aid began roughly a generation ago. Most of those immunizations were paid for by the World Health Organization (WHO) or by the contractor agencies it dispatches to small villages across the globe; WHO draws most of its money from donor countries in the West. It took about 150 years, from 1800 to 1950, for typical European life expectancy to advance from four decades to six. In most of the developing world, it has taken just 40 years, from 1930 to 1970. Helping most of the world achieve, in 40 years, the life-span improvement that took Europe 150 years is a spectacular achievement—partly attributable to the advent of antibiotics and partly to international aid.

Next, aid has brought birth control and reproductive-health care to the world's impoverished. The International Planned Parenthood Federation, funded mainly by Western aid, has established family-planning clinics throughout the developing world. Its work, and that of other groups with Western support (including about $300 million annually from the United States), is a primary reason the global population runaway, predicted as recently as the '70s, has not come to pass. In the '50s developing-world women averaged six live births; now the figure is three and still declining.

Or consider agriculture. In the mid-'60s India and Pakistan experienced wrenching national famines; Ehrlich predicted that hundreds of millions in the subcontinent would starve to death. A doomsday book called *Famine 1975!*, by Paul and William Paddock, forecasting imminent global starvation, was a bestseller in the late '60s. International aid organizations—some government, some private (mainly the Rockefeller and Ford Foundations)—launched an ambitious initiative to introduce high-yield crops into Pakistan and India, shipping trucks of high-yield seeds from a research institute in Mexico to the subcontinent on an emergency basis. In two years in the mid-'60s conversion to these high-yield crop strains stopped the famines that had been widespread in India and Pakistan for several years. Rather than starving as predicted, Pakistan and India have used these high-yield crops to become self-sufficient in food production despite population growth. Pakistan, which harvested roughly 3.4 million tons of wheat annually in the early '60s, was harvesting 18 million tons by the late '90s; India went from 11 million tons to about 60 million.

Would the Green Revolution have come to developing nations regardless of aid? Perhaps eventually, but many millions would have died needlessly in the interim. China researched high-yield rice strains on its own and dramatically increased rice production without Western help. But the increase did not become significant until the '80s, almost two decades after productivity rose dramatically in the subcontinent, with huge numbers of Chinese suffering malnourishment as their nation went it alone in adopting Green Revolution farming. As Norman Borlaug, the Nobel Peace Prize-winning Iowa plant breeder who oversaw the effort in Pakistan and India, puts it: "What was accomplished in India and Pakistan could not have happened without aid support money."

Foreign aid's success in promoting macroeconomic—as opposed to social and agricultural—development is decidedly mixed. But for every Tanzania, Zambia, and Congo, there is a South Korea. Destitute after Japanese occupation and the Korean War, South Korea has lifted its per capita GDP to $16,000 since the '50s. Of course, Seoul's export-led economic model played a major role in that success. But developmental aid was also an important part— particularly in improving the country's infrastructure and educational system. In their book about foreign aid, *A Half Penny on the Federal Dollar*, Brookings Institution analysts Michael O'Hanlon and Carol Graham cite South Korea as a prime example of development aid translating directly into development.

And South Korea isn't alone. Ideologically charged debates often center around whether aid or trade conquers poverty. But in countries that have followed the "Tiger" model of export growth—Indonesia, Bolivia, Ghana, Vietnam, Chile, Botswana—foreign aid has often buttressed market-oriented development strategies. Indeed, foreign aid is one of the levers donors have used in recent years to push developing-world governments to liberalize their economies. Donors, for instance, have helped Vietnam start a thriving coffee-export trade; Chad is on the verge of tapping its petroleum reserves via an oil pipeline built with World Bank support.

Uganda, another African country that has opened itself to market economics, owes about half of its national budget to Western aid. Progress there has been remarkable: Poverty has declined by one-third in the last decade; spiraling inflation has been stopped; the percentage of children in school is sharply increasing; and Uganda also leads Africa in progress against AIDs. Americans spent about 29 cents each on Uganda last year and, for that tiny amount, made a palpable difference in people's lives.

It is true, of course, that not all foreign aid has been this successful. In the '60s and '70s, for instance, European donors pushed on developing countries a statist, socialism-lite that mirrored their own domestic economic organization. And the United States often funneled aid as a reward to corrupt and repressive cold war allies like Mobutu. But there is at least some reason to believe foreign aid is becoming more effective—that we will see more Ugandas and fewer Tanzanias. A group of World Bank analysts concluded in their 1998 study *Assessing Aid* that the main thread running through positive foreign aid projects was association with positive social trends. Give money to entrenched, anti-reform governments, and you shouldn't expect much; fund new or reformist regimes, and good things might happen. Worry less about exactly which project to pick—no one can ever be sure what will work—and more about whether the governments being supported are progressive and accountable. Easterly, in *Elusive Quest for Growth*, adds that it is essential that all aid projects contain incentives. On the local level, individuals in the project area must participate or otherwise get something tangible, however small, granting them an incentive to cooperate. At the national level, only reformist governments should be funded, giving leaders the right incentive. Those principles now form the basis of most International Monetary Fund (IMF) and World Bank aid—and they were the basis for Bush's March speech as well. Foreign aid, which for all its foibles has done tremendous good, may do even more good in the years to come.

But for all its merits, foreign aid lacks a stable, powerful American political constituency. Long scorned by the libertarian right, it is now increasingly scorned by the anti-globalization left as well.

Conservative opposition to foreign aid is nothing new. Some of it is based on stinginess; some on distaste for developing-world anti-Americanism; and some on the view, best elucidated by the late British economist Peter Bauer, that aid actively backfires by keeping corrupt, incompetent developing-world elites in power. But to view foreign aid with disdain, as many conservatives do, deprives the West of credit for one of its signal contributions to the world. It also dishonors America's spiritual heritage, as all the monotheistic faiths impose on the favored an obligation to help the less-well-off. After declaring that the United States should fund a crash program to stop African AIDS, the 81-year-old Jesse Helms gave the suggestion that he had dropped his long-standing opposition to increased foreign aid because he is aware that he will soon meet God. This is clear thinking.

Traditional conservative objections to foreign aid have been joined in recent years by left-wing complaints, via the anti-globalization movement—leaving foreign aid politically stranded. Anti-globalizers don't oppose assistance per se; they oppose aid that comes with reformist strings, such as IMF loans with "structural adjustment" clauses that require recipient governments to adopt free-market and free-trade policies. The leftish Institute for Policy Studies protests that IMF requirements mean the bank "holds virtual neocolonial control over developing nations." Yet though anti-globalizers don't want Western governments imposing terms on aid, *they* are happy to do so. One is that money be spent only on projects deemed "sustainable." In recent years anti-globalizers and various foundation-grant trendies have so barraged the World Bank and similar organizations with the buzz term "sustainability" that donors are increasingly leery about funding anything much beyond village solar projects or similar ideas that sound small-scale and low-input.

Very little in Western economies meets the left's shaky definition of "sustainability," which more or less translates as no fossil fuels, no built structures, and no packaging. Indeed, foisting this standard on the developing world is a formula for keeping living standards low. Even the environmental rationale for "sustainability" doesn't make much sense: No current resource is in short supply, not even petroleum, and no important resource—other than groundwater in China and freshwater in a few other regions—seems likely to fall into short supply in the near future, based on present information. In turn, some industries that could be sustained indefinitely—like logging—are amongst the ones anti-globalizers hate most.

The anti-globalization movement may be politically marginal, but its street protests have proved effective means of political pressure. The movement, for example, has successfully mau-maued the World Bank and other donor institutions into withdrawing essentially all support for the construction of hydro dams, such as the Narmada River complex that India is now struggling to complete without Western financing. To be sure, dam construction in the developing world often displaces the poor, usually on rough terms, and this is a serious problem; the West should not support developing-world dam construction without ensuring fair treatment of the displaced. But the dams themselves provide pollution-free energy, clean drinking water, and irrigation for agriculture. Not only do nearly all developing nations need more electricity generation to improve living standards—India currently has 6 percent as much electric power per capita as the United States—but to improve human health. The number of children in developing nations who die each year from respiratory diseases caused by *indoor* air pollution, mainly from indoor fires for cooking and heating, exceeds the number of Americans of all ages who die of all causes. Anti-globalizers oppose the dams, power plants, water reservoirs, and other big aid-backed projects that could change that, and their views have gained surprising influence among the Western donor institutions that now live in terror of PC criticism.

SIMILAR PRESSURE INHIBITS Western donors from exporting high-yield agricultural techniques to Africa, the one part of the world where subsistence farming remains ubiquitous. Environmentalists and anti-globalizers have, for example, fought Western proposals to send fertilizer to Africa, which struggles with some of the world's poorest soils. Antiglobalizers romanticize farming with animal-drawn plows as "appropriate," but most wouldn't last a day at the dehumanizing toil this form of existence imposes and rewards only with meager survival. The United States could do its next great service to the world by teaching Africa our granary-filling methods of farming. Instead, trendy complaints from well-fed anti-globalizers and enviros keep donor institutions from supporting an African Green Revolution. It is a strange and terrible irony that the left's influence over the World Bank and IMF, and the American right's influence in Congress, are conspiring to deny needed and worthwhile foreign aid at the very moment, as Stiglitz notes, that such aid is becoming more honest and effective.

As a result, the United States today gives shockingly little to the world's poor. In the years immediately after World War II, when the United States was funding the successful reconstruction of Western Europe under the Marshall Plan, 15 percent of the federal budget went to foreign aid. Today polls show that most Americans *think* 15 percent of the federal budget goes to foreign aid—and they consider that too much. According to polls, most Americans think foreign aid should consume between 1 percent and 5 percent of the federal budget.

If only it did. In fact, the United States devotes about half of 1 percent of federal spending to foreign aid. The roughly $11 billion for foreign aid in Bush's latest budget request is only about one-seventh as much, in inflation-adjusted dollars, as the United States spent on foreign aid during the Marshall Plan years, when the United States had many more domestic problems that required government money. The aid total in Bush's current fiscal year budget is lower in absolute dollars—not inflation-adjusted—than was spent on aid during most years of the Ronald Reagan administration. And the real figure is even smaller than that, since about $4 billion of the $11 billion goes to Israel and Egypt in what is officially called development aid but is for all intents and purposes an annual payment to ensure observation of the treaty that ended the 1973 Israel-Egypt war. By contrast, all of Africa—excluding Egypt—wracked by poor public health, illiteracy, and malnutrition, receives a mere $1.3 billion annually in U.S. development support, a small fraction of the annual amount Congress just added to U.S. agriculture subsidies.

All industrial nations have reduced aid over the last decade, as "donor fatigue" has set in and rising social-welfare obligations to retirees have increasingly dominated the balance sheets of Western governments. But America's reduction has been the most stark, with the country now spending just 0.1 percent of GDP on aid—less a share of national income than that spent by *Portugal*—versus an average of 0.24 percent for the OECD nations. True, other OECD nations—like Japan, usually the world leader in total international development assistance, and Denmark, the world leader in assistance as a share of GDP—do not have significant military obligations, mainly because the United States shields them. But while the United States must protect the globe, the United States is also the wealthiest nation on Earth. To trail Portugal in an important measure of moral responsibility to the world should keep even foreign aid skeptics awake at night.

Before the O'Neill-Bono road show, Bush pledged the United States to a roughly $1.5 billion per year increase in foreign aid. That is admirable but would still leave the country well behind what it was spending in the '50s, when the United States was far less affluent. The administration has proposed that the defense budget rise by $48 billion this year, and the increase is justified. But various estimates, including those from WHO and from Columbia University aid advocate Jeffrey Sachs, hold that increasing America's spending against international AIDs by around merely $2.5 billion per year could dramatically improve African public health, perhaps saving millions of lives. Two and a half billion dollars is about what it costs to build a new attack submarine. Can anyone seriously argue that the same amount to help stop a continent-wide epidemic wouldn't be money well spent?

In the aftermath of September 11 some have suggested the United States should increase foreign aid to improve its image abroad and thus make us safer. This would probably not work—after all, most of the 9/11 murderers came from Saudi Arabia, a wealthy nation that receives no aid; and in any case all were fanatics. But most of the men and women of the developing world mean no harm to anyone. They are struggling to survive and appeal to us for aid. Aid we must send them even though we know some will be stolen or wasted. We must do this because it is right.

The Cartel of Good Intentions

The world's richest governments have pledged to boost financial aid to the developing world. So why won't poor nations reap the benefits? Because in the way stands a bloated, unaccountable foreign aid bureaucracy out of touch with sound economics. The solution: Subject the foreign assistance business to the force of market competition.

By William Easterly

The mere mention of a "cartel" usually strikes fear in the hearts and wallets of consumers and regulators around the globe. Though the term normally evokes images of greedy oil producers or murderous drug lords, a new, more well-intentioned cartel has emerged on the global scene. Its members are the world's leading foreign aid organizations, which constitute a near monopoly relative to the powerless poor.

This state of affairs helps explain why the global foreign aid bureaucracy has run amok in recent years. Consider the steps that beleaguered government officials in low-income countries must take to receive foreign aid. Among other things, they must prepare a participatory Poverty Reduction Strategy Paper (PRSP)—a detailed plan for uplifting the destitute that the World Bank and International Monetary Fund (IMF) require before granting debt forgiveness and new loans. This document in turn must adhere to the World Bank's Comprehensive Development Framework, a 14-point checklist covering everything from lumber policy to labor practices. And the list goes on: Policymakers seeking aid dollars must also prepare a Financial Information Management System report, a Report on Observance of Standards and Codes, a Medium Term Expenditure Framework, and a Debt Sustainability Analysis for the Enhanced Heavily Indebted Poor Countries Initiative. Each document can run to hundreds of pages and consume months of preparation time. For example, Niger's recently completed PRSP is 187 pages long, took 15 months to prepare, and sets out spending for a 2002–05 poverty reduction plan with such detailed line items as $17,600 a year on "sensitizing population to traffic circulation."

Meanwhile, the U.N. International Conference on Financing for Development held in Monterrey, Mexico, in March 2002 produced a document—"the Monterrey Consensus"—that has a welcome emphasis on partnership between rich donor and poor recipient nations. But it's somewhat challenging for poor countries to carry out the 73 actions that the document recommends, including such ambitions as establishing democracy, equality between boys and girls, and peace on Earth.

Visitors to the World Bank Web site will find 31 major development topics listed there, each with multiple subtopics. For example, browsers can explore 13 subcategories under "Social Development," including indigenous peoples, resettlement, and culture in sustainable development. This last item in turn includes the music industry in Africa, the preservation of cultural artifacts, a seven-point framework for action, and—well, you get the idea.

It's not that aid bureaucrats are bad; in fact, many smart, hardworking, dedicated professionals toil away in the world's top aid agencies. But the perverse incentives they face explain the organizations' obtuse behavior. The international aid bureaucracy will never work properly under the conditions that make it operate like a cartel—the cartel of good intentions.

ALL TOGETHER NOW

Cartels thrive when customers have little opportunity to complain or to find alternative suppliers. In its heyday during the 1970s, for example, the Organization of the Petroleum Exporting Countries (OPEC) could dictate severe terms to customers; it was only when more non-OPEC oil exporters emerged that the cartel's power weakened. In the foreign aid business, customers (i.e., poor citizens in developing countries) have few chances to express their needs, yet they cannot exit the system. Meanwhile, rich nations paying the aid bills are clueless about what those customers want. Non-governmental organizations (NGOs) can hold aid institutions to task on only a few high-visibility issues, such as conspicuous environmental

The Aid Cartel's Golden Oldies

Many of the "new" themes that the international aid agencies emphasize today have actually been around for several decades.

Donor Coordination	"[Foreign aid] should be a cooperative enterprise in which all nations work together through the United Nations and its specialized agencies." (U.S. President Harry Truman, 1949)	"Aid coordination... has been recognized as increasingly important." (World Bank, 1981)	"We should improve coherence through better coordination of efforts amongst international institutions and agencies, the donor community, the private sector, and civil society." (World Bank President James Wolfensohn, 2002)
Aid Selectivity	"Objective No. 1: To apply stricter standards of selectivity... in aiding developing countries." (President John F. Kennedy, 1963)	"The relief of poverty depends both on aid and on the policies of the recipient countries." (Cassen Development Committee Task Force, 1985)	"[The International Development Association] should increase its selectivity... by directing more assistance to borrowers with sound policy environments." (International Development Association, 2001)
Focus on Poverty	"[The aid community must] place far greater emphasis on policies and projects which will begin to attack the problems of absolute poverty." (World Bank President Robert McNamara, 1973)	"The Deputies encouraged an even stronger emphasis on poverty reduction in [the International Development Association's] programs." (Former World Bank Managing Director Ernest Stern, 1990)	"The Poverty Reduction Strategy Paper aims at... increasing the focus of... assistance on the overarching objective of poverty reduction." (International Development Association, 2001)
African Reforms	"Many African governments are more clearly aware of the need to take major steps to improve the efficiency... of their economies." (World Bank, 1983)	"African countries have made great strides in improving policies and restoring growth." (World Bank, 1994)	"Africa's leaders... have recognized the need to improve their policies, spelled out in the New Partnership for African Development." (World Bank, 2002)

Sources: William Easterly, "The Cartel of Good Intentions: Bureaucracy vs. Markets in Foreign Aid" (Washington: Center for Global Development, 2002); James Wolfensohn, "Note From the President of the World Bank" (April 12, 2002)

destruction. Under these circumstances, even while foreign aid agencies make good-faith efforts to consult their clients, these agencies remain accountable mainly to themselves.

The typical aid agency forces governments seeking its money to work exclusively with that agency's own bureaucracy—its project appraisal and selection apparatus, its economic and social analysts, its procurement procedures, and its own interests and objectives. Each aid agency constitutes a mini-monopoly, and the collection of all such monopolies forms a cartel. The foreign aid community also resembles a cartel in that the IMF, World Bank, regional development banks, European Union, United Nations, and bilateral aid agencies all agree to "coordinate" their efforts. The customers therefore have even less opportunity to find alternative aid suppliers. And the entry of new suppliers into the foreign assistance business is difficult because large aid agencies must be sponsored either by an individual government (as in the case of national agencies, such as the U.S. Agency for International Development) or by an international agreement (as in the case of multilateral agencies, such as the World Bank). Most NGOs are too small to make much of a difference.

Of course, cartels always display fierce jostling for advantage and even mutual enmity among members. That explains why the aid community concludes that "to realize our increasingly reciprocal ambitions, a lot of hard work, compromises and true goodwill must come into play." Oops, wait, that's a quote from a recent OPEC meeting. The foreign aid community simply maintains that "better coordination among international financial institutions is needed." However, the difficulties of organizing parties with diverse objectives and interests and the inherent tensions in a cartel render such coordination forever elusive. Doomed attempts at coordination create the worst of all worlds—no central planner exists to tell each agency what to do, nor is there any market pressure from customers to reward successful agencies and discipline unsuccessful ones.

As a result, aid organizations mindlessly duplicate services for the world's poor. Some analysts see this duplication as a sign of competition to satisfy the customer—not so. True market competition should eliminate duplication: When you choose where to eat lunch, the restaurant next door usually doesn't force you to sit down for an extra meal. But things are different in the world of foreign aid, where a team from the U.S. Agency for International Development produced a report on corruption in Uganda in 2001, unaware that British analysts had produced a report on the same topic six months earlier. The Tanzanian government churns out more than 2,400 reports annually for its various donors, who send the poor country some 1,000 missions each year. (Borrowing terminology from missionaries who show the locals the one true path to heaven, "missions" are visits of aid agency staff to developing countries to discuss desirable government policy.) No wonder, then, that in the early 1990s, Tanzania was implementing 15 separate stand-alone health-sector projects funded by 15 different donors. Even small bilateral aid agencies plant their flags everywhere. Were the endless meetings and staff hours worth the effort for the Senegalese government to receive $38,957 from the Finnish Ministry for Foreign Affairs Development Cooperation in 2001?

By forming a united front and duplicating efforts, the aid cartel is also able to diffuse blame among its various members when economic conditions in recipient countries don't improve according to plan. Should observers blame the IMF for fiscal austerity that restricts funding for worthy programs, or should they fault the World Bank for failing to preserve high-return areas from public expenditure cuts? Are the IMF and World Bank too tough or too lax in enforcing conditions? Or are the regional development banks too inflexible (or too lenient) in their conditions for aid? Should bilateral aid agencies be criticized for succumbing to national and commercial interests, or

should multilateral agencies be condemned for applying a "one size fits all" reform program to all countries? Like squabbling children, aid organizations find safety in numbers. Take Argentina. From 1980 to 2001, the Argentine government received 33 structural adjustment loans from the IMF and World Bank, all under the watchful eye of the U.S. Treasury. Ultimately, then, is Argentina's ongoing implosion the fault of the World Bank, the IMF, or the Treasury Department? The buck stops nowhere in the world of development assistance. Each party can point fingers at the others, and bewildered observers don't know whom to blame—making each agency less accountable.

THE $3,521 QUANDARY

Like any good monopoly, the cartel of good intentions seeks to maximize net revenues. Indeed, if any single objective has characterized the aid community since its inception, it is an obsession with increasing the total aid money mobilized. Traditionally, aid agencies justify this goal by identifying the aid "requirements" needed to achieve a target rate of economic growth, calculating the difference between existing aid and the requirements, and then advocating a commensurate aid increase. In 1951, the U.N. Group of Experts calculated exactly how much aid poor countries needed to achieve an annual growth rate of 2 percent per capita, coming up with an amount that would equal $20 billion in today's dollars. Similarly, the economist Walt Rostow calculated in 1960 the aid increase (roughly double the aid levels at the time) that would lift Asia, Africa, and Latin America into self-sustaining growth. ("Self-sustaining" meant that aid would no longer be necessary 10 to 15 years after the increase.) Despite the looming expiration of the 15-year aid window, then World Bank President Robert McNamara called for a doubling of aid in 1973. The call for doubling was repeated at the World Bank in its 1990 "World Development Report." Not to be outdone, current World Bank President James Wolfensohn is now advocating a doubling of aid.

By forming a united front and duplicating efforts, the foreign aid community is able to diffuse blame among its members when economic conditions in poor countries fail to improve.

The cartel's efforts have succeeded: Total assistance flows to developing countries have doubled several times since the early days of large-scale foreign aid. (Meanwhile, the World Bank's staff increased from 657 people in 1959–60 to some 10,000 today.) In fact, if all foreign aid

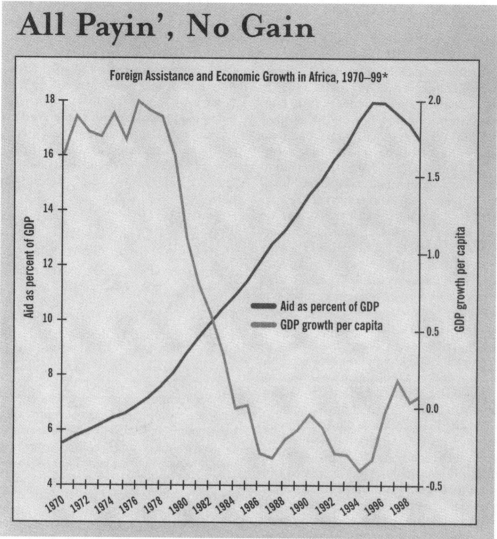

All Payin', No Gain

Foreign Assistance and Economic Growth in Africa, 1970–99*

Aid as percent of GDP

GDP growth per capita

Aid as percent of GDP

GDP growth per capita

*The data for each year represents the average per capita GDP growth rate and the average rate of aid (as a percentage of GDP) over the previous 10 years.

Source: *World Development Indicators* (Washington: World Bank, 2001); calculations by author

given since 1950 had been invested in U.S. Treasury bills, the cumulative assets of poor countries by 2001 from foreign aid alone would have amounted to $2.3 trillion. This aid may have helped achieve such important accomplishments as lower infant mortality and rising literacy throughout the developing world. And high growth in aid-intensive countries like Botswana and Uganda is something to which aid agencies can (and do) point. The growth outcome in most aid recipients, however, has been extremely disappointing. For example, on average, aid-intensive African nations saw growth decline despite constant increases in aid as a percentage of their income.

Aid agencies always claim that their main goal is to reduce the number of poor people in the world, with poverty defined as an annual income below $365. To this end, the World Bank's 2002 aid accounting estimates that an extra $1 billion in overseas development assistance would lift more than 284,000 people out of poverty. (This claim has appeared prominently in the press and has been repeated in other government reports on aid effec-

tiveness.) If these figures are correct, however, then the additional annual aid spending per person lifted out of poverty (whose annual income is less than $365) comes to $3,521. Of course, aid agencies don't follow their own logic to this absurd conclusion—common sense says that aid should help everyone and not just target those who can stagger across the minimum poverty threshold. Regrettably, this claim for aid's effect on poverty has more to do with the aid bureaucracy's desperate need for good publicity than with sound economics.

A FRAMEWORK FOR FAILURE

To the extent that anyone monitors the performance of global aid agencies, it is the politicians and the public in rich nations. Aid agencies therefore strive to produce outputs (projects, loans, etc.) that these audiences can easily observe, even if such outputs provide low economic returns for recipient nations. Conversely, aid bureaucrats don't try as hard to produce less visible, high-return out-

"Do Everything" Development

In September 2000, representatives of 189 countries met at the U.N. Millennium General Assembly in New York and adopted the Millennium Declaration concerning peace, security, and development issues. The Millennium Development Goals (MDGs), listed below, emerged from this gathering. Since then, virtually all the leading aid institutions have endorsed the MDGs, including the World Bank, International Monetary Fund, Organisation for Economic Co-operation and Development, and the Inter-American Development Bank.

Goal 1: Eradicate extreme poverty and hunger

Halve, between 1990 and 2015, the proportion of people whose income is less than $1 a day. Halve, between 1990 and 2015, the proportion of people who suffer from hunger.

Goal 2: Achieve universal primary education

Ensure that, by 2015, children everywhere, boys and girls alike, will be able to complete a full course of primary schooling.

Goal 3: Promote gender equality and empower women

Eliminate sender disparity in primary and secondary education preferably by 2005 and in all levels of education no later than 2015.

Goal 4: Reduce child mortality

Reduce by two-thirds, between 1990 and 2015, the under-five mortality rate.

Goal 5: Improve maternal health

Reduce by three-quarters, between 1990 and 2015, the maternal mortality ratio.

Goal 6: Combat HIV/AIDS, malaria, and other diseases

Have halted by 2015 and begun to reverse the spread of HIV/AIDS. Have halted by 2015 and begun to reverse the incidence of malaria and other major diseases.

Goal 7: Ensure environmental sustainability

Integrate the principles of sustainable development into country policies and programmes and reverse the loss of environmental resources. Halve, by 2015, the proportion of people without sustainable access to safe drinking water. Have achieved, by 2020, a significant improvement in the lives of at least 100 million slum dwellers.

Goal 8: Develop a global partnership for development

Develop further an open, rule-based, predictable, non-discriminatory trading and financial system.... Address the special needs of the least developed countries.... Address the special needs of landlocked countries and small island developing states.... Deal comprehensively with the debt problems of developing countries.... In cooperation with developing countries, develop and implement strategies for decent and productive work for youth. In cooperation with pharmaceutical companies, provide access to affordable, essential drugs in developing countries. In cooperation with the private sector, make available the benefits of new technologies, especially information and communications.

Source: United Nations Development Programme and World Bank

puts. This emphasis on visibility results in shiny showcase projects, countless international meetings and summits, glossy reports for public consumption, and the proliferation of "frameworks" and strategy papers. Few are concerned about whether the showcase projects endure beyond the ribbon-cutting ceremony or if all those meetings, frame-works, and strategies produce anything of value.

This quest for visibility explains why donors like to finance new, high-profile capital investment projects yet seem reluctant to fund operating expenses and maintenance after high-profile projects are completed. The resulting problem is a recurrent theme in the World Bank's periodic reports on Africa. In 1981, the bank's Africa study concluded that "vehicles and equipment frequently lie idle for lack of spare parts, repairs, gasoline, or other necessities. Schools lack operating funds for salaries and teaching materials, and agricultural research stations have difficulty keeping up field trials. Roads, public buildings, and processing facilities suffer from lack of maintenance." Five years later, another study of Africa found that "road maintenance crews lack fuel and bitumen... teachers lack books... [and] health workers have no medicines to distribute." In 1986, the Word Bank declared that in Africa, "schools are now short of books, clinics lack medicines, and infrastructure maintenance is avoided." Meanwhile, a recent study for a number of different poor countries estimated that the return on spending on educational instructional materials was up to 14 times higher than the return on spending on physical facilities.

And then there are the frameworks. In 1999, World Bank President James Wolfensohn unveiled his Comprehensive Development Framework, a checklist of 14 items, each with multiple subitems. The framework covers clean government, property rights, finance, social safety nets, education, health, water, the environment, the spoken word and the arts, roads, cities, the countryside, microcredit, tax policy, and motherhood. (Somehow, macroeconomic policy was omitted.) Perhaps this framework explains why the World Bank says management has simultaneously "refocused and broadened the development agenda." Yet even Wolfensohn seems relatively restrained compared with the framework being readied for the forthcoming U.N. World Summit on Sustainable Development in Johannesburg in late August 2002, where 185 "action recommendations"—covering everything from efficient use of cow dung to harmonized labeling of chemicals—await unsuspecting delegates.

Of course, the Millennium Development Goals (MDGs) are the real 800-pound gorilla of foreign aid frameworks. The representatives of planet Earth agreed on these goals at yet another U.N. conference in September 2000. The MDGs call for the simultaneous achievement of multiple targets by 2015, involving poverty, hunger; infant and maternal mortality, primary education, clean water, contraceptive use, HIV/AIDS, gender equality, the environment, and an ill-defined "partnership for development". These are all worthy causes, of course, yet would the real development customers necessarily choose to spend their scarce resources to attain these particular objectives under this particular timetable? Economic principles dictate that greater effort should be devoted to goals with low costs and high benefits, and less effort to goals where the costs are prohibitive relative to the benefits. But the "do everything" approach of the MDGs suggests that the aid bureaucracy feels above such tradeoffs. As a result, government officials in recipient countries and the foreign aid agency's own frontline workers gradually go insane trying to keep up with proliferating objectives—each of which is deemed Priority Number One.

A 2002 World Bank technical study found that a doubling of aid flows is required for the world to meet the U.N. goals. The logic is somewhat circular, however, since a World Bank guidebook also stipulates that increasing aid is undoubtedly "a primary function of targets set by the international donor community such as the [Millennium] Development Goals." Thus increased aid becomes self-perpetuating—both cause and effect.

FOREIGN AID AND ABET

Pity the poor aid bureaucracy that must maintain support for foreign assistance while bad news is breaking out everywhere. Aid agencies have thus perfected the art of smoothing over unpleasant realities with diplomatic language. A war is deemed a "conflict-related reallocation of resources."

Countries run by homicidal warlords like those in Liberia or Somalia are "low-income countries under stress." Nations where presidents loot the treasury experience "governance issues." The meaning of other aid community jargon, like "investment climate," remains elusive. The investment climate will be stormy in the morning, gradually clearing in the afternoon with scattered expropriations.

Another typical spin-control technique is to answer any criticism by acknowledging that, "Indeed, we aid agencies used to make that mistake, but now we have corrected it." This defense is hard to refute, since it is much more difficult to evaluate the present than the past. (One only doubts that the sinner has now found true religion from the knowledge of many previous conversions.) Recent conversions supposedly include improved coordination among donors, a special focus on poverty alleviation, and renewed economic reform efforts in African countries. And among the most popular concepts the aid community has recently discovered is "selectivity"—the principle that aid will only work in countries with good economic policies and efficient, squeaky-clean institutions. The moment of aid donors' conversion on this point supposedly came with the end of the Cold War, but in truth, selectivity (and other "new" ideas) has been a recurrent aid theme over the last 40 years.

Unfortunately, evidence of a true conversion on selectivity remains mixed. Take Kenya, where President Daniel arap Moi has mismanaged the economy since 1978. Moi has consistently failed to keep conditions on the 19 economic reform loans his government obtained from the World Bank and IMF (described by one NGO as "financing corruption and repression") since he took office. How might international aid organizations explain the selectivity guidelines that awarded President Moi yet another reform loan from the World Bank and another from the IMF in 2000, the same year prominent members of Moi's government appeared on a corruption "list of shame" issued by Kenya's parliament? Since then, Moi has again failed to deliver on his economic reform promises, and international rating agencies still rank the Kenyan government among the world's most corrupt and lawless. Ever delicate, a 2002 IMF report conceded that "efforts to bring the program back on track have been only partially successful" in Kenya. More systematically, however, a recent cross-country survey revealed no difference in government ratings on democracy, public service delivery, rule of law, and corruption between those countries that received IMF and World Bank reform loans in 2001 and those that did not. Perhaps the foreign aid community applies the selectivity principle a bit selectively.

DISMANTLING THE CARTEL

How can the cartel of good intentions be reformed so that foreign aid might actually reach and benefit the world's poor? Clearly, a good dose of humility is in order, considering all the bright ideas that have failed in the past. Moreover, those of us in the aid industry should not be so arrogant to think we are the main determinants of whether low-income countries develop—poor nations must accomplish that mainly on their own.

Antiglobalization protesters are largely on target when it comes to the failure of international financial institutions to foment "adjustment with growth" in many poor countries.

Still, if aid is to have some positive effect, the aid community cannot remain stuck in the same old bureaucratic rut. Perhaps using market mechanisms for foreign aid is a better approach. While bureaucratic cartels supply too many goods for which there is little demand and too few goods for which there is much demand, markets are about matching supply and demand. Cartels are all about "coordination," whereas markets are about the decentralized matching of customers and suppliers.

One option is to break the link between aid money and the obligatory use of a particular agency's bureaucracy. Foreign assistance agencies could put part of their resources into a common pool devoted to helping countries with acceptably pro-development governments. Governments would compete for the "pro-development" seal of approval, but donors should compete, too. Recipient nations could take the funds and work with any agency they choose. This scenario would minimize duplication and foster competition among aid agencies.

Another market-oriented step would be for the common pool to issue vouchers to poor individuals or communities, who could exchange them for development services at any aid agency, NGO, or domestic government agency. These service providers would in turn redeem the vouchers for cash out of the common pool. Aid agencies would be forced to compete to attract aid vouchers (and thus money) for their budgets. The vouchers could also trade in a secondary market; how far their price is below par would reflect the inefficiency of this aid scheme and would require remedial action. Most important, vouchers would provide real market power to the impoverished customers to express their true needs and desires.

Intermediaries such as a new Washington-based company called Development Space could help assemble the vouchers into blocks and identify aid suppliers; the intermediaries could even compete with each other to attract funding and find projects that satisfy the customers, much as venture capital firms do. (Development Space is a private Web-based company established last year by former World Bank staff members—kind of an eBay for foreign aid.) Aid agencies could establish their own intermediation units to add to the competition. An information bank could facilitate transparency and communication, posting news on projects searching for funding, donors searching for projects, and the reputation of various intermediaries.

Bureaucratic cartels probably last longer than private cartels, but they need not last forever. President George W. Bush's proposed Millennium Challenge Account (under which, to use Bush's words, "countries that live by these three broad standards—ruling justly, investing in their people, and encouraging economic freedom—will receive more aid from America") and the accompanying increase in U.S. aid dollars will challenge the IMF and World Bank's near monopoly over reform-related lending. Development Space may be the first of many market-oriented endeavors to compete with aid agencies, but private philanthropists such as Bill Gates and George Soros have entered the industry as well. NGOs and independent academic economists are also more aggressively entering the market for advice on aid to poor countries. Globalization protesters are not well informed in all areas, but they seem largely on target when it comes to the failure of international financial institutions to foment "adjustment with growth" in many poor countries. Even within the World Bank itself, a recent board of directors paper suggested experimenting with "output-based aid" in which assistance would compensate service providers only when services are actually delivered to the poor—sadly, a novel concept. Here again, private firms, NGOs, and government agencies could compete to serve as providers.

Now that rich countries again seem interested in foreign aid, pressure is growing to reform a global aid bureaucracy that is increasingly out of touch with good economics. The high-income countries that finance aid and that genuinely want aid to reach the poor should subject the cartel of good intentions to the bracing wind of competition, markets, and accountability to the customers. Donors and recipients alike should not put up with $3,521 in aid to reduce the poverty head count by one, 185-point development frameworks, or an alphabet soup of bureaucratic fads. The poor deserve better.

William Easterly is senior fellow of the Center for Global Development and the Institute for International Economics in Washington, D.C., and former senior advisor of the development research group at the World Bank. He is the author of The Elusive Quest for Growth: Economists' Adventures and Misadventures in the Tropics *(Cambridge: MIT Press, 2001).*

[Want to Know More?]

The Web site of the **World Bank** contains many of the documents cited in this article, as well as relevant works such as **"World Development Report 2002: Building Institutions for Markets"** (Washington: World Bank, 2001) and **"Assessing Aid: What Works, What Doesn't, and Why"** (Washington: World Bank, 1998). Visit the Web site of the **United Nations** to find the **"Monterrey Consensus"** documents from the March 2002 U.N. International Conference on Financing for Development in Mexico. **The Poverty Reduction Strategy Papers** are available on the Web site of the **International Monetary Fund**.

Insightful books on foreign assistance include Judith Tendler's classic *Inside Foreign Aid* (Baltimore: Johns Hopkins University Press, 1975), James Morton's *The Poverty of Nations: The Aid Dilemma at the Heart of Africa* (London: I.B. Tauris Publishers, 1996), and Nicolas van de Walle's *African Economies and the Politics of Permanent Crisis: 1979–99* (Cambridge: Cambridge University Press, 2001). Other useful works include *Tropical Gangsters* (New York: Basic Books, 1991) by Robert Klitgaard and *The World Bank: Its First Half Century, Volume I* (Washington: Brookings Institution Press, 1997) by Devesh Kapur, John P. Lewis, and Richard Webb. In his recent book *On Globalization* (New York: PublicAffairs, 2002), philanthropist George Soros suggests market mechanisms for foreign aid.

Enduring works on bureaucracy include William A. Niskanen Jr.'s *Bureaucracy and Representative Government* (Chicago: Aldine-Atherton, 1971) and James Q. Wilson's *Bureaucracy* (New York: Basic Books, 1989). Lant Pritchett and Michael Woolcock assess the problems of bureaucracy in economic development in **"Solutions When the Solution Is the Problem: Arraying the Disarray in Development"** (Washington: Center for Global Development, 2002).

FOREIGN POLICY has a long history of covering economic development and foreign aid, including Samuel Huntington's **"Foreign Aid: For What and for Whom"** (Winter 1970-71), appearing in *FP*'s inaugural issue. Also see **"Development: The End of Trickle Down?"** (Fall 1973) by James Grant, **"The Third World: Public Debt, Private Profit"** (Spring 1978) by Albert Fishlow, et al., **"Funding Foreign Aid"** (Summer 1988) by David R. Obey and Carol Lancaster, and **"The IMF: A Cure or a Curse?"** (Summer 1998) by Devesh Kapur. More recent *FP* coverage includes Ricardo Hausmann's **"Prisoners of Geography"** (January/February 2001), Dani Rodrik's **"Trading in Illusions"** (March/April 2001), Stephen Fidler's **"Who's Minding the Bank?"** (September/October 2001), and William Easterly's **"Think Again: Debt Relief"** (November/December 2001).

This article is based on a longer research paper by Easterly, **"The Cartel of Good Intentions: Bureaucracy vs. Markets in Foreign Aid"** (Washington: Center for Global Development, 2002). Fox a comprehensive treatment of foreign aid and the problems of economic development, see Easterly's *The Elusive Quest for Growth: Economists' Adventures and Misadventures in the Tropics* (Cambridge: MIT Press, 2001).

For links to relevant Web sites, access to the *FP* Archive, and a comprehensive index of related FOREIGN POLICY articles, go to **www.foreignpolicy.com**.

From *Foreign Policy*, July/August 2002, pp. 40-49. © 2002 by William Easterly.

Nongovernmental organizations (NGOS) and third world development: An alternative approach to development

The growth of nongovernmental organizations (NGOS) is due to the changing attitudes of donor agencies about development assistance in Third World Countries. NGOs hold a great promise to provide self-help solutions to problems of poverty in many Third World societies.

J Wagona Makoba

The phenomenal growth of nongovernmental organizations (NGOs) at both international and national levels is due to the changing attitude of donor agencies about development assistance and the increased demand for NGO services in Third World countries.[1] NGOs are non-membership support organizations involved in relief, rehabilitation, or community development work in developed and, especially, developing or Third World countries.[2] Considered part of the civil arena in society which also includes trade unions, people's associations and membership organizations, cooperatives and religious-based charities, NGOs provide a third approach to development between market-led and state-led strategies.[3] In the post-Cold War era, governments in Third World countries are experiencing a steady decline in both fiscal support and public credibility.[4] On the other hand, markets globally are on the ascent in terms of ideological and resource support, while those in the Third World are still nascent or in decline.

As I argue later, the prevalence of weak states and declining markets in the Third World inevitably have development-oriented NGOs as the only alternative to promote grassroots development. Hence development-oriented NGOs are not simply located somewhere between the state and market in terms of institutional space, but are emerging as a critical "third or middle sector" fostering the development of marginalized segments of the population. As one observer recently pointed out, developmental NGOs are "needed to cater for those groups whose place at the state or market table is not reserved."[5]

All NGOs share several characteristics including dependency on donor funding; the need for self-financing, transparency or accountability to donors and clients; and targeting the needy or marginalized segments of the population or operating in various sectors of society depending on the needs to be met as well as resources available in the local community.[6]

This paper is organized in three sections. The first section discusses the major reasons behind the changing attitude of the international donor community towards development assistance and its impact on NGO growth or expansion. The second critically examines how the failure of both governments and markets in Third World countries to deliver economic development has led to an increasing reliance on the NGOs to fill the "void" created. The final section analyzes the emerging significance of NGOs as an alternative approach to grassroots development in the Third World.

THE CHANGING ATTITUDE OF THE INTERNATIONAL DONOR COMMUNITY AND NGO GROWTH

The phenomenal growth of NGOs on the world scene has been aptly described as "a global associational revolution."[7] The number of development-oriented NGOs registered in countries of the industrialized North "grew from 1,600 in 1980 to

2,970 in 1993."[8] It is also reported that "over the same period, the total spending of these NGOs rose from US $2.8 billion to US $5.7 billion in current prices…"[9] By 1993 there were an estimated 28,900 international NGOs worldwide, with approximately 20,000 of these in Third World countries.[10] The growth of the NGOs on the global scene is associated with the changing attitude of the international donor community about development assistance. Such a change of attitude has been influenced by the end of the Cold War, the dwindling development aid from leading industrialized countries, and a new emphasis on targeting aid to benefit marginalized segments of society.

With the end of the Cold War, "development policy and aid transfers have come to be dominated by…a new policy agenda."[11] Such a policy agenda "is driven by…liberal economics and liberal democratic theory."[12] Also, recent developments in economic and political thinking about roles of markets and states in promoting economic development have been heavily influenced by neoliberal economic and democratic theory. As a result of such thinking, markets and private sector initiatives are considered the most efficient mechanisms for achieving economic growth and providing most services, including social services (via privatization) to most people. In contrast, governments whose role within the economy is being reduced, are expected to create an "enabling environment" or a legal and policy framework for the private sector provision of goods and services. Under the New Policy Agenda, NGOs are seen as vehicles for democratization as well as for providing goods and services in Third World countries where markets are inaccessible to the poor or where governments lack capacity or resources to reach them. In the eyes of the international donor community, NGOs are both cost-effective in reaching the poor and are considered "the preferred channel for service provision, in deliberate substitution for the state."[13] For most Western industrialized countries including the United States, the end of the Cold War has meant an end to using foreign aid to "buy" allies in the Third World to support it against the former Soviet Union. As a result, the strategic and military importance of development aid has diminished.

In addition to the post-Cold War considerations, the declining aid levels to the Third World have forced the donor community "to ensure that the developmental impact of scarce aid resources is maximized."[14] According to the 1998 World Bank Report: "Net official development finance, consisting of loans and grants from government agencies and multilateral institutions, has declined by almost 50 percent in real […] terms since the early 1990s."[15] The decline in aid has been due to several factors including "the pressing need for fiscal consolidation in most [industrialized] countries, the declining strategic and military importance of development aid since the end of the Cold War, and weak public support for aid in some major donor countries, due in part to skepticism about its effectiveness."[16] Increasingly, many donors "want to fund projects and programs that have a greater chance of being sustainable and of standing on their own."[17] As a result, capacity building and sustainability are now the watchwords for donor-supported development projects.

In order to maximize the impact of scarce development aid, many donors are channeling an increasing share of their overseas development aid (ODA) through NGOs. For example, in 1980, funding from the international donor community "accounted for less than 10 percent of NGO budgets, [but] by the 1990s their share had risen to 35 percent."[18] As a result of increased donor funding, NGOs in some African countries "now provide or implement more than a fifth of total aid flows, compared with less than one percent fifteen years ago."[19] Increasingly, a large number of NGOs in the Third World are funded by a small number of donors such as the World Bank, the United States Agency for International Development (USAID), and the United Nations Development Program (UNDP). The World Bank not only encourages member governments to work with NGOs on development projects, but also directly funds the NGO projects. It is reported that, "from 1973 to 1988, NGOs were involved in about 15 [World] Bank projects a year. By 1990 that number had jumped to 89, or 40 percent of all new projects approved."[20] And in 1997, approved World Bank projects in Third World countries involving NGOs were: 84 percent in South Asia, 61 percent in Africa, and 60 percent in Latin America and the Caribbean.[21] USAID is said to funnel 20 percent of its funds through NGOs.

The Inter-American Foundation Survey of North and Latin America "revealed that most of the multilateral banks, the United Nations, USAID, and advocacy organizations in the United States and Europe favored a restucturing of aid away from traditional government-to-government approaches toward people-to-people programs."[22] Furthermore, the public sentiment in the United States and Europe seems to be "in favor of NGO participation in the development process and against governmental foreign aid."[23]

The shift toward people-oriented programs and NGO participation in the development process is a result of declining foreign aid and the failure of many large-scale development projects funded by multilateral agencies and Third World governments. Major donors now view NGOs as potentially effective agents of development assistance, especially aid targeted to benefit the poor, women, and children. The donor community's concern with people-centered development projects has led to considerable rethinking about cost-effectiveness and impact on the beneficiaries.

As a result, donor agencies now favor new and innovative development strategies such as the "minimalist cost-effective approach" (used by microfinance institutions), "assisted self-reliance" or "participatory development." NGOs appear well suited to adapt the use of new and innovative strategies because of their small scale, flexibility and wide-ranging capacity to mobilize resources and to organize people to solve their own problems. The new development strategies, which are considered to be "people-friendly" engage grassroots energies, resources and talents and perceive people as active participants of their own development. Finally, the objectives of the new development models can be achieved much more cost-effectively through NGOs. This is primarily because NGOs tend to be flexible vehicles for meeting a variety of human needs including: self-expression or empowerment, promoting equity, self-help, participation and mutual assistance. Increasingly, donors believe that many of their aid priorities in the post-debt and post-Cold War era can be effectively implemented through a variety of NGOs found in virtually all Third World countries. The aid priorities of the international donor community include: "…developing the private sector, alleviating poverty, encouraging more equitable income distribution, supporting women, and promoting participatory development approaches and modalities that stress individual action and collective initiatives."[24]

Despite donor interest in channeling development aid through NGOs, critics contend that funds from such powerful donors as the World Bank or USAID are likely "to compromise the independence and effectiveness of NGOs in achieving their social goals."[25]

FAILURE OF ECONOMIC DEVELOPMENT IN THE THIRD WORLD AND NGO GROWTH

As noted in the previous section, donor agencies increasingly support NGOs in providing services to the poor in Third World countries where markets are inaccessible and where governments lack capacity or resources to reach the poor. In most Third World countries including those in Africa, both states and markets are weak or in decline. In Africa, the persistence of the dual crisis of weak states and nascent or declining markets poses a classic dilemma for proponents of either market- or state-led economic development. The failure of both markets and governments in Africa to deliver economic development has contributed to the rapid growth and expansion of NGOs on the continent.

Evidence accumulated over the past three decades shows "the inability of the African State to deliver on its development promise."[26] In fact, the African State is now perceived as "the inhibitor of social, economic, and political development."[27] The demise of the African State has inevitably given rise to the ascendancy of NGOs to fill up the "development vacuum" that has been created. The expansion of the NGO sector in Africa is most clearly reflected at the country level. For example, in Kenya there are about 500 NGOs and in Uganda there are more than 1,000 registered foreign and indigenous NGOs. Similarly, other African countries have a large number of active NGOs. These countries include: "Zambia with 128, Tanzania with 130, Zimbabwe with 300, and Namibia with over 55."[28] The growing role of NGOs in all sectors of development is an indication of the decreasing capacity of the African state to undertake meaningful development. Besides increases in NGO numbers, the amount of development resources they receive or handle for development purposes has grown over the years. It is estimated that "official aid to Kenyan NGOs amounts to about US $35 million a year, which is about 18 percent of all official aid received by Kenya annually [and]…in Uganda, NGOs disburse an estimated 25 percent of all official aid to Uganda."[29]

The weakening financial situation of Uganda and Kenya, like that of other African countries, is due to a combination of huge external debts, corruption and the effects of structural adjustment programs imposed by the International Monetary Fund (IMF). In particular, the structural adjustment programs have "strained the ability of the African states to provide services and has attracted more NGOs to cushion the adverse short-term effects of adjustment programs, such as by providing affordable healthcare services."[30] Given the prevailing political and economic conditions in Uganda and Kenya, as well as elsewhere in Africa, the role and contribution of NGOs to the development process is expected to increase.

Donor agencies increasingly funnel development assistance through NGOs and other non-state institutions because the states in Sub-Saharan Africa are considered both inefficient and corrupt. As Dicklich observes, the "failure of the [African] state to provide for basic services has led to many official donors to use NGOs rather than the local state to provide services."[31] In Uganda, a succession of inefficient, violent and corrupt regimes since 1971 has contributed to the emergence of over 1,000 indigenous NGOs to provide self-help solutions to the poor. Most "ordinary Ugandans have had to fend for themselves, relying on organizations outside of the state rather than on the state itself to provide basic necessities."[32] In general, most service-oriented NGOs have generally "moved into service provision where the state has moved out."[33] No doubt, NGOs have been necessary in Uganda and other African countries to fill up the "developmental gaps" caused by the weak post-independence state.

While African States have become increasingly weak, formal markets have steadily declined and in some cases have been replaced by informal or parallel markets. According to Callaghy, most African economies are faced with

"declining or negative rates and stagnating or falling per capita income figures; balance of payments and debt problems (which have become more severe (since) the 1980s, requiring IMF and the World Bank programs with their attendant conditionality packages and consequences. Many (export) commodity prices remain low while most import prices remain high. In many countries, agricultural production is falling while aid levels stagnate. Health and nutrition levels are falling while informal or magendo economies (have) become more important as states weaken and formal markets decline. 'Socialist' states have performed poorly and 'capitalist' ones are not significantly better. Hopes for economic growth and development have shriveled on all sides."[34]

In Uganda and other African countries, authoritarian regimes "induced an 'exit' from the formal economy [as well as] a general avoidance of state institutions by a wide range of groups and occupations."[35] Furthermore, economic restructuring due to structural adjustment programs and privatization contributed to the retreat of African states from their responsibilities of promoting economic development and providing "basic social services such as health care, education, sanitation and basic security,…"[36] Given the weak private sector and the state withdrawal from the provision of basic economic necessities and social services, "many NGOs are being pressurized into dealing with poverty alleviation (not eradication), and the provision of basic social services...."[37] Thus, NGOs increasingly fill in social and economic spaces created by weak markets or retreating states. As a result, "NGOs have been heralded as...new agents with the capacity and commitment to make up for the shortcomings of the state and market in reducing poverty."[38] Some critics of NGO participation in economic development contend that such involvement provides legitimacy and support to governments that have failed to deliver economic development or provide basic social services to their citizens. Other critics charge that NGOs save "donors money and allow them to avoid addressing implementation difficulties, while also allowing them [the donors] to retain ultimate control over activities."[39]

The absence of viable states or markets in most Third World countries including African states has left NGOs as the most important alternative for promoting economic development. Thus, the failure or inability of both states and markets to meet the basic needs of the majority of the people in the Third World has given rise to the growing importance of the NGO sector in the development process.[40] Such inability has also exposed the inherent limitations of the state or private sector as major agents of promoting economic development in the Third World.

NGOS AS AN ALTERNATIVE APPROACH TO DEVELOPMENT IN THE THIRD WORLD

The rapid growth and expansion of NGOs worldwide attest to their growing critical role in the development process. At the international level, NGOs are perceived as vehicles for providing democratization and economic growth in Third World countries. Within Third World countries, NGOs are increasingly considered good substitutes for weak states and markets in the promotion of economic development and the provision of basic services to most people.

NGOs are seen by their proponents as a catalyst for societal change because they are responsive to the needs and problems of their clients, usually the poor, women and children. Because of targeting and being responsive to marginalized groups in society, NGOs are being heralded as "important vehicles for empowerment, democratization and economic development."[41] In fact, some NGOs are "driven by strong values and...interests...,geared toward empowering communities that have been traditionally disempowered."[42] International donor agencies see NGOs as "having the capacity and commitment to make up for the shortcomings of the state and market in reducing poverty."[43] Perhaps the greatest potential NGOs have is to generate self-help solutions to problems of poverty and powerlessness in society. This is based on the view of NGOs as independent, "efficient, less bureaucratic, grassroots oriented, participatory and contributing to sustainable development in grassroots communities."[44] But for NGOs to remain independent of donor or elite control and achieve their social and economic goals, they have to work diligently toward capacity building and financial sustainability."[45]

NGOs are increasingly playing an important role in the development process of most Third World countries as discussed in section three of this paper. The growing importance of NGOs in the development process is attributed to the fact that they are considered suitable for promoting participatory grassroots development and self-reliance, especially among marginalized segments of society—namely, the poor, women and children. In fact, some NGOs seek to organize and involve the marginalized groups in their own development. And sometimes, they try to link their clients to the powerful segments of society by providing access to resources that are normally out of reach to the poor. For example, within development-oriented NGOs, microfinance institutions (MFIs) try to contribute to the economic improvement of the poor by: "... bringing in new income from outside the community, preventing income from leaving the community, providing new [self] employment opportunities and stimulating backward and forward linkages to other community enterprises."[46]

As stated earlier, most development-oriented NGOs in the Third World use new and innovative development strategies such as the "minimalist costeffective approach (favored by microfinance institutions/poverty lending programs), "assisted self-reliance" or "participatory development." Overall, NGOs appear well suited to adapt the use of such innovative strategies because of their small scale of operations, flexibility and great capacity to mobilize resources and to organize people to solve their own problems. The new development strategies perceive people as active participants in their own development. These "bottom-up" development strategies stand in sharp contrast to the "topdown" capitalist and state socialist models of development. Both the capitalist model based on "trickle-down" and the state socialist model of "egalitarian development" based on central/state planning of economic activity have failed to meet basic needs of the poor, women and children and have not helped these marginalized groups to solve their own problems. Both models offer no real choice to the poor about immediate local problems faced or needs. Both forms of institutional "top-down" directed development discourage popular citizen participation and de-emphasize people-centered development activity. And finally, both models stress large-scale, capital intensive projects that are susceptible to elite control, corruption, and inefficiency.

CONCULSION

As we start the 21st century, NGOs hold a great promise to provide self-help solutions to problems of poverty and powerlessness in many Third World societies. They are increasingly making up for the shortcomings of the state and market in reducing poverty in Third World countries. Furthermore, their future role in development is expected to increase precisely because of favorable international donor support. In fact, since the launching of the Structural Adjustment Participatory Review Initiative (SAPRI) in 1997, NGOs are increasingly influencing economic and social development policy in Third World countries. SAPRI provides a framework for joint evaluation of economic reform by the World Bank, Third World governments, and one thousand civil society organizations including NGOs.[47] But despite their growing role and expected contribution to Third World development, NGOs should neither be considered a panacea nor the "magic bullet" for solving the problems of development.

NOTES

1. Michael Edwards and David Hume (eds.). *Beyond the Magic Bullet: NGO Performance and Accountability in the Post-Cold War World* (Westford, CT: Kumarian Press, 1996), pp. 2–4.

2. Anthony Bebbington and John Farrington, "Governments, NGOs and Agricultural Development: Perspectives on Changing Inter-Organizational Relationships," *The Journal of Development Studies,* 29:2, January 1993, p. 201.

3. In this paper, NGOs are considered part of the "third or middle sector," wedged between the state/public and the market/private sectors. In this sense, NGOs are neither part of the public nor private sector—even though they may receive resource or ideological support from either or both. This view of NGOs contrasts sharply with that expressed by Norman Uphoff (1996: 23–27) who contends that "NGOs are best considered a sub-sector of the private sector [as] this is implied by the synonym used for [northern] NGOs-Private Voluntary Organizations (PVOs)."

4. Norman Uphoff, "Why NGOs are not a Third Sector: A Sectoral Analysis with some Thoughts on Accountability, Sustainability, and Evaluation," Chapter One, pp. 23–29, in Michael Edwards and David Hume, *Beyond the Magic Bullet: NGO Performance and Accountability in the Post-Cold War World,* p. 23.

5. Goran Hyden, "Civil Society, Social Capital and Development: Dissection of a Complex Discourse," in *Studies in Comparative International Development,* 32:7, Spring 1997, p. 27.

6. E. A. Brett, "Rebuilding Survival Structures for the Ugandan Poor: Organizational Options for Reconstruction and Development in the '90s," *Institute of Development Studies,* Brighton, United Kingdom, December, 1990, pp. 7–8.

7. Lester M. Salamon, "The Rise of the Nonprofit Sector," *Foreign Affairs*, 73:4, July/August, 1994, p. 109.

8. Michael Edwards and David Hume (eds.). *Beyond the Magic Bullet: NGO Performance and Accountability in the Post-Cold War World,* p. 1.

9. Ibid.

10. Salamon, "The Rise of the Nonprofit Sector," p. 10.

11. Edwards and Hume, *Beyond the Magic Bullet: NGO Performance and Accountability in the Post-Cold War World,* p. 2.

12. Ibid.

13. Ibid., and Anthony Bebbington and John Farrington, "Governments, NGOs and Agricultural Development: Perspectives on Changing Inter-Organizational Relationships," p. 202.

14. OECD, Development Assistance Committee, *New Directions in Donor Assistance to Microenterprises*, OECD/OCDE, Paris, 1993, p. 8.

15. World Bank, *Global Development Finance Report* (Washington, D.C.), p. 49.

16. Ibid.

17. Roger C. Riddell, "The End of Foreign Aid to Africa? Concerns About Donor Policies," *African Affairs,* 98, 1999, p. 321.

18. World Bank, *Global Development Finance Report*, p. 51.

19. Nicolas Van de Walle, "Aids Crisis of Legitimacy: Current Proposals and Future Prospects;" *African Affairs*, 98, 1999, p. 345.

20. Marguerite Michaels, "Retreat from Africa:" *Foreign Affairs*, 72:1, 1993, p. 103.

21. World Bank, *Progress Report*, (Washington, D.C., 1997), p. 49.

22. Stephen G. Vetter, "The Business of Grassroots Development," *Grassroots Development*, 19:2, 1995, p. 6.

23. Ibid.

24. OECD, Development Assistance Committee, p. 7.

25. Marguerite Michaels, "Retreat from Africa," p. 103 and Nicolas Van de Walle, "Aids Crisis of Legitimacy: Current Proposals and Future Prospects," p. 346.

26. Stephen N. Ndegwa, *The Two Faces of Civil Society: NGOs and Politics in Africa,* (Westford, CT: Kumarian Press, 1996), p. 15.

27. Ibid.

28. Ibid., p. 20.

29. See Ibid. and Susan Dicklich, "Indigenous NGOs and Political Participation," in Holdger B. Hansen and Michael Tweedle (eds.), *Developing Uganda*, (Oxford: James Curry, 1998) p. 148. The 25 percent in Uganda represents the annual average expenditure by NGOs. During particular fiscal years, NGO expenditures may be higher. For example, it is reported that during the 1992/93 fiscal year, the expenditure of foreign and indigenous NGOs "was US $125 million,… almost equal to the expected World Bank contribution to the rehabilitation and Development plan for the same year," (Dicklich, 1998a: 148).

30. Stephen N. Ndegwa, *The Two Faces of Civil Society: NGOs and Politics in Africa,* p. 2.

31. Susan Dicklich, *The Elusive Promise of NGOs in Africa: Lessons from Uganda,* (New York: St. Martin's Press, 1998b), p. 6.

32. Ibid., pp. 22–3.

33. Ibid., p. 6.

34. Thomas M. Callaghy, "State, Choice and Context: Comparative Reflections on Reform and Intractability," in David E. Apter and Carl G. Rosberg (eds.), *Political Development and The New Realism in Sub-Saharan Africa*, (Charlottesville and London: University Press of Virginia, 1994), p. 201.

35. Dicklich, *The Elusive Promise of NGOs in Africa: Lessons from Uganda*, p. 145.

36. Ibid.

37. Ibid., p. 149.

38. Dicklich, *The Elusive Promise of NGOs in Africa: Lessons from Uganda*, p. 3.

39. Nicolas Van de Walle, "Aids Crisis of Legitimacy: Current Proposals and Future Prospects," p. 346.

40. Daniel C. Levy, *Building the Third Sector*, (Pittsburgh, PA: University of Pittsburgh Press, 1996), pp. 1–3; Vetter, "The Business of Grassroots Development, P.2; and Salamon, "The Rise of the Nonprofit Sector," p. 116.

41. Dicklich, *The Elusive Promise of NGOs in Africa: Lessons from Uganda*, p. 2.

42. Ibid., p. 8.

43. Ibid., p. 3.

44. Ndegwa, *The Two Faces of Civil Society: NGOs and Politics in Africa*, p, 20.

45. Van de Walle, "Aids Crisis of Legitimacy: Current Proposals and Future Prospects," p. 346.

46. Management Systems International Report, "Assessing the Impact of Microenterprise Interventions: A Framework for Analysis:" March 1995, p. ii.

47. World Bank, *Progress Report*, 1997, pp. 49–50.

J. Wagona Makoba is Associate Professor of Sociology at the University of Nevada, Reno. He is the author of Government Policy and Public Enterprise Performance in Sub-Saharan Africa: The Case Studies of Tanzania and Zambia, 1964–1984. (Lewiston: The Edwin Mellen Press, 1998) and several articles on political and economic reform in Sub-Saharan Africa.

From *Journal of Third World Studies*, Spring 2002, pp. 53-63. © 2002 by Journal of Third World Studies. Reprinted by permission.

Fishermen on the net

The digital revolution is helping the poor, too

THE global great and good are obsessed with the "digital divide". Half the people in the world, they fret, have never made a telephone call. Africa has less international bandwidth than Brazil's city of Sao Paulo. How, ask dozens of inter-governmental task-forces, can the poor get connected? Amid all the attention paid to developing countries' lack of Internet access, some people feel that more fundamental problems are being ignored. Ted Turner, an American media boss, observed last year that there was no point in giving people computers when they had no electricity.

He may be wrong. Indian scientists recently produced a prototype of a battery-powered device called the Simputer—short for "simple computer"—that is expected to cost only $200 a unit. Even at that price it may be too expensive for the truly poor. But computers can be shared. And the time may come when they will pay for themselves. Information and communication technology (ICT) may be overhyped, but it does matter. True, people in poor countries need food and medicine before they need Internet access, but ICT could help them lay hands on both of these more easily.

In several fishing villages on the Bay of Bengal, for example, an Internet link-up allows a volunteer to read weather forecasts from the US Navy's public website and broadcast them over a loudspeaker. For fishermen who work from little wooden boats, knowing that a storm is looming can mean the difference between life and death. The Internet also lets them know the market price for their catch, which helps them haggle with middlemen. And they can download satellite images revealing where fish shoals are.

Communication, as everybody knows, is getting cheaper. Any task that can be digitised can now be done at a distance, which creates all sorts of opportunities for developing countries. Fixing software for a London firm does not require Indians to travel to Britain. They can do it from Bangalore. That is why India's software industry has grown from almost nothing ten years ago into the most dynamic business on the subcontinent, employing 400,000 people and generating more than $8 billion in sales last year. The country has almost as many fluent English speakers as England, and universities that turn out hordes of computer-literate graduates. Charging a small fraction of what Californians demand, Indian programmers fixed a large chunk of the world's millennium-bug troubles, and take on a larger share of western companies' back-room operations each year.

And they are not the only ones. Dial a helpline for an American bank, and you may find yourself talking to someone in the Philippines or Puerto Rico. A computer screen tells the telephonist what the weather is like in whichever city the customer is calling from, and suggests appropriate responses to the customer's grumbles about the local baseball team's performance last night. Creepy? Maybe, but it lowers costs for consumers and creates jobs where none existed before.

Accurate, timely information is useful in almost any field. Take health care. The Internet is the quickest and cheapest way yet devised of disseminating medical research. Using websites such as Healthnet, doctors in poor countries can easily and cheaply keep up to speed with the latest developments in their field. In Bangladesh, the local Medinet system provides access to hundreds of expensive medical journals for less than $1.50 a month. Throughout Africa, outbreaks of meningitis are tracked over the Internet so that epidemics can be stopped early.

Yuppie toys in rural Bangladesh

Granted, these health-care schemes mostly depend on charity for their funding. And the lack of Internet content in languages other than English is a problem. But new websites in other languages pop up every day. And as the cost of ICT falls, it may not be long before poor people start clubbing together to buy time online with their own money. It has already happened with mobile telephones.

For many Bangladeshi women, the mobile phone provides upward mobility. Women in rural areas borrow money from Grameen Bank, a microlender, to buy a GSM phone. Handsets can cost a year's income for the whole household, but more than 90% of borrowers are able to service their loans because they earn money charging other villagers to make calls.

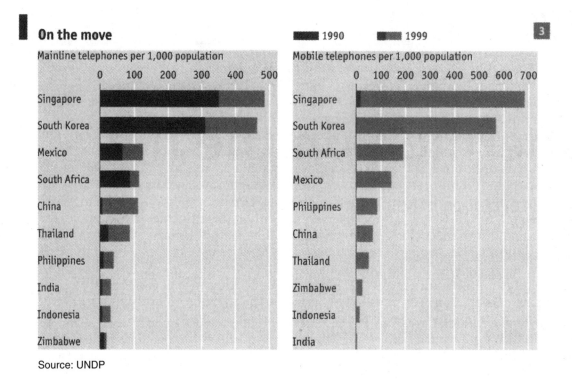

On the move

■ 1990   ■ 1999

Mainline telephones per 1,000 population

Singapore
South Korea
Mexico
South Africa
China
Thailand
Philippines
India
Indonesia
Zimbabwe

Mobile telephones per 1,000 population

Singapore
South Korea
South Africa
Mexico
Philippines
China
Thailand
Zimbabwe
Indonesia
India

Source: UNDP

Everyone benefits. In Bangladesh as a whole, 97% of homes lack a telephone. In rural areas, practically no one has a phone. So when a Grameen-backed phone entrepreneur sets up shop, the whole village is suddenly connected to the outside world. Parents can call the nearest city, for example, to find out what has happened to the remittance from their son working on a construction site in the Gulf. A telephone call can take the place of a long and expensive trip: one study found it typically saved between 3% and 10% of the caller's monthly household income.

The service also turns a profit. The callers may be poor, but because there are so many of them, rural phones in Bangladesh generate more revenue than urban ones. Grameen Phone, a Grameen Bank offshoot founded to provide services for the poor, now sells to everyone and has become the largest provider of mobile-phone services in Bangladesh.

Mobile phones are popular everywhere, but in poor countries they have extra advantages. It is quicker and cheaper to sell a poor person a mobile telephone than to install a payphone in a remote village. A phone in someone's house is constantly guarded, which makes it unlikely to be vandalised.

Mobiles are user-friendly, too. Fixed-line telephone services in most developing countries are provided by awful state-owned monopolies. Ask for a telephone line to be installed in your home in Zimbabwe, and you can choose between bribing someone or waiting for several years. But if you walk into a mobile-phone shop, your handset will be up and running in five minutes.

Nor is lack of a credit history a problem. Poor people are barred from opening accounts with traditional tele-phone companies because no one trusts them to pay their bills. With mobiles, however, they can buy pre-paid cards. When they have used up all their minutes, they buy another. There is no chance that they will receive an unmanageable bill at the end of the month.

Because the mobile-phone companies get paid in advance, they waste no time or money on chasing bad debts, as fixed-line firms do. So their cashflow is better and they are able to expand their networks faster. In many poor countries the number of mobile users has overtaken the number of land-line users in less time than it takes to get a land line installed. In Uganda, for example, it took two years for MTN, a private mobile-phone firm, to outstrip UTL, the clumsy state-owned telecoms firm. Before it was privatised last year, UTL in a typical year lost 40% of revenues in bad debts, and another 30% because of "audit adjustments". In 1997–99, the number of land-line users in Uganda inched up from 54,000 to 59,000. Over the same period, the number of mobile users exploded from 7,000 to 87,000.

This still means that only one Ugandan in 240 has a mobile phone, but it's a start. And ICT is spreading faster than any technology in the whole of human history. In 1900, 24 years after the telephone was invented, only 5% of homes in America were hooked up to the telephone network. In most other countries, the figure was negligible. Compare this with the spread of the Internet. In 2000, 18 years after the creation of a rudimentary Internet and 11 years after the beginning of the World Wide Web, 6.7% of the world's population were logging on. In other words, the Internet is spreading around the whole world faster than the telephone spread around its richest country a century ago.

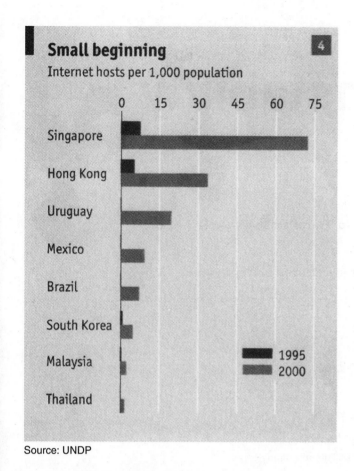

Small beginning

Internet hosts per 1,000 population

4

Singapore
Hong Kong
Uruguay
Mexico
Brazil
South Korea
Malaysia
Thailand

■ 1995
■ 2000

Source: UNDP

The poor are catching up

And poor countries, for once, are not missing out on this revolution. In 1998, according to the UNDP, only 12% of Internet users were in non-OECD (ie, less developed) countries. By 2000 this proportion had almost doubled, to 21%, when the cake itself had more than doubled in size over the same period.

The Internet is not only a marvellous technology; it is a tool that helps developing countries to adopt outside technology faster, and sometimes to develop their own. The makers of the Simputer used free open-source software, which they could not have downloaded without the Internet. Indeed, without the Internet, open-source software would not exist, as it depends on large numbers of volunteer programmers swapping ideas online.

In developing countries scientists are thin on the ground. A decade ago, if a researcher from a poor country wanted to bounce ideas off lots of other experts every day, he probably had to move to a rich country. Now he simply logs on. Cheaper communications mean more north-south collaboration, and indeed more south-south collaboration. In 1995–97, American scientists co-wrote papers with colleagues from 173 other countries. Kenyans published papers with scientists from 81 other nations.

The Internet also allows more collaboration within countries. Brazilian biologists work with colleagues at opposite ends of the Amazon river and on both sides of the equator. Some of them surprised the world last year by sequencing the genome of a bacterium that shrivels oranges. It was the first time that a free-living plant pathogen had been sequenced anywhere, a breakthrough for southern-hemisphere science. According to Andrew Simpson, one of the project's leaders and an old Brazil hand, this would not have been possible without the Internet. "It's a godsend," he says. "It gives us confidence that what we are doing is as good as what is going on in America, because we can see what is going on in America."

From *The Economist*, November 10, 2001, Survey 8-10. © 2001 by The Economist, Ltd. Distributed by the New York Times Special Features. Reprinted by permission.

UNIT 3

Conflict and Instability

Unit Selections

Key Points to Consider

- What factors account for the emergence of Islamic extremism?

- How might Muslim definitions of terrorism differ from those of the West?

- What are the causes of the most recent round of tension between India and Pakistan?

- Why has the war in the Democratic Republic of Congo received so little attention?

- How have President Mugabe's policies brought Zimbabwe to the brink of instability?

- What accounts for the continuing violence between Palestinians and Israelis?

- How have the September 11, 2001, attacks affected the Palestinian-Israeli conflict?

- What are the causes and effects of the breakdown of Colombia's peace efforts?

 **Links: www.dushkin.com/online/**
These sites are annotated in the World Wide Web pages.

The Carter Center
http://www.cartercenter.org
Center for Strategic and International Studies (CSIS)
http://www.csis.org/
Conflict Research Consortium
http://www.Colorado.EDU/conflict/
Institute for Security Studies
http://www.iss.co.za
PeaceNet
http://www.igc.org/peacenet/
Refugees International
http://www.refintl.org

The sources of conflict and instability in the developing world are deep, varied, and complicated. Some of the factors that fuel this have contributed to the anger and resentment behind the September 11, 2001, attacks against the United States.

Conflict and instability stem from a combination of sources. Among these are ethnic and religious diversity, nationalism, the struggle for state control, competition for resources, and the cold war's legacy. In many cases, boundaries that date from the colonial era encompass diverse groups. A state's diversity can increase tension among groups competing for scarce resources and opportunities. When some groups benefit or are perceived as enjoying privileges at the expense of others, ethnicity can offer a convenient vehicle around which to organize and mobilize. Moreover, ethnic politics lends itself to manipulation both by regimes that are seeking to protect privileges, maintain order, or retain power and those that are challenging existing governments. In a politically and ethnically charged atmosphere, conflict and instability often result as groups vie to gain control of a state apparatus that can extract resources and allocate benefits. While ethnicity has certainly played a role in many recent conflicts, conflicts over power and resources may be mistakenly viewed as ethnic in nature. Recent conflicts in Africa and Asia have increasingly centered around valuable resources. Conflicts like the war in the Democratic Republic of Congo generate economic disruption, population migration, environmental degradation, and may also draw other countries into the fighting.

Early literature on modernization and development speculated that as developing societies progressed from traditional to modern, primary attachments such as ethnicity and religious affiliation would fade and be replaced by new forms of identification. Clearly, however, ethnicity remains a potent force, as does religion. Ethnic politics and the emergence of religious radicalism demonstrate that such attachments have survived the drive toward modernization.

Initially inspired and encouraged by the theocratic regime in Iran, Islamists have challenged a number of governments, seeking to install governments based on Islamic law. These Islamic movements may reflect a combination of disillusionment with Western cultural influences, but also represent one of the few ways to challenge entrenched authoritarian regimes. Radical Islamic groups that advocate a more rigid and violent interpretation of Islam have increasingly challenged more mainstream Islamic thought. These radicals are driven by hatred of the West and the United States in particular. They were behind the 1993 New York City World Trade Center bombing, the United States embassy bombings in Kenya and Tanzania in 1998, and the devastating attacks that destroyed the World Trade Center and damaged the Pentagon in September 2001. Although there is a tendency to equate Islam with terrorism, it is a mistake to link the two. A deeper understanding of the various strands of Islamic thought is required to separate legitimate efforts to challenge repressive regimes and forge an alternative to Western forms of political organization from radicals who pervert Islam and use terrorism.

There is no shortage of tension and conflict around the world. Prospects for peace in the Middle East have faded in the midst of Palestinian suicide bombings and Israeli retaliation. Although the Taliban and its Al Qaeda allies seem to have been ousted in Afghanistan, tensions remain. The Afghan vice president and a cabinet minister have been assassinated, there was an attempt on the life of President Karzai, and regional warlords continue to compete for power. Tensions continue between India and Pakistan over the disputed territory of Kashmir, a situation that threatens not only war between two countries armed with nuclear weapons but also has implications for the war on terrorism. Within India, religious nationalism is behind periodic clashes between Hindus and minority Muslims and Christians. Efforts by lower castes to gain more political power have also resulted in violence. Parts of Africa also continue to be conflict-prone. A tenuous peace prevails in the Democratic Republic of Congo where the fighting drew neighboring countries into the conflict, resulting in what has been called Africa's "first world war." Angola's long, brutal civil war finally ended with the death of rebel leader Jonas Savimbi in February 2002, but rebel troops must still be reintegrated into society, always a potentially dangerous process. Cote d'Ivoire, which had been stable and successful, was rocked by a coup in 1999, and a recent rebel uprising threatens to plunge the country into a conflict similar to those that took place in neighboring Sierra Leone and Liberia. In South America, the Colombian government continues to face a serious challenge by leftist rebels, a conflict made more complicated by ties between the rebels and drug traffickers as well as the activities of right-wing paramilitaries.

The threats to peace and stability in the developing world remain complicated and dangerous and clearly have the potential to affect the Western industrialized countries. These circumstances require a greater effort to understand and resolve conflicts, whatever their source.

Fight to the Finish

Has a 'clash of civilizations' threatened 'the end of history'?

By Jay Tolson

Americans now have at least an idea of who the enemy is: a network of terrorists who may or may not have the support of one or more nations. But there is another question, with potentially greater implications for U.S. foreign policy: What are we fighting?

Terrorism alone is a complicated challenge, of course. Defeating it, experts say, will require the same sustained commitment needed for fighting a conventional war, along with a shrewd appreciation of how it differs from a classic military operation. Without drawing these subtle distinctions, says Martha Crenshaw, a professor of government at Wesleyan University and a terrorism specialist, "you are likely to create expectations… that might be impossible to meet: What is victory? And when does it arrive?" To balance the desire for revenge and justice with the need for lasting security, two objectives should be utmost, Crenshaw says: "In the short run, we have to crack down on the networks; in the long run, we have to drain the swamps that spawned them."

And to drain those swamps, Washington has to know what it is fighting—a question that goes to the heart of the two most provocative theories on global politics after the Cold War, published in articles by Francis Fukuyama ("The End of History?") and Samuel Huntington ("The Clash of Civilizations?").

Writing in the same year as the fall of the Berlin Wall, Fukuyama, then a State Department official, argued that the end of the Cold War signaled the approaching end of "mankind's ideological evolution and the universalization of Western liberal democracy as the final form of human government." In other words, the West had won. There would be blips on the screen, he allowed, and temporary setbacks in different parts of the world, but the path to an essentially harmonious global political economy was set.

Not too long after Fukuyama outlined his scenario, Huntington, a professor of government at Harvard, issued a far more ominous forecast. The passing of the Cold War had brought an end to serious competition among nation-states, but it had also launched an era of growing competition among the world's major civilizations. Where once ideologies had been the points of conflict, now religion, ethnicity, and cultural values were. "Most important," he wrote, "the efforts of the West to promote its values of democracy and liberalism as universal values, to maintain its military predominance, and to advance its economic interests engender countering responses from other civilizations." The stage for epic struggles—the West versus the rest—had been set. And none loomed more menacingly than the one between the West and the Islamic world.

So, then, in the still-smoky aftermath of the horrors, has Huntington been vindicated and Fukuyama disproved? Huntington has so far declined to comment, but Fukuyama remains guardedly hopeful. "In general, the Muslim world has had the most problems with modernizing," he says, "but even there the modernizing trends are strong." At the same time, Fukuyama cautions, "A lot depends on how the administration responds."

To many Islamic specialists, even suggesting that the current conflict is a clash between civilizations is handing victory to the terrorists. To Osama bin Laden, who in 1998 declared that his struggle was a continuation of the struggle against infidel Crusaders of the Middle Ages, nothing sounds sweeter than President Bush using the word "crusade" to describe America's intentions. It confirms, in his own eyes and those of his followers, bin Laden's sense of himself as a modern-day Saladin—the sultan of Egypt and Syria who captured Jerusalem and defended it during the Third Crusade.

Islam versus Islam. Some scholars say the battle lines are different: "It's not Islam versus the West as much as it is Islam versus Islam," says Akbar Ahmed, former Pakistani ambassador to the United Kingdom and professor of Islamic studies at American University in Washington, D.C. Ahmed calls for what he terms a more Islamic response to the radical fringe. Many Islamic extremists (like many American commentators) equate globalism with Americanism and Westernization, for example, ignoring the Koran's own words: "There is neither East nor West for God." They also fail to acknowledge, Ahmed says, the long history of Muslim interaction with disparate religious cultures. Other scholars go even further, saying that extremism denies the spiritual richness and diversity within Islam, ignoring the Prophet's own call for tolerance.

While there are many variants of extreme Islam, most share qualities with a movement launched by an 18th-century scholar, Mohammed ibn Abd al-Wahhab, who lived in central Arabia and found a receptive ear in the Saud family, which in 1932 became the ruling dynasty of the new kingdom of Saudi Arabia. Claiming to return to original Islam, Wahhabism rejected all innovations, stressed literal belief in the Koran and hadith (the traditions of Mohammed), and called for the creation of a state run strictly according to Islamic law. Most threateningly, says Sheik Hisham Kabbani, chairman of the Islamic Supreme Council of America, the fundamentalists "considered traditional Muslims apostates," a judgment that the Prophet had said no mortal should make. The Prophet had foreseen that there would be many divergent practices within Islam, Kabbani notes, but said only God could judge which version is true.

Islamic civilization continued to spread and diversify. But the European nations' surrender of their colonies during the first half of the 20th century left the newly independent Islamic nations throughout North Africa and the Middle East with great challenges. Without democratic traditions, all acquired autocratic regimes of one kind or another. Their intellectuals and elites, meanwhile, tended to embrace the secularism of their former colonizers in their quest for modernity and progress. Religious practice and scholarship were neglected, and the rulers increasingly cracked down on their critics. Into this era of ferment came the Jewish state of Israel, which arose in the center of the Levant. One of its primary supporters, the United States, became a major player in the region, more than once having its way by supporting or installing unsavory regimes.

Social justice. Thus the region became ripe for movements (particularly fundamentalist ones) promising social justice and ethnic solidarity. And all the better if they painted Uncle Sam as the Great Satan, the wicked meddler in local affairs. But the notion that the United States is the decisive factor behind the rise of fanatical Islam—or of the terror it inspires—is absurd, as Richard Cohen noted in the Washington Post. Even as Washington attempted to broker a settlement between Israel and the Palestinian Authority, bin Laden's associates carried out their attacks on U.S. embassies and the USS Cole.

If there is a failing within America's foreign-policy establishment, says Charles Fairbanks, a professor at the Johns Hopkins School of Advanced International Studies, it is that it needs to figure religion more seriously into the geopolitical equation: "It's in the interest of the United States and other countries to try to encourage the recovery of traditional Islamic religious learning." Sheik Kabbani concurs, adding that it's the fanatics who "now have the mike." But to be helpful, Americans need a rudimentary knowledge of Islamic traditions and teaching so that they will know who should get aid, and who should not. At present, says Kabbani, some extremists "are getting support even from the regimes they are trying to destroy." If the United States and other nations do not help the traditionalist Muslims win the theological struggle within Islamic civilization, then the West may face what the extremists want: a real and possibly cataclysmic clash between civilizations.

The Crisis Within Islam

Who speaks for Islam? The events of September 11, which left
the world waiting for a decisive repudiation of terrorism by
Islam's leaders, give that question fresh urgency. Answering it
will require the resolution of a century-old crisis that has
silenced many of those who speak for the Muslim majority.

by Richard W. Bulliet

Islam is a religion of peace, President George Bush has declared. The imam at the local mosque has likely offered the same
assurance, as has your Muslim neighbor or coworker. Yet many
in the West remain suspicious that Islam is not at all a peaceful
faith, and that the conflict sparked by the September 11 attacks
is not just a war against terrorism but a "clash of civilizations."

It's not hard to understand why. Osama bin Laden, who became the world's best-known Muslim during the 1990s, declared that there is no path open to a believing Muslim except
jihad, or holy war, against the United States. Islamic authorities
who refuse to join him, bin Laden said, are betraying the faith.
At the same time, the few prominent Muslims who have dis-
owned the terrorism perpetrated in Islam's name on September
11 and actively affirmed its peaceful character have been
drowned out by the silence of the many others who have not, or
who have in their confusion failed to condemn unequivocally
bin Laden's acts.

This strange silence does not reflect the attitude of traditional
Islam but is a painful manifestation of a crisis of authority that
has been building within Islam for a century. It is this crisis that
allowed bin Laden, despite his lack of a formal religious education or an authoritative religious position, to assume the role of
spokesman for the world's Muslims. The crisis has undermined
the traditional leaders who should be in a position to disqualify
or overrule a man who does not speak—or act—for Islam.

Today's crisis grows in part out of the structure of Islam itself—a faith without denominations, hierarchies, and centralized institutions. The absence of such structures has been a
source of strength that has permitted the faith to adapt to local
conditions and win converts around the world. But it is also a
weakness that makes it difficult for Muslims to come together
and speak with one voice on important issues—to say what is
and what is not true Islam.

Islam's structural weakness has been immeasurably magnified by a series of historical forces that have gradually compromised the authority of its traditional religious leaders in the
Middle East and elsewhere. The imams and muftis (legal
scholars) who once shaped the worldviews of ordinary Muslims
and confidently articulated the meaning of the faith have been
overshadowed by more innovative and often radical figures
with much shallower roots in tradition. Hundreds of millions of
ordinary Muslims feel that they understand their religion per-
fectly well, and that it provides no justification for the murderous crashing of airliners into the World Trade Center and the
Pentagon. But until Islam's crisis of authority is resolved, these
people will have no voice, and public confusion about what
Islam really stands for will persist.

The crisis has three related historical causes: the marginalization of traditional Muslim authorities over the past century
and a half; the rise of new authorities with inferior credentials
but greater skill in using print and, more recently, electronic
media; and the spread of mass literacy in the Muslim world,
which made the challengers' writings accessible to vast new
audiences.

The deepest roots of the crisis go back to the early 19th century, when the Muslim world was forced to begin coming to grips with the challenge of European imperialism. Governments in these countries responded by embracing a variety of reforms based on European models. This response began in Egypt and the Ottoman Empire (which both escaped the imperial yoke) in the early 19th century; spread to Iran, Tunisia, and Morocco by the end of the century; and was then embraced in many countries during the era of decolonization after World War II. In subject lands—including India, Indonesia, Malaysia, Algeria, and West Africa—European colonial governments imposed similar reforms from above.

Strongly influenced by the example of European anticlericalism, which seemed to 19th-century leaders in Egypt and the Ottoman Empire to be an essential element in the making of European might, these leaders moved to strip traditional Muslim religious authorities of their institutional and financial power. Later, popular leaders such as Mustafa Kemal Ataturk (1881–1938) in Turkey and Hafez al-Assad (1928–2001) in Syria, continued the attack in the name of secular nationalism. By secularism, however, they meant not separation of church and state but suppression of the church by the state.

For centuries, the traditional religious authorities had interpreted and administered the law in Muslim lands. The reformers replaced Islamic sharia with legal codes of European inspiration, and lawyers trained in the new legal thinking took the place of religiously trained judges and jurisconsults in new European-style courts.

The 19th-century Egyptian and Ottoman reformers also established new schools to train military officers and government officials. These elite institutions, which were to serve as models for most mass school systems in the Middle East after World War II, taught modern subjects such as science and foreign languages—though, signicantly, little in the way of liberal arts—and worked to instill a secular outlook in their students. The traditional Islamic schools were discontinued, downgraded, or stripped of funding.

Another traditional element that lost prominence in 19th-century Muslim society due to the opposition of reformist governments was the ubiquitous Sufi brotherhoods—mass religious organizations that held out the promise of a mystical union with God. The secular leaders of the modernizing nations feared that the Sufi sheiks, with their otherworldly perspectives and intellectual independence, might become a significant source of resistance to reform. But the decline of Sufism left a spiritual vacuum that nationalist zeal ultimately fell far short of filling.

In many parts of the Islamic world after 1800, governments took control of the financial endowments that mosques, seminaries, and other religious institutions had amassed over the years from the contributions of the faithful. Many of these endowments were considerable, and in Egypt, Iran, and other countries had had the effect of gradually concentrating a significant share of the national wealth under religious control. Confiscating this resource, as Egypt did early in the 19th century, or centralizing its administration in a government ministry, the later Ottoman practice, put financial control in the hands of the state. Mosque officials, teachers, and others employed in many religious institutions now were subject to government pressure.

This slow but persistent assault on the foundations of religious authority diminished the stature and influence of traditional religious leaders in public life. Many ordinary Muslims grew to distrust the pronouncements of their religious leaders. Were their views shaped by religious conscience and learning, or by the need to curry favor with the government officials who controlled their purse strings? By the 1930s the sun clearly was setting on the old authorities.

Even as governments in the Middle East and elsewhere were hammering at the sources of traditional religious authority, a powerful technological revolution struck a second blow. Printing technology, which had begun to transform European society in the 15th century, had its first impact in the Islamic religious world only in the second half of the 19th century (though government and the technical fields were affected somewhat earlier). For centuries, the lines of religious authority within Islam had been formed by personal links between teachers and their disciples. Now this traditional mode of preserving, refining, and transmitting ideas faced competition from writers, editors, and publishers with little or no formal religious training and few ties to established teachers. They became authorities simply by virtue of putting their words into print. A Muslim in Egypt could become a devoted follower of a writer in Pakistan without ever meeting him or anyone else who had met him.

Al-Manar (The Minaret), a magazine published in Cairo by Rashid Rida between 1898 and 1935, provides a typical example of how this new trade in religious ideas worked. Rida had studied in both an Ottoman state school with a "modern" curriculum and an Islamic school, but he wielded his influence as a writer and editor. In the pages of *Al-Manar* thousands of Muslims around the world first encountered the modernist ideas of Rida's mentor, Muhammad Abduh, an advocate of Islam's compatibility with modern science and of greater independence in Muslim thought. But Rida soon took the magazine in another direction, advocating Arab nationalism and eventually embracing the religious conservatism of Saudi Arabia.

By tradition, a Muslim teacher's authority rested on his mastery of many centuries of legal, theological, and ethical thought. But as lawyers, doctors, economists, sociologists, engineers, and educators spewed forth articles, pamphlets, and books on the Islamic condition, this ancient view lost force. After World War II, the most popular, innovative, and inspiring thinkers in the Islamic world boasted secular rather than religious educational backgrounds. (This is still the case. Bin Laden, for example, was trained as an engineer; his associate Ayman al-Zawahiri was a surgeon; and their ideological predecessor Sayyid Qutb was an Egyptian schoolteacher.)

Because radio and television were under strict government control in most Muslim countries, these new thinkers ex-

pounded their ideas in print—at least until the advent of audio-
and videocassettes made other mediums possible. The Islamic
Revolution of 1979 in Iran brought worldwide prominence not
only to Ayatollah Khomeini, an authority of the old type who
used books and audiotapes to spread his views, but also to the
sociologist Ali Shariati, whose writings and spellbinding ora-
tory galvanized Iran's university students, and the economist
Abolhasan Bani Sadr, who was elected president of the new Islamic Republic in 1981. In Sudan, lawyers Mahmoud Muhammad Taha and Hasan Turabi gained large followings; the
philosophers Hasan Hanafi in Egypt, and Muhammad Arkoun
in Algeria both propounded influential interpretations of Islam.

The new thinkers of the past half-century have offered a wide
variety of ideas. Some have called for a return to life as it was
lived in Muhammad's time (though they often disagree about
what seventh-century life was like) and disparaged the teachings of scholars from later centuries. Others have joined bin
Laden in preaching terrorist violence as the solution to Islam's
problems. Still others, such as Rashid Ghannushi in Tunisia and
Abbassi al-Madani in Algeria, have called for the creation of Islamic political parties and for their open competition with other
parties in free and democratic elections. In Iran, President Muhammad Khatami leads a powerful, democratically oriented reform movement.

It is also true, however, that some of the leaders who capitalized on the new media to build large followings were both extremists and formally trained religious figures. Khomeini is the
most obvious example; Egypt's Sheik Umar Abdur-rahman,
who is languishing in an American prison since being convicted for his role in the 1993 World Trade Center bombing, is
another.

The final element in the making of today's crisis was the decision by the newly independent states of the post-World War II
era to pursue compulsory education and mass literacy. The
young Muslims who came of age in the developing world
during the l960s thus had the tools to read what the new authorities were writing. Because their schooling included minimum
exposure to the traditional religious curriculum and texts—and
in many cases admonitions by their government teachers not to
put too much stock in religious scholarship—they did not feel
obliged to follow the dictates of the old authorities. And they
appreciated the contemporary vocabulary and viewpoints of the
new Islamic writers. So long as nationalism offered them the
promise of a better future, they remained loyal to their political
leaders and governments. But when the nationalists' dreams
failed and the future dimmed, as it did in most Muslim countries
during the 1970s, people looked elsewhere for hope and inspiration, and they didn't have to look far.

Traditional Islam is far from dead. Many Muslims still stand
firmly by the legal opinions (fatwas) and moral guidance of traditionally educated muftis and the orthodox teachings of the
imams at their local mosques. But the momentum seems to be
with the new authorities. This has created an unusual dynamic

within the Muslim world. While the new authorities seldom de-
fer to the old, the old feel compelled to endorse some of their
rivals' ideas in order to seem up to date and retain influence.
The locus of debate thus has been steadily shifting in favor of
the new authorities.

Local imams and other religious officials are also dependent
(in a way their rivals usually are not) on their national government. They are caught in a three-way squeeze between government interests, their religious training, and the popular
teachings of their rivals. This helps explain the strange silence
that has prevailed since September 11. Some traditional religious figures have chosen to say nothing. Some have tacitly admitted the evil of terrorism while denying that Islam and
Muslims had anything to do with the attacks. Some have resorted to anti-American rhetoric. And some have condemned
the terrorist acts but stopped short of recognizing and condemning the instigators.

This failure of the traditional leadership has left Muslims everywhere in a quandary. They know what their faith means to
them, and they think this meaning should be obvious to everyone. They do not pray five times a day, fast during Ramadan,
make the pilgrimage to Mecca, and live modest, peaceful, hardworking lives for the secret purpose of destroying Western civilization and slaughtering Americans. They find the association
of such violent ideas with their religion odious and prepos-
terous-and threatening if they happen to live in the United
States. Yet nobody seems to speak for them.

This is not to suggest that giving voice to the feelings of ordinary Muslims would somehow release a hidden reservoir of
support for America's global preeminence and its policies in the
Middle East and other regions. Many, if not most, Muslims are
highly critical of these policies. Those with the strongest anti-
American feelings applauded the events of September 11 and
praised bin Laden for launching them—even, in some cases,
while shuddering at the thought of living in a world governed
by his religious vision. But these supporters of terror, though
prominently featured on television, do not represent the Muslim
majority. Indeed, a good number of the Muslim world's apolo-
gists for terror are not themselves religious people.

In any event, opposition to U.S. policies is hardly restricted
to the Islamic world. No one should mistake political views for
religious ones—millions of non-Muslims (including some
Americans) voice similar criticisms of the United States. For
Americans to want Muslims to repudiate terrorism and disown
its authors is reasonable. To want them to agree wholeheartedly
with everything America does in the world is unrealistic.

What Muslims lack in this moment of crisis is a clear, decisive, and unequivocal religious authority able to declare that the
killing of innocents by terrorist attacks is contrary to Islam and
to explain how Muslims can stand firmly against terrorism
without seeming to embrace the United States and its policies.
When authority itself is in question, the middle gives way.

History suggests that Islam will overcome its current crisis of
authority, just as it has overcome a number of other crises in its
past. The first of these arose soon after the prophet Muham-

mad's death in A.D. 632. Later in the seventh century, as the generation that had personally known Muhammad died off, the Muslim community split over several issues, particularly the proper line of succession to the caliphate that had been established after Muhammad's death. (It was from this crisis that the Sunni-Shiite split grew.) Civil wars erupted. The crisis of authority was temporarily resolved by the consolidation of a military state, the Umayyad Caliphate, and the suppression of dissent. The caliphate shifted the seat of power from Medina, in Arabia, to Damascus, and quickly extended its rule over a vast empire that stretched from Spain in the west to what is now Pakistan in the east.

In the middle of the ninth century, as the conversion of non-Arab peoples brought into Islam people bearing the traditions of Christianity, Judaism, Zoroastrianism, Buddhism, and Greek philosophy, Islam again entered a period of uncertainty. The caliphate had passed into the hands of the Abbasids, so named because they claimed descent from the Prophet's uncle Abbas. The caliphate, its seat now in Baghdad, flourished—this period was in many ways the apex of Arab civilization. But when a new religious challenge arose, the caliph's resort to force failed. Against him was arrayed a new class of religious scholars who maintained that Muslims should follow the tradition of the prophet Muhammad, as preserved in a multitude of sayings and anecdotes, rather than the dictates of a caliph in Baghdad. Today's declining Islamic authorities date the beginnings of their power to this confrontation. Under the leadership of the scholar Ahmad ibn Hanbal and others who followed him, it was eventually agreed that Muslims would look to a consensus of scholars—in theory, throughout the Muslim lands, but in practice within each locality—for guidance on how to live moral lives. (Ahmad ibn Hanbal himself was founder of one of the four main schools of Islamic law within the Sunni tradition.)

A fresh crisis of authority arose, however, as it became evident that the sayings of the Prophet were too numerous and internally contradictory for all of them to be true. A new group of scholars set out to establish rules for determining which sayings were most likely to be true, and they gradually collected the most reliable of them into books. Nevertheless, several centuries elapsed before these books of "sound" traditions won recognition as the sole authoritative guides to Muslim behavior.

The key to this recognition was the spread during the 12th and 13th centuries of *madrasas*, Islamic seminaries that had first appeared in Iran in the 10th century. Institutions such as al-Azhar in Cairo, the Zaituna Mosque in Tunis, the Qarawiyin Madrasa in Fez, and clusters of seminaries in Mecca and in Ottoman Istanbul and Bursa gained particular eminence. The *madrasas* adopted the authoritative compilations of prophetic traditions as a fundamental part of their curricula, along with instruction in the Koran and the Arabic language. Other collections were gradually forgotten. The Muslim religious schools of today, whether grand edifices like al-Azhar and the Shiite seminaries at Qum in Iran, or the myriad humble *madrasas* of Pakistan and *pesantrens* of Indonesia, have roots in the resolution of this crisis of authority that arose more than 800 years ago.

Even as the *madrasas* were being established, a new upheaval was beginning. It grew out of the feeling of many common people—including those in late-converting rural areas of the Middle East and more recently Islamized lands in West Africa, the Balkans, and Central, South, and Southeast Asia—that Islam had become too legalistic and impersonal under the guidance of the scholars and *madrasas*. Religious practice, these Muslims felt, had become a matter of obeying the sharia and little else. The rise of Sufi brotherhoods beginning in the 13th century was a response to this popular demand for a more intense spiritual and communal life. Born in the Middle East, Sufism spread quickly throughout the Muslim world. The Sufis made room for music, dancing, chanting, and other manifestations of devotion that were not permitted in the mosque. But Sufi practices did not supersede conventional worship; the sheiks who led the Sufi brotherhoods provided religious guidance that paralleled rather than opposed the authority exercised by the established scholars and seminaries.

One can see in this capsule history of Islamic religious development a demonstration of the fact that a faith with no central institution for determining what is good or bad practice is bound to experience periodic crises of authority. But this history also demonstrates that the Muslim religious community has overcome every crisis it has confronted.

How will it overcome this one? There is no way to rebuild religious authority on the old foundations. The modern state, the modern media, and the modern citizen must be part of any solution. Islam's history suggests that any new institutions that grow out of the current crisis will not supplant those already in place. Seminaries will continue to impart to their students a mastery of fundamental legal and interpretive texts, and their graduates will continue to issue weighty legal opinions. Because Muslims retain a historical memory of being unified under a caliphate—a powerful state predicated on Islamic teachings—the dream of Islamic political unity will not disappear.

Any response to the current crisis must appeal to the many Muslims whose spiritual, moral, and intellectual needs have not been met by the faith's traditional institutions. Fortunately, the violent, totalitarian philosophy of bin Laden and his allies represents only one of the possible responses. Others are more promising.

Throughout the Muslim world organizations modeled (consciously or unconsciously) on the ancient Sufi brotherhoods but expounding this-worldly interpretations of Islam have been able to attract thousands of members. (A revival of Sufism itself seems to be underway in Iran, Central Asia, and other areas.) In some ways resembling political parties, but dedicated as well to the pursuit of social welfare programs, these fraternal organizations often present themselves as prototypes of a modern, nonclerical form of Islamic government. The Muslim Brotherhood in Egypt, the Islamic Salvation Front in Algeria, and the Hezbollah (Party of God) in Lebanon differ widely in their interpretations of Islam, but they share a willingness to exist in a modern political world of participatory institutions. The Islamic Salvation Front actually triumphed in the first round of Algeria's 1991 parliamentary elections and failed to take power only because the Algerian military stepped in. The country has been convulsed by violence ever since.

No one can safely predict whether the participation of such groups in an electoral system would further the spread of democracy or simply give them a platform for preaching noxious doctrines. Hezbollah leader Sheik Muhammad Fadlallah, for example, has embraced the concept of a secular, multiparty political system in Lebanon, even at the cost of alienating some of the support within Iran for his Shiite group. But Hezbollah originally rose to prominence in Lebanon through violence during the country's years of civil war (and it has continued its campaign against Israel). Still, the fact that such groups formally advocate participatory governing institutions—and that the Islamic Republic of Iran has developed such institutions—does give reason for hope.

Another set of possibilities for change within Islam is provided by educational and research institutions that exist independently of both traditional seminaries and formal government educational systems. These institutions provide venues for modern Muslim intellectuals to develop new ideas about contemporary issues. They are as likely to be found in London, Paris, and Washington as in Cairo and Istanbul—London's Institute of Islamic Political Thought and the Institute of Islamic and Arabic Sciences in America, outside Washington, D.C., are leading examples—and the thinkers they host often provide valuable guidance for the growing population of Muslims living outside the Muslim world.

In some Muslim countries, governments now sponsor educational institutions devoted to teaching about Islam from the perspective of the contemporary world. The Institutes of Higher Islamic Studies in Indonesia are a notable example. Some of these institutes may soon become full-fledged universities offering both religious and secular courses.

Iran may seem an unlikely quarter in which to look for encouragement, but it too may provide some clues to the future direction of Islam. There, an avowedly Islamic state is pursuing a unique experiment integrating elections and other modern political elements into an Islamic framework of government. Though Iran may prove to be the first and only enduring Islamic republic, the intellectual trends that have developed there, sometimes to the dismay of conservative religious leaders with seminary backgrounds, encourage Muslims to think that a lively intellectual life and engagement with worldwide currents of thought can survive and flourish in a religious environment. Iran remains far from a model republic, but the trajectory that has taken it from being a country bent on the export of revolution to one with a sizable electoral majority favoring liberalization is encouraging.

Finally, another source of innovation may be the substantial numbers of secular Muslims who — contrary to the Western stereotype — live not only outside the traditional boundaries of the Islamic world but within them. Secular Muslim thinkers have been elaborating the idea of *turath* (heritage) as a point of intersection between the past and a present in which the particulars of religious law and practice seem irrelevant. In engaging the "modern" Muslim intellectuals, these secularists are striving to create legitimacy for nonobservant forms of Islam.

Although these modernizers within contemporary Islam seem to work at cross purposes as much as they work in concert, some sort of fusion among them seems the most likely route to resolving today's crisis of authority. There is little possibility that nonobservant Muslim intellectuals, ideologues of Islamic political parties, thinkers attached to centers and institutes, and teachers in government-sponsored religious schools will ever see eye to eye on everything. But in the past, discord within Islam was often resolved when Muslim leaders agreed to respect divergent views while recognizing a common interest in the welfare of the global Muslim community. Muhammad himself declared, in one of his most often-cited sayings, "The difference of opinion in my community is a divine mercy."

But more immediate action is needed than the development of long-term concord within Islam. The ugly alternative is a "clash of civilizations" like the one envisioned by Harvard University political scientist Samuel Huntington and echoed in the propaganda of bin Laden and other extremists. Polarizing the world between Islam and the West would serve the interest of the people who fly airliners into skyscrapers; it would spell tragedy for everybody else. Even if Islam's uncertain authorities, new and old, cannot agree on issues that might imply support for American foreign policy, they should be able to recognize an oncoming catastrophe and take measures to avoid it.

Islam's leaders must act. The heads of Islamic centers and institutes around the world, along with leading Muslim intellectuals of every persuasion, must clarify the meaning of their faith. Non-Muslims in the United States and other countries are eager for signs of leadership in the Muslim world. They await an affirmation that the vision of a peaceful, fraternal world embodied in Islam's past and in the hearts of most ordinary Muslims still guides the people who claim to speak in Islam's name. The crisis of September 11 can be the crucible in which the tools for resolving Islam's own crisis of authority are forged. The lessons of the past encourage hope that Islam will find a path out of its confusion of voices. We listen with hope in our hearts.

RICHARD W. BULLIET *is a professor of Middle Eastern history at Columbia University and directed Columbia's Middle East Institute for 12 years. He is the author of* Islam: The View from the Edge *(1994) and editor of* The Columbia History of the Twentieth Century *(1998). Copyright (c) 2002 by Richard W. Bulliet.*

From *The Wilson Quarterly*, Winter 2002, pp. 11-19. © 2002 by Richard W. Bulliet. Reprinted by permission.

Mixed Message: The Arab and Muslim Response to 'Terrorism'

Mustafa Al Sayyid

Many Arab and Muslim countries sympathized with the victims of September 11 and offered valuable support to the United States in its campaign against Osama bin Laden's organization and the Taliban regime in Afghanistan. Yet, large sections of the Arab and Muslim public, as well as many of their governments, cannot offer the United States full support in its fight against terrorism because they do not share with the United States the same definition of terrorism and suspect a hidden agenda behind the future phases of this campaign. The general public in the West, particularly in the United States, may not realize that the earliest victims of armed groups claiming to be inspired by certain interpretations of Islam were themselves Muslims—intellectuals, senior officials of government, ordinary citizens, and security forces. These people lived in Muslim countries such as Algeria, Egypt, Yemen, and Saudi Arabia years before a small group of alleged members of an Islamic organization launched its deadly attacks of September 11 in New York City and Washington, D.C. Arab and Muslim countries, therefore, did not need any particular preaching on the part of Washington to join an international campaign against terrorism because many of them had long been involved. Arab people have learned, however, that terrorism cannot be defeated if those who fight it rely exclusively on military force.

Islam and Terrorism

To start this story at its inception, reflection on any possible link between Islam and terrorism is important. Because some Western media tend to label those individuals involved in terrorist actions "Muslim terrorists," the positions taken by Arab and Muslim states on the White House's "war on terrorism" should be analyzed. One should consider that the U. S. state Department's list of terrorist organizations mostly includes organizations that are active in Muslim countries, which becomes the focus of media reporting, while ignoring organizations in non-Muslim countries, such as Spain, Northern Ireland, and Latin America, and reinforcing a perception in Western public opinion that terrorism is exclusively Islamic. Western media also uses jihad to convey the notion of an armed struggle launched by Muslims against people of other religions in order to compel them to renounce their religions and adopt Islam. Based on these observations, examining if there is an unbroken link between Islam and terrorism is necessary.

Popular Western media tends to misconstrue the relationship between Islam and terrorism significantly. Attributing a particular policy position on the use of armed struggle for political ends to any religion as a whole is difficult. Among Christians, one can find militant priests in Latin America, such as Colombia's Camillo Torres, who justify armed struggle in terms of a theology of liberation. Islam, as any other world religion, can be interpreted in various ways.

More importantly, those who believe that Islam can guide and inspire a political order do not necessarily seek to establish that political order using peaceful methods. The Rafah Party in Turkey is one; others can be found in Jordan, Yemen, Lebanon, Indonesia, Malaysia, Pakistan, and Bangladesh. Islamic organizations such as the Muslim Brothers in Egypt, the Islamic Salvation Front in Algeria, and the Nahda Party in Tunisia accept a pluralist political system and an electoral path to political power. The governments in each of these countries bans these parties, however, because they constitute serious contenders for power that the ruling groups, who are reluctant to accept any transfer of power through the ballot box, reject.

Furthermore, the notion of forced conversion is alien to Islam. Although explaining their religion to others is a duty incumbent on Muslims, Islam considers the question of faith a personal matter. Most Muslims recognize and respect the religions of Christians and Jews and consider their holy books sacred texts for Muslims as well. Indeed, the Prophet Muhammad married a Coptic Christian from Egypt. When the notion of jihad was applied during the early days of the Prophet Muhammad, it meant armed struggle against the enemies of the new faith who were launching war against it. Once the new faith triumphed with Muhammad's entry into Mecca, jihad acquired a new meaning. In Muhammad's words, the "greater jihad" meant an inner struggle to suppress one's evil desires and elevate one's soul.

According to the most authoritative statements on the Islamic theory of international relations, the so-called division of the world into the realm of war (*dar al-harb*) and the realm of Islam (*dar al-islam*) does not hold in the modern world because reciprocal commitments to maintain friendly relations tie Muslim states to other countries. The mere establishment of diplomatic relations with other nations signals that the other country has become part of the realm of commitment (*dar al-'ahd*).[1]

Yet, political Islam—just like Arab nationalism and Marxism—can be interpreted in several ways. Some versions would call for the use of exclusively peaceful methods of political action. Other versions of the same ideologies would justify and legitimize armed struggle against those domestic and foreign powers that seem to pose an obstacle to the realization of the political strategy inspired by these ideologies. Thus, some Muslim activists would interpret verses of the Qur'an or traditions of the Prophet Mohammed to serve their own political ends, however they conceive them. Nevertheless, just as no one in his or her right mind would charge all Protestants or Catholics of being terrorists because certain Protestant or Catholic groups in Northern Ireland resort to armed action, by the same logic, the presence of certain terrorist groups that call themselves Islamic does not make Islam and all its adherents potential terrorists and a threat to the rest of humanity.

As this introduction conveys, Arab and Muslim states do not feel that they bear any special responsibility in the battle against terrorism. Even if one accepts the claims bin Laden made in his televised statements—as well as the charges of the U. S. government that the perpetrators of the tragic and condemnable attacks on September 11 are all Arabs and Muslims—terrorism has nothing to do with Islam and Arabism because these individuals cannot, by any stretch of the imagination, be considered representative of approximately one billion Muslims.

Not reflecting on the reasons that would lead people to commit such acts, however, would be completely irresponsible. Inquiring about the motivations is not an attempt to justify or excuse what happened on September 11, as commonly assumed in the United States, but just the opposite. If the causes of such acts are not understood, the victims of the tragic events of Sep-

tember 11, as well as the victims of any future terrorist acts, will be disserved. Those who do not learn from history are bound to pay a bitter price in the future.

Arab and Muslim governments also do not feel that they bear more responsibility than other countries in the fight against terrorism. Not only have a number of these governments been engaged in the fight against terrorism for many years, but they have also not received much support from the governments of the United States and the United Kingdom. At different times during the 1980s, Egypt, Algeria, Jordan, Syria, Saudi Arabia, and even Iraq and Libya were engaged in a fight against Islamic organizations that were using armed struggle in an attempt to overthrow or at least destabilize their governments. The extent of the armed insurrection has varied from country to country—sporadic in Jordan, Libya, Iraq, and Tunisia; more serious in Saudi Arabia; quite protracted in Egypt; and bloodiest, but short-lived, in Syria and Algeria. Pointing the finger at these countries and accusing them of not doing enough to curb terrorism would simply be ignoring well-established facts.

Moreover, leaders of these countries, particularly President Hosni Mubarak of Egypt, have called on Western governments not to provide easy asylum to well-known key figures in militant Islamic organizations who are wanted for trial for their involvement in terrorist acts. For example, Shaykh Omar Abdel-Rahmanthe—Spiritual leader of the Islamic Group, or *Al-Jama'a al-Islamiyyah,* who calls on members to launch armed attacks to topple the Egyptian government—was given an entry visa to the United States, where he stayed and continued to agitate against the Egyptian government until he was arrested for his role in the 1993 attempt to blow up the World Trade Center. Members of the Islamic Group, including two of the shaykh's sons, joined the International Islamic Front led by bin Laden in Afghanistan. The United Kingdom gave political asylum to Yasser Al-Sirri, suspected master-mind of the assassination of Ahmed Shah Mas'oud, the former leader of Afghanistan's Northern Alliance. Many leaders of Islamic organizations in Tunisia and Algeria, such as Rashid Al-Ghanoushi, have also found asylum in the United Kingdom. Habib Al-Adli, the Egyptian minister of interior, pointed out that, of all Western countries that were warned about the presence of Egyptian terrorists on their territories, only Italy was willing to offer some cooperation.[2]

Reactions of Arab and Muslim Countries to September 11

As with any broad discussion, examining "the" Arab and Muslim response to terrorism in general, and the September 11 attacks in particular, is difficult. The positions of Arab and Muslim countries on these issues cover a large spectrum. These positions, moreover, have not been consistent from the beginning of the crisis with the attacks of September 11 through the U. S. military operations that led to the fall of the Taliban regime in Afghanistan and the installment of a provisional administration in late December 2001. In addition, Arab governments' formal positions have not always truly reflected public opinion in these countries. Indeed, a major dilemma for

U. S. commentators has been not just the indifference but the open hostility with which the public in a number of Arab and Muslim countries met the U. S. war against terrorism.[3]

When news of the September 11 events reached the capitals of Arab and Muslim countries, their reactions were varied. On the whole, all Arab and Muslim governments condemned the attacks and expressed sympathy for the American people. Holding ranks with other Arab governments, Iraq's President Saddam Hussein, Deputy Prime Minister Tariq Aziz, and Permanent Representative to the United Nations (UN) Mohammed al-Douri all deplored these attacks and expressed sympathy with their victims. Aziz in particular categorically rejected any link between Iraq and the perpetrators of the attacks.[4] When U. S. authorities first revealed the identities of those suspected of hijacking the planes, public opinion in Arab countries was generally skeptical of the reliability of such information, particularly because some of the names were of people who were either still alive in Saudi Arabia and the United Arab Emirates or who had died years earlier.

Numerous Arab commentators believed it to be unlikely that bin Laden, whom U. S. authorities declared a prime suspect in the attacks, could have been capable of masterminding such an elaborate, well-timed, and well-synchronized attack from his hideout in the mountains of Afghanistan. Many people in Arab countries remained perplexed about why bin Ladin declared his support for the attacks in later televised statements relayed to the Arab world through Qatar's Al Jazeera television network. Even when he suggested in another televised videotape that he knew in advance of the attacks and its details, Arab public opinion continued to be divided between those who believed that he masterminded the attacks and condemned him and those who thought that the attacks were the acts of the enemies of Arabs and Muslims who wanted to drive a wedge between the Arabs and Muslims and Western people. The conversation on that televised tape, for them, was simply fake.[5]

When the U. S. government resolved to undertake military action against Al Qaeda—the organization believed to be led by bin Laden and supported by the Taliban regime in Afghanistan, where bin Ladin has been living—most Arab and Muslim governments reluctantly joined what the U. S. government called "a worldwide coalition against terror." The degree of support varied, however, from total support (Kuwait, Bahrain, Qatar, and Jordan), to verbal support (Tunisia, Algeria, and Morocco), and support coupled with criticism (Saudi Arabia and Egypt).

Other than Pakistan, Uzbekistan, Turkey, and Jordan, few Arab and Muslim governments adopted a position of total support for U. S. military action in Afghanistan. Pakistan, which had been the Taliban's major source of economic, military, and diplomatic aid, shifted its posture completely—first, by attempting to persuade Afghanistan's Taliban leaders to agree to hand bin Laden over to U. S. authorities; later, by offering logistical and intelligence support to the U. S. military; and finally, by withdrawing diplomatic recognition of the Taliban regime following its retreat from Kabul in November when it seemed that its days were numbered. Pakistan's alignment with the United States was a major risk for its head of state, General Pervez Musharraf, who faced difficult foreign and domestic policy choices. On the one hand, India, Pakistan's traditional rival, was offering the United States varied military assistance in the war against the Taliban regime. Had Pakistan denied the United States the same assistance, it would have incurred the United States' wrath, enabling India to gain a diplomatic advantage. On the other hand, if Pakistan aided the United States, Musharraf's government would face serious domestic opposition and would be sacrificing a friendly neighboring regime in Afghanistan. Musharraf decided to take the risk of increased domestic tension to gain some U. S. diplomatic and economic support, deny India its monopoly of U. S. favors in the subcontinent, and hopefully ensure a voice for Pakistan in deciding the future of Afghanistan.

Uzbekistan also took a major risk by allowing the United States to use its air bases to launch operations in northern Afghanistan, to provide areas from which U. S. Special Forces could undertake underground operations into northern Afghanistan in the early phases of the war, and to assist Northern Alliance troops. Uzbekistan's strategic location and the Northern Alliance's early successes, gaining control over Mazar-e Sharif as well as Kabul, reduced the strategic importance of Pakistan in the later stages of the U. S. military effort in Afghanistan. Other countries of the Commonwealth of Independent States, such as Tajikistan, Turkmenistan, Kyrgyzstan, and Kazakhastan, offered logistical services, later acknowledged by Secretary of State Colin Powell in his tour of these countries following the fall of the Taliban regime in December. Some of them faced internal opposition by Islamic movements that the Taliban regime supported; all were hoping to benefit from U. S. generosity after the war.

Turkey's support for the United States surprised no one. The Turkish government offered to send a small contingent of troops to train Northern Alliance forces and expressed its readiness to send troops to participate in a future peacekeeping force in Afghanistan. The Turkish government's position is consistent with the domestic and foreign policy it has pursued in recent years, which the country's military establishment firmly supports. The Turkish government has been adopting an extremely secular stance in its relations toward its domestic Islamic parties, to the extent that it has outlawed even those Islamic parties that abide by the rules of democratic politics. In foreign policy, the Turkish government consider alignment with U. S. policies the best way to get the Bush administration to use its leverage to persuade European countries to relax the requirements for Turkey's admission into the European Union. Moreover, Turkey views a military pressence in Afghanistan as a way to strengthen its position as a key player in the regional politics of Central Asia, a region in which it shows great interest.

Thus far, the Jordanian government has been the only other Arab government to offer to send troops to partake in a future peacekeeping force in Afghanistan. Jordan's King Abdullah, standing next to President George W. Bush at the White House in mid-September 2001, was eager to announce Jordan's total solidarity with the American people as well as its willingness to offer all the aid it could under the present circumstances. The young king has been careful to pursue the policy that his father, who tried to safeguard the independence of his country by allying Jordan with the dominant powers in the Middle East and in the world, had laid down. King Abdullah wants to maintain cordial relations with the United States and avoid deteriorating relations with Israel, despite the intransigent policies of the government of Prime Minister Ariel Sharon. Algerian president Abdul-Aziz Bouteflika's forceful position in support of U. S. actions should occasion no surprise.

Many other countries in the Arab and Muslim world support the United States in its political and military campaign against both bin Laden and Al Qaeda as well as the Taliban regime, as clearly indicated in statements issued by the Organization of the Islamic Conference in its Doha summit in October 2001.[6] Their support, however, is limited to backing the U. S. effort diplomatically, sharing intelligence, and freezing funds that individuals and organizations suspected of sympathizing with terrorists allegedly used, steps that follow UN Security Council Resolution 1373.

Another group of Arab and Muslim countries did not hesitate to condemn the terrorist attacks on the United States or to share intelligence with U. S. authorities, but these states—Egypt and Saudi Arabia, in particular—were critical of certain aspects of the U. S. response to the attacks. The *Washington Post* and the *New York Times,* among other U. S. newspapers, have published editorials that were quite critical of the positions of these two governments, accusing them of not providing the United States with sufficient support in its battle against terrorism.

Before the United States began military operations in Afghanistan, Mubarak, who is skeptical about using massive force to deal with terrorism, suggested that terrorism is a complex problem, one that is rooted in frustrations caused by a lack of progress in resolving issues affecting Arab peoples and one that requires an international solution. He reiterated a proposal he made for the first time in 1995 following a failed attempt on his life by Egyptian Islamists in Addis Ababa, Ethiopia. Mubarak attenuated his reservations against the use of force to deal with the problem of terrorist attacks on the United States, but he emphasized that no Egyptian troops would be sent to fight in Afghanistan. He added that Egypt would not participate in a peacekeeping force in Afghanistan. Mubarak was careful to point out, however, that the Egyptian government supported the U. S. military effort. He even expressed delight that the problem of bin Laden would finally vanish.[7]

Some Saudi officials were more outspoken in their disapproval and concern about the suffering of the civilian population in Afghanistan as a result of the U. S. military campaign. The Arab media, especially Al Jazeera, which is widely watched in Saudi Arabia, kept its viewers well informed about reports of the extent of civilian casualties in Afghanistan, a particularly sensitive issue for Saudi Arabia not only because of the country's claims to leadership in the Muslim world but also because of Saudi Arabia's close relationship with the Taliban regime. Saudi Arabia was one of three countries (along with Pakistan and the United Arab Emirates) that had recognized and maintained diplomatic relations with the Taliban government. (Only Pakistan maintained diplomatic relations with the Taliban regime after the fall of Kabul, allegedly on advice from the United States, as a way to maintain a channel of communication with the Taliban leadership.) Some Saudi officials also ex-

pressed displeasure that some of the personalities and organizations whose assets were frozen at the request of the UN Security Council committee in charge of implementing the resolution on fighting terrorism were included on that list on very shaky and suspicious grounds. Prince Nayef, the minister of interior, reiterated several times that U. S. authorities did not provide any convincing evidence about the involvement of Saudis in the September 11 attacks; the Saudi government revoked bin Laden's Saudi nationality some time ago. Although some Saudis criticized the civilian casualties in Afghanistan, U. S. officials nevertheless declared their appreciation of the cooperation they received from Saudi authorities in the war.

Other voices in the Arab and Muslim world were critical of the U. S. military campaign, but this reaction came from quarters that are critical of U. S. policy in general. Iran's President Muhammad Ali Khatami, for example, did not like the two choices that Bush put to the rest of the world—"Either you are with the United States, or you are with the terrorists."—arguing that Iran was neither with the United States nor with terrorists. The Iranian government also condemned the use of military force against the Afghan people. Nevertheless, the Iranian government has been careful not to take any practical steps to thwart U. S. military action in Afghanistan.

Public Opinion: Divided or Critical?

Despite the overall supportive positions taken by governments in Arab and Muslim countries toward the United States in its campaign against terrorism, reading a U. S. newspaper would give one the impression that public opinion in those countries is quite critical of the United States and even sympathetic toward bin Laden and his followers. Because most of these countries have no reliable polls to gauge public opinion, one can only attempt to do so by looking at media coverage in these countries. Some judgments can be made by examining a sampling of what has appeared in the mainstream media in the Arab world.

Public opinion in many Arab countries has been quite divided on the issue of what happened in New York City and Washington, D.C., on September 11, as well as the U. S. response to these events and the impact of the attacks on Arab and Muslim countries. Some have been concerned that these events would have a negative impact on relations between Arab as well as Muslim countries on the one hand and Western countries on the other, particularly the United States. Others were concerned about the projected images of Islam and Muslims, which certain media and prominent Western politicians immediately linked with terrorism and the absence of freedom. Some in the business sector were apprehensive that they might be punished twice—first by the economic forces of a deepened recession, then as a result of prejudices against Muslims that would be the natural outcome of the association some in the West try to establish between Islam and terrorism. Finally, some political strategists believed that Muslim countries had a great deal to gain from the West by aligning completely on the side of the United States at the moment when it most needed their moral support. Turkish, Pakistani, and Jordanian government supporters were among these groupings, as were those who called on the Egyptian government to endorse the U. S. position more forthrightly.

Nevertheless, many people in Arab and Muslim countries presumably were not convinced of any of the evidence that the U. S. government presented to prove that bin Laden and his organization organized the September 11 events. Many did not see any reason to insist that 25 million Afghan people pay the price for something that was allegedly done by an individual or a small group of people. The release of two of bin Laden videotapes, in which he came close to admitting knowledge of and even inspiration for the September 11 attacks, did not change these initial positions. According to one school of thought, no matter how distasteful the Taliban regime was, the war that the United States launched in Afghanistan has disrupted the lives of millions of Afghans who should have been spared that heavy price—notwithstanding the airdrops of food packages in the same areas that U. S. warplanes had bombed. For these segments of Arab and Muslim public opinion, the belief that an end to terrorism could be sought through the use of military means is quite naive. The use of force by the United States would breed more terrorism in the future, fed by the frustrations of victims of the U. S. attacks, many of whom perhaps had not been involved with the Taliban regime but suffered as a result of the so-called collateral damage of U. S. military operations.

Moreover, these same segments of Arab and Muslim public opinion consider the U. S. administration's definition of terrorism to be rather selective. One can understand that the United States considers bin Laden and members of Al Qaeda—who have acknowledged their responsibility for attacks on the U. S. embassies in Nairobi and Dar es Salaam as well as on U. S. military bases and a warship in Saudi Arabia and the Persian Gulf region—to be terrorists. The same standard would no doubt apply to those who perpetrated the September attacks. Arabs, Muslims, and many others around the world, however, question the U. S. criteria that categorize groups that are fighting against foreign military occupation as terrorist organizations, even though U. S. State Department experts know fully well that international law recognizes the legitimate right of all people to self-defense. Thus, many Middle Easterners feel that the White House adapts its concept of terrorism to whatever suits its needs at the moment and considers those fighting Israeli military occupation of the Palestinian territories in the West Bank and Gaza Strip and what remains in southern Lebanon to be terrorists.

From this viewpoint, the U. S. administration prefers to ignore real terrorist practices when its closest friends, foremost Israel, conduct them. The Israeli government not only has persisted in its military occupation of Arab territories in Palestine, the Golan Heights, and the Sheb'a farms in southern Lebanon, but it also has consolidated this occupation by building settlements in defiance of several UN resolutions that the U. S. government itself supported. Reneging on its promise to negotiate an end to occupation, the Israeli government has declared the Oslo accords dead and, in the face of Palestinians' legitimate resistance to the occupation, has proceeded to carry out a policy of siege, closure of Palestinian towns, economic blockade, and

targeted assassination of Palestinian leaders. Any objective observer would find that state terrorism is the only name that can be used to describe this Israeli policy. The U. S. administration's reaction has been at best a mild reproach of Israel that rarely comes at the worst moment of Israeli atrocities and is usually pronounced by a junior member of the administration. In the eyes of the Arab and Muslim public, this course of action amounts to a double standard.

Another major cause for concern among Arab and Muslim people has been the mystery the U. S. administration has fostered about its plans for the duration of the campaign against terrorism. Bush has stated a number of times that the campaign against terrorism would be a long one that would continue for years and involve other countries and organizations. Initial reports in the U. S. press and later statements by senior U. S. officials suggested that some Arab countries were indeed targeted for future phases of the campaign, the prime candidate being Iraq. Bush himself threatened the use of force if Hussein refused to allow UN arms inspectors to return to Iraq. Powell said on December 5 that the United States had not yet decided whether to launch a strike against Iraq.

Such statements have been so alarming that many world leaders, including the German chancellor and French president as well as Egyptian and Saudi senior officials, have warned that striking at Iraq would be a mistake. It is doubtful that any of those leaders have much sympathy for the Iraqi regime, but implicating Iraq in this war against terrorism casts serious doubts on U. S. intentions underlying the whole enterprise. In fact, immediately after the September 11 attacks, some U. S. officials suggested that Iraqi intelligence services had had contact with some of the people who hijacked the planes that struck the World Trade Center, but the officials did not offer any convincing evidence and did not repeat the allegation. More importantly, talk of toppling Saddam was an open secret in Washington in the spring of 2001, months before the September attacks. Large segments of the Arab and Muslim world, including some policymakers, believe that the U. S. administration is using the campaign against terrorism as a pretext to carry out its own hidden agenda, which has nothing to do with the war against terrorism.

Where We Go from Here

In the wake of the fall of the Taliban regime, the vast majority of Arab and Muslim countries are abiding by the Security Council resolution on fighting terrorism and cooperating with the established committee to implement the different measures included in that resolution. Four Muslim countries reportedly expressed interest in participating in the peacekeeping force in Afghanistan, namely Jordan, Turkey, Bangladesh, and Indonesia. Saudi Arabia is participating in the financial effort to help rebuild Afghanistan. Arab and Muslim and other concerned countries are expected to continue sharing intelligence on terrorist organizations.

Continued talk by U. S. officials of other phases in the war, however, causes much concern in the Arab and Muslim world for two reasons. First, the U. S. definition of terrorism does not distinguish between those launching a just armed struggle against the illegal occupation of their land and those using force against elected, legitimate governments. Accordingly, the U. S. administration considers Palestinians fighting Israeli troops inside the occupied West Bank and Gaza Strip to be terrorists. Second, the U. S. administration does not condemn state terrorism. U. S. spokespersons did not recognize the Israeli government's siege of Palestinian towns and villages and targeted assassination of Palestinian officials by Israeli secret services, which provoked suicide attacks on the part of Palestinian groups, as terrorist actions. That the United States provides Israel with the economic and military means through which it executes such policies but protects Israel from any UN condemnation is even more frustrating for Arabs. For example, in early December the United States thwarted a resolution supported by 12 members of the UN Security Council to send international observers to monitor the safety of the civilian population in the West Bank and Gaza Strip.

Under these conditions, for most Arab and Muslim governments to extend more support to the United States for future phases of its war against terror becomes difficult. Of course, Yemen, concerned that it might become a possible target in a second phase in this so-called war, did begin its own internal efforts against alleged members of Al Qaeda, with the help of special forces trained by the United States and under the personal guidance of President Ali Abdallah Saleh's son. Somalia, another possible target in a future phase in this war, offered to cooperate with the United States while denying any presence of Al Qaeda members on its territory.

These two examples demonstrate that, although the governments of a few countries had incomplete control over their territories, such as Afghanistan, Yemen, and Somalia, no Muslim government encouraged terrorist activities against U. S. citizens or officials. Many Arab and Muslim governments did face problems with their own terrorists. They spared no effort fighting them and did not always receive the timely support they expected from the international community. Ironically, the U. S. government played a role in encouraging the groups of young people who went to Afghanistan in the 1980s to fight the Soviet military presence there and who ended up fighting the United States.

Even before the events of September 11, a gulf of mistrust separated the United States from large sections of the Arab and Muslim public and policymakers. What the U. S. administration is contemplating at present—launching a second phase of this war against other Arab and Muslim countries while doing very little to stop state terrorism practiced, for example, by Israel—risks widening this gulf even further and does not bode well for friendly relations between the United States and Arab and Muslim nations in the future.

Apart from exchanging intelligence, participating in a peacekeeping force in Afghanistan, and contributing to the rebuilding of Afghanistan, the best service that Arabs and Muslims can offer to the U. S. administration and to the U. S. people is its advice for the United States to cooperate with Arab and Muslim peoples and governments to find nonmilitary solutions for the profound causes that push some young people to take up arms against their own governments and those of other countries. This approach could perhaps be the best way to promote the ideals of the American Revolution; life, liberty, and the pursuit of happiness, by all people.

Notes

1. Majid Khadduri, "The Islamic Theory of International Relations and Its Contemporary Relevance," in Harris Proctor, ed., *Islam and International Relations* (London and Dunmow, England: Pall Mall Press, 1964), pp. 26–27, 35–38.
2. *Mussawwar,* December 28, 2001, p. 8.
3. See in particular articles by Thomas Friedman in the *New York Times* and several editorials in the *Washington Post* during October 2001.
4. *Iraqi News Agency,* October 21, 2001; *Al-Ahram,* September 18 and October 7, 2001.
5. For examples of press reactions in the Arab world, see "Slalma Ahmed Salama," *Al-Ahram,* December 2001.
6. Statement of the Secretary General of the Organization of the Islamic Conference, October 11, 2001.
7. Mubarak became more forthcoming in his support of the U. S. preparation to launch military action in Afghanistan following his visit to Paris and his meeting with President Jacques Chirac on September 24, 2001. Prior to that date, he was critical of both the notion of an international alliance against terrorism, arguing that it would be divisive, and of the use of massive military force against terrorist organizations. His statements were reported in *Al-Ahram* and *Al-Hayat* newspapers.

Mustafa Al Sayyid is a professor of political science and director of the Center for the Study of Developing Countries at Cairo University. The author would like to thank his assistants Ingy Abdel-Hamid, Sodfa Mahmoud, and Karam Khamis for the valuable research they contributed to this paper.

Back to Brinksmanship

How India and Pakistan arrived at a nuclear standoff

BY SUMIT GANGULY

INDIA AND PAKISTAN STAND ONCE AGAIN ON THE BRINK OF war. The moment is a precarious one and the stakes are high, not just for the region but potentially for the world. The United States has burgeoning interests in the subcontinent since the war in Afghanistan, and renewed Indo-Pakistani conflict could divert needed resources from the effort to stamp out terrorism. Incautious statements from both Indian and Pakistani leaders have also raised fears that a nuclear exchange may be in the offing. The consequences would be far-reaching and devastating. Nonetheless, only three years after their last confrontation prompted frantic U.S. diplomatic overtures and direct personal intervention by President Bill Clinton, these two nuclear-armed adversaries have, since the beginning of this year, been staring each other down across their shared border.

The trigger for the current crisis was an incident last December, when operatives of two Pakistan-based insurgent groups attacked the Indian parliament. Security guards managed to keep the terrorists away from legislators, but in the shootout that followed, six Indians were killed along with the five attackers. Pakistan's president, General Pervez Musharraf, condemned the attack, but his principal military spokesman suggested that India had assaulted its own parliament in an effort to implicate Pakistan. Under intense pressure from India and the United States, Musharraf banned the two groups responsible for the attacks and promised to squelch the activities of other terrorists operating from inside Pakistan. He refused, however, to hand over 20 individuals whom the Indian government accuses of involvement in a range of terrorist activities on Indian soil. In the intervening months, it turns out, Musharraf has also failed to end his country's support for terrorism in Kashmir, even while he has supported the U.S. effort to root out al-Qaeda and Taliban fighters along his western border with Afghanistan.

Indeed, although he initially cracked down on several of Pakistan's militant Islamic organizations, Musharraf looked the other way when the groups' members resumed activities under new names. In response, India has adopted a strategy of coercive diplomacy, massing close to half a million troops along the India-Pakistan border and the so-called Line of Control that divides the disputed state of Jammu and Kashmir. India's leaders have made clear that for New Delhi to reverse the military buildup, the infiltration of terrorists from Pakistan into India must end.

Indo-Pakistani relations have a long and troubled history, of which the current crisis is merely the latest chapter. Since both independent states emerged from the detritus of the British empire in 1947, they have fought four wars (1947–48, 1965, 1971, and 1999). Their most intractable conflict is the one over Kashmir, the mostly Muslim state whose Hindu ruler chose to join his lands to India in 1947. Pakistan contested that arrangement and invaded the territory, touching off the first Indo-Pakistani war. By the end, Pakistan controlled about one-third of Kashmir. The status of the state has remained unresolved ever since.

The Indo-Pakistani conflict lay mostly dormant for several decades. During the 1970s and 1980s, the Indian government sought to win the hearts and minds of the Kashmiris by investing in education, mass media, and social welfare. Yet at the same time, the authorities engaged in considerable political chicanery, as they attempted to prevent a secessionist elite from taking power through the electoral process. By 1989, these policies, combined with fundamental social changes within Indian-controlled Kashmir, had helped spark an ethno-religious insurgency in the fabled Kashmir Valley.

Pakistan's political and military leadership saw a vital opportunity in Kashmir's brimming reservoir of discontent with Indian misrule. Over the next several years, Pakistan's military intelligence organization, the Inter-Services Intelligence Agency, provided Kashmiri rebels with military training, logistical support, and physical sanctuaries. The Pakistani authorities also brought in disaffected Afghans, radical Arabs, and Pakistani jihadis to support and extend the uprising. By the mid-1990s, a spontaneous and largely disorganized uprising had been transformed into a well-orchestrated insurgency. The principal local insurgent organization, the Jammu Kashmir Liberation Front (JKLF), found itself caught in a vicious vise: It faced relentless military pressure from the Indian security forces at the same time as it suffered routine depredations at the hands of Pakistan-sponsored militant Islamic organizations. By the mid-1990s, the JKLF had eschewed violence as a political strategy for fear of being destroyed on the battlefield.

As Pakistan-sponsored, nonindigenous groups came to dominate the insurgency, Kashmiri support for it subsided. That moment was not lost on New Delhi, which conducted a successful election for the state's legislature in 1996. An unprecedented number of Kashmiris turned out to vote, and foreign and domestic observers concluded that the election was mostly free of fraud. Many Kashmiris greeted the emergence of a popularly elected government with considerable optimism: After more than half a decade of political turmoil and civil violence, perhaps some modicum of law and order might soon return to their disputed state. Indeed, by the late 1990s, the insurgency was clearly fading.

But when both India and Pakistan tested nuclear devices in May 1998, Kashmir would feel the aftershocks. The hawkish Indian home minister, L. K. Advani, cavalierly announced that Pakistan's ability to foment mischief in Kashmir was now effectively constrained. That statement, designed to instill fear in the minds of risk-prone Pakistani decision makers, revealed Advani's myopic understanding of the strategic significance—as well as the military limitations—of nuclear weapons. For although nuclear weapons could dramatically reduce the likelihood of full-scale war, they could also create permissive conditions for more low-level conflict—a situation that political science scholars refer to as the "stability-instability paradox."

Between May and June of 1999, the subcontinent saw the first test of this paradox. During the preceding winter, units of Pakistan's Northern Light Infantry had penetrated Indian territory in a successful surprise attack at three points along the Line of Control. The waning of Kashmir's insurgency had led India's military circles to grow complacent and vulnerable. On the Pakistani side, tensions between the civilian regime of Prime Minister Nawaz Sharif and the military had made reviving the flagging insurgency politically attractive. Moreover, Pakistani decision makers surmised that given the nuclear risks, their Indian counterparts would be loath to punish Pakistan by expanding the scope of the conflict. They were right: Out of fear of nuclear escalation, India kept the conflict confined to the areas of incursion. Ultimately, significant Indian military pressure, combined with forceful U.S. intercession, persuaded Sharif to withdraw his troops in late July 1999.

That debacle proved fatal for Sharif. Three months later, Musharraf, the chief of army staff and the architect of the incursion, seized power in a bloodless coup. A renewed burst of Pakistani support for the Islamic militants soon followed. Terrorist attacks increasingly expanded outside the Kashmir Valley to neighboring regions of India, as the December attack on the parliament building in New Delhi so brazenly demonstrated.

Last month, India's frustrations with Musharraf's regime reached their apex after a May 14 terrorist attack killed 34 Indians, including a number of wives and children of military personnel in the Kashmiri city of Jammu. Within days, militants also killed Abdul Ghani Lone, a 70-year-old moderate Kashmiri separatist leader who had indicated a willingness to begin talks with New Delhi. Musharraf publicly condemned these attacks while also insinuating that both episodes were the handiwork of al-Qaeda forces. But despite the strong urging of the United States and other Western powers, the Pakistani military leadership has evinced little willingness to curb the terror emanating from its lands. India's prime minister, Atal Behari Vajpayee, has consequently assumed a more bellicose posture.

Pakistan's persistent dissembling on the question of military support to terrorists, and India's growing impatience and belligerence, have stoked fears of a conventional war between these two long-standing foes. Concern that any war between India and Pakistan could escalate to the nuclear level has prompted calls for restraint from all corners of the world. Rightly so: A nuclear war in South Asia would produce horrific human loss and a humanitarian crisis of unprecedented magnitude. It would also breach the unspoken post-Nagasaki taboo on the use of nuclear weapons. The rupture of this fire wall would make the world a far, far more dangerous place.

War between India and Pakistan would also hobble the U.S.-led effort to eviscerate the remnants of al-Qaeda and Taliban forces, many of whom have taken refuge in the poorly administered, trackless reaches of Pakistan's Northwest Frontier Province. As Pakistan's army gets increasingly drawn into a conflict with India, its ability to cooperate with the United States will inevitably dissipate. Meanwhile, the United States may find itself in the singularly unenviable position of having to choose between an uncertain but necessary ally, Pakistan, and a long-term potential strategic partner and democracy, India.

The most immediate interest of the United States, clearly, is to forestall and ideally prevent another war between India and Pakistan. In all likelihood, U.S. pressure on both capitals will lead the two states to step away from the brink. Then the United States must do two things: It must forcefully persuade Pakistan to eschew support for the Islamic militants in Kashmir and simultaneously convince India that a lasting peace can emerge only if the genuine grievances of the Muslim population in the Kashmir Valley are adequately addressed. Adopting these two negotiating principles will be neither easy nor painless. But for India and Pakistan, there is no other path away from the precipice.

SUMIT GANGULY *is the author of* Conflict Unending: India-Pakistan Tensions Since 1947, *and a professor of Asian studies and government at the Univeristy of Texas at Austin.*

A report from Congo

Africa's great war

Congo's civil war may have claimed as many as 3m lives. The country is so inaccessible that its horrors are rarely reported

BUKAVU, KINSHASA AND KISANGANI

IN THE chocolate waters of the Congo river, a mutilated corpse rolls by. The rebels' "minister for children" shivers. How is he going to explain this to the horrified UN peace envoys from the capital, Kinshasa, who are at that moment stepping on to the quay to meet him? Not by telling the truth, obviously, which was that his rebel group had slaughtered 150 people in the town of Kisangani on May 14th–15th, then pitched their disembowelled bodies into the river with stones crammed into their bellies. Instead, he smiles, accepts the envoys' offerings of food aid, and talks chummily of other things.

Over the past four years, Congo's war has claimed more lives than any other. The International Rescue Committee, an American aid agency, says that by the middle of last year, 2.5m people had died because of the war in eastern Congo alone. Some were shot or hacked to death; many more succumbed to starvation or disease as nine national armies and a shifting throng of rebel groups pillaged their country. By now, the death toll is probably over 3m, although this is the roughest of estimates. As one UN worker puts it: "Congo is so green, you don't even see the graves."

Western powers seem barely to have noticed the catastrophe. This is partly because, unlike the Middle East, Congo has no strategic importance. But it is also because it is two-thirds the size of Western Europe, thickly-forested, incredibly dangerous and has hardly any paved roads or working telephones. Simply finding out what is happening in Congo is a challenge, as your correspondent discovered while accompanying militiamen on patrol by the shore of Lake Kivu last week, when he was forced to hide in a bush to avoid 200 hostile Rwandan soldiers passing by.

And yet there is hope. In April, the Congolese government signed a peace deal with most rebel groups. By offering rebel leaders a share of power, the government won nominal control

Room for improvement	
Democratic Republic of Congo, 2000	
GDP per person, $	90
Population per doctor	24,319
Telephone lines per 1,000 people	<1
Life expectancy at birth, years*	51
Access to safe water, %	45
Undernourishment, %†	61
Sources: UNDP; World Bank, Oxfam	*1999 †1998

of 70% of the country (see map). This part of Congo is now relatively peaceful, and slowly recovering economically. But one important rebel group refused to sign. The Rwandan-controlled Rally for Congolese Democracy (RCD), whose minister for children brazened out the watery corpse, still holds most of eastern Congo.

Here, the war blazes on. Or rather, dozens of overlapping micro-wars flicker, in which almost all the victims are civilians. A typical village can expect to be looted by several different bands of armed men. Under what amounts to Rwandan occupation, eastern Congo is arguably the most miserable place on earth.

A rough neighbourhood

The story of Congo's war began in 1994, with the genocide in Rwanda, Congo's tiny neighbour. A government dominated by Rwanda's Hutu tribe tried to exterminate the Tutsis, a prosperous minority. In 100 days, 800,000 Tutsis, and Hutus who refused to co-operate, were murdered. The slaughter stopped when an army of exiled Tutsis invaded from Uganda, and drove the killers into Congo (which was then called Zaire). The new, Tutsi-dominated government in Rwanda was afraid that the *genocidaires* would regroup and return to finish the job. So when Mobutu Sese Seko, Zaire's dictator, gave them succour, Rwanda engineered a rebellion that toppled him.

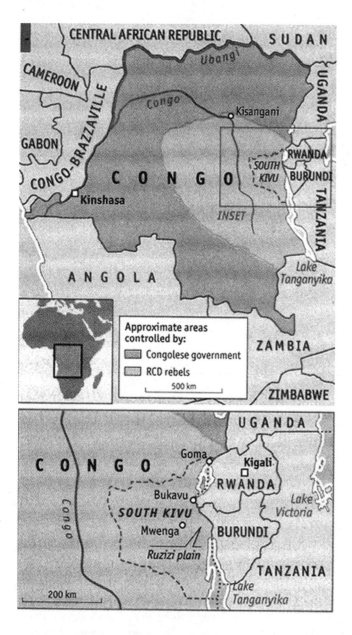

In his place, the Rwandans installed a guerrilla leader and drunk, Laurent Kabila. They hoped he would do their bidding. Instead, he rearmed the *genocidaires*. So Rwanda tried to topple him, too. With help from Uganda and Burundi, it nearly succeeded. Kabila was saved by five friendly nations, of which the most effective were Angola and Zimbabwe. Most of the countries that intervened had legitimate interests in Congo. Rebels from most neighbouring states were using Congo's lawless forests as a base from which to launch cross-border raids. The failure of the Kabila government to curb these rebels prompted Rwanda, Uganda and Angola to enter the war. Zimbabwe, which shares no border with Congo, sent troops for different reasons: to satisfy the power-broking pretensions of its president, Robert Mugabe, and his army's appetite for loot.

Before long, the war reached a stalemate and the miscellany of armies settled down to the serious business of plunder. Zimbabwe bagged diamond seams in the south. Angola joined the Congolese government in an oil venture. Rwanda and Uganda began digging for diamonds and coltan (a mineral used in mobile telephones), harvested timber and ivory, and even emptied schools of desks and chairs. Though supposedly allies, Rwandan and Ugandan troops have occasionally clashed over the spoils. But in general, the more all the armies plundered, the less willing they became to fight each other (as opposed to unarmed peasants).

A younger, thinner, nicer Kabila

Then, in January last year, one of Kabila's bodyguards shot him once in the gullet and twice in the guts. No one knows who gave the order, because the assassin was himself killed almost immediately. One rumour is that his allies were so sick of his double-crossing that they had him killed.

Kabila's legacy is not a happy one. At his mausoleum in Kinshasa, a bronze statue shows the dead president clutching a book, which is an odd memorial for a man who halted book-keeping at the Congolese finance ministry. In the parts of Congo he controlled, Kabila ruled as despotically as his predecessor, but less competently. He scared off investors by jailing foreign businessmen and demanding million-dollar ransoms. He banished aid by insulting foreign diplomats. By printing money while enforcing price controls, he caused Soviet-style shortages. By the end, petrol was so scarce that the state oil firm resorted to flushing its main pipeline with river water to force out the dregs.

Kabila was succeeded by his son, Joseph. Now 31, Mr Kabila junior is doing a better job. He endorsed a peace plan that his father rejected. A ceasefire followed, policed, where possible, by the UN. Angola withdrew to its border, and Uganda pulled out some troops, leaving others to mind its businesses in northern Congo.

Zimbabwe stayed, at the government's request. The young Mr Kabila wants to be protected from the Rwandans, who refuse to leave. The Rwandan government says it will not withdraw its army, which occupies a slice of Congo 27 times bigger than its own country, until the last Hutu fighter has been captured or killed.

Rwanda reckons that 55,000 Hutus, still intent on genocide, continue to lurk in Congo's jungles or have joined the Congolese army under false names. The estimate of the International Crisis Group, a think-tank, is half that number. All the same, to prevent the Hutus from regrouping and invading Rwanda, the UN has proposed setting up a Congo-based border force. Mr Kabila's government accepts this. To show goodwill, it has already released 2,000 Rwandan Hutus from its army and invited the international genocide tribunal to investigate. But Rwanda says no.

No one is suggesting that Rwanda's fears are unfounded. Yet many observers say that Rwanda and its pet rebels are no longer pursuing the *genocidaires* with much passion. Refugees from

Mwenga, a bush-town in South Kivu, a mineral-rich province, complain that their homes were attacked by Hutu militiamen a dozen times in the past two months, but that the town's Rwandan garrison did nothing. The townspeople say they asked the Rwandan commander why. He is said to have replied: "These are our brothers, do you think we can kill them?"

Another example: the Hutus' camps on South Kivu's Ruzizi plain are well known to the Rwandan troops nearby, say UN officials, but they are seldom, if ever, attacked. A local Hutu commander claims that the Rwandans regularly supply him with arms. Elsewhere, according to a senior Rwandan officer who recently sought asylum in Belgium, the army hires Hutu militiamen to work its coltan mines. He told the Belgian senate that they could otherwise be eradicated "in less than a month".

Besides securing its own borders, Rwanda claims to be protecting Congo's ethnic Tutsis, the foot-soldiers of its two rebellions, from being murdered by Hutu militias or other Congolese. But even this group is now rising against Rwanda's occupation. Early this year, a brigade of Congolese Tutsis mutinied and returned to their tribal homeland, in the hills above Lake Tanganyika. Around 1,000 rebel fighters have deserted to join them. Rwanda swiftly sent 8,000 troops to crush the revolt, and is now reportedly herding Tutsi villagers into a barren camp, to stop them feeding the deserters. Tutsi against Tutsi, this is now the fiercest battle of the war.

With mid-ranking rebels meanwhile crossing to the government side, and senior rebels expected to follow, the RCD could be on the verge of disintegration. Rwanda has tried to deter desertions by force, of which the massacre in Kisangani was an example. But if the RCD crumbles, Rwanda will doubtless find an alternative cover for its occupation of eastern Congo. Indeed, it has already started looking: in June, Rwanda deposed the RCD "government" of South Kivu, and replaced it with the leaders of a more pliable militia.

The players change; but the game remains the same. This is bad news for the hapless people of eastern Congo. Everywhere they look, they see men with guns. Besides the Hutu militias, the Rwandans and their rebel allies, there are the Mai-Mai, a warrior cult, and myriad gangs of bandits. All prey on the poor.

Across Kivu, a land of postcard-perfect green hills, villages are half-deserted, fields neglected and livestock a fond memory. In the forests around Mantu, near Bukavu, villagers have dug waist-deep trenches and covered them with branches. When Hutu marauders come, as they do most weeks, they hide in these holes.

The peasants of Ramba Chitanga, a village too tiny to appear on any map, tell a grisly tale. When the RCD left, Hutus moved in, and accused the villagers of feeding their enemies. Then the Mai-Mai attacked. During the ensuing battle, the Hutus hacked off 29-year-old Janet Vumilia's hands. Now, with her skittle-like stumps, she ticks off the relatives they killed: her parents-in-law, her brother-in-law, her pregnant sister, her niece.

Villagers say they can distinguish different factions by their actions. The Hutus, they say, are more vicious than the Mai-Mai, while the rebels are more likely than the Rwandans to abduct children. But sometimes the distinctions become blurred. Francine, a 14-year-old new mother, says she thinks her baby's father was an RCD rebel. But he could have been a Mai-Mai; men from both groups raped her. When her father objected, the Mai-Mai slit his throat.

In Kivu, cattle are now scarce, yet livestock prices have plummeted. Locals calculate that if they buy a cow, armed men will take it, so they don't buy cows. The rebels levy a "security tax" of $1 per hut, but payment does not seem to reduce the likelihood that, sooner or later, killers will come in the night.

In Walungu hospital, near Bukavu, 1,200 patients compete for 300 beds, and the attention of three doctors. About half the inmates are relatively healthy, but too terrified to go home. Murals in the hospital depict black doctors in white coats peering into microscopes, recalling the lost hopes of the 1960s. Now, the wards are full of black children with blonde hair, a sign of malnutrition. The hospital's 32-year-old head of nursing says she has received only three months' salary in her whole career. Why does she carry on? "It's our country," she shrugs. "It's painful, but there it is."

If Rwanda had been a benevolent occupier, it might have turned a profit without stealing. As Congo's infrastructure crumbled over the past 40 years, people in the east of the country strengthened their commercial links with Rwanda. This trend accelerated when war closed the Congo river, the country's main highway, severing eastern Congo from Kinshasa. Had the Rwandans built a decent road or two linking eastern Congo with Kigali, their own capital, they could have exploited this.

Instead, they chose to murder and plunder. "We couldn't believe the things these people did during the genocide, until they came and started doing them to us," said a market-woman in Bukavu, jumbling together Hutu killers and Tutsi invaders.

Heroic optimism

In Kinshasa, meanwhile, stinking bank notes suggest that things are looking up. The new government printed no money in the second half of last year, and lifted price controls. Inflation is currently below 5%, down from 511% in 2000. On June 13th, the World Bank approved a $450m loan, part of a multilateral aid package that could total $1.7 billion. The Bank's new man in Kinshasa, the first posted there in a decade, says that Congo could be about to see surprisingly rapid development.

This is heroic optimism. Congo's institutions are as dirty as its bank notes. Corrupt parastatals disgorge no revenue to the central bank. Their past thieving has saddled Congo with $13 billion in debt. Civil servants make ends meet by demanding bribes or doing other jobs on the side. In the offices of the immigration ministry, for example, not only visas but also sexual favours are for sale. Respect for human rights has improved under the new regime, but only a bit. Political troublemakers, and journalists, are regularly arrested. Policemen and soldiers, often drunk, in mirrored sunglasses, freelance as muggers. During a random "visa inspection", *The Economist*'s correspondent was bundled into a black Mercedes and relieved of $450 by four men claiming to be security officers. It could have been worse: an American aid worker was recently beaten senseless after overtaking a general on the boulevard.

Whether Joseph Kabila could control these abuses is unclear. He is popular throughout Congo, because people regard him as the best hope for peace. But he has some ugly people around him, including some of the worst crooks and thugs of his father's administration, and of Mobutu's too. The people of Congo tell anyone who asks that they want their country to be united, invader-free and democratic. The prospects for this are woeful, in the short term at least.

Mr Kabila has promised to hold free and fair elections as soon as the country is reunited. But no Congolese leader has ever kept such a promise before, and some of Mr Kabila's allies have cause to try to prevent the young president from keeping his. For example, Mwenze Kongolo, who is Mr Kabila's uncle and Mr Mugabe's good friend, is associated with the worst abuses of the old Kabila regime, and so cannot expect anyone to vote for him. He is however popular with certain army brigades.

If Congo's different factions agree about anything, it is that white foreigners are responsible for their troubles. Given a history of western meddling—King Leopold of the Belgians plundered Congo in the 19th century, and America first propped up Mobutu and then helped overthrow him—such a view is understandable. But it is hopelessly out of date. Since the cold war ended, western governments have seen no reason to care much about Congo.

Mostly, they ignore it, considering its problems too tangled to solve. The UN's ceasefire-monitoring mission consists of only 442 military observers. Efforts to disarm the militias in eastern Congo have been crippled by the RCD's refusal to grant the UN free passage.

Since 1994, western donors have put a bigger effort into Rwanda than Congo. In part, this stems from guilt at having failed to stop the genocide. It is also because bite-sized Rwanda's problems seem more manageable, and its efficient government is refreshing to deal with. Throughout Congo's war, aid has poured into Kigali.

All the same, Rwanda's behaviour in Congo may soon cost it some friends. In August, a UN Security Council panel will deliver its third report on the looting of Congo's resources, focusing on the humanitarian impact of Rwanda's plunder in the east of the country. Sanctions, or at least a reduction of aid, are possible. Britain, Rwanda's biggest bilateral donor and keenest apologist, is expected to oppose such moves. But France will probably be in favour. And the Americans may be willing to go along with some sort of censure of Rwanda, even if it is only a reprimand.

It will take more than a reprimand to eject Rwanda from Congo. Its leaders—the Tutsi fighters who stopped the genocide—have turned their country into one of the most militarised in Africa. By plundering Congo, some have found a way of enriching themselves without upsetting their own people. Meanwhile, war gives Rwanda's president, Paul Kagame, an excuse to squelch dissent and muffle the press. It also keeps at least 20,000 unruly, mostly-Hutu soldiers from coming home and stirring fresh trouble in Rwanda, where it all began.

Zimbabwe:
The Making of an Autocratic "Democracy"

"Two decades after independence, the fruits of President Robert Mugabe's rule are a rapidly declining economy, the systematic dismantling of constitutional government, growing political violence, a costly war in Congo, and international condemnation."

ROBERT B. LLOYD

Zimbabwe's elections this March pitted a seasoned leader from the independence era against an upstart from the trade-union movement. In the early stages of the campaign, many compared the contest to Zambia's 1991 election, which saw long-serving President Kenneth Kaunda yield power peacefully to opposition union leader Frederick Chiluba. But in Zimbabwe this spring, President Robert Mugabe employed harsh means to maintain his 22-year hold on power in the face of increasing domestic and international challenges to his rule. The final results of the highly charged election showed the effectiveness of these measures. Excluding spoiled ballots, Robert Mugabe won 56 percent of the total votes cast. His opponent, Morgan Tsvangirai, won 42 percent.

Both the election process and the final tally were disputed. Tsvangirai, Western governments, and international organizations bluntly accused Mugabe of voter intimidation and outright election fraud. Southern African governments—sensitive to questions about their own legitimacy, fearing turmoil from Zimbabwe spilling into their own countries, and preferring private diplomacy with an African colleague—generally supported the legitimacy of the election. Nevertheless, despite the electoral outcome, the days of the nearly 80-year-old president clearly are numbered. Mugabe has damaged the fabric of both the Zimbabwean state and civil society. Two decades after independence, the fruits of President Robert Mugabe's rule are a rapidly declining economy, the systematic dismantling of constitutional government, growing political violence, a costly war in Congo, and international condemnation.

OF VIOLENCE BORN

The roots of Zimbabwe's current crisis are seen in the political and economic circumstances of its birth. Zimbabwe, unlike most countries in Africa, was born in violent struggle. The English who settled in Zimbabwe—then called Rhodesia—were attracted by cheap land and a mild tropical climate. These set-

tlers, with the use of local labor, transformed a well-watered savanna terrain into an agricultural cornucopia. Exports of tobacco and mineral commodities provided cash for the development of a first-class infrastructure. During the 1960s, Rhodesia successfully diversified into manufacturing.

Despite the country's economic success, the political and economic disenfranchisement of a majority of the population by the English settlers ultimately proved to be Rhodesia's undoing. Grievances against colonial government in Rhodesia and throughout Africa grew louder. Black nationalist organizations became increasingly successful in challenging the legitimacy and power of colonial rule. As the European decolonization of the continent began in the late 1950s, Rhodesia felt increasing international and domestic pressure to yield to this continent-wide movement to independent African states. But the settlers said they would not negotiate, and in November 1965 declared independence from Great Britain. While the British government refused to accept this independence, it did not mount any military action to force the white Rhodesians to submit to British rule.

The subsequent failure of white Rhodesians to negotiate with the black Zimbabwean nationalist movement led to the growing radicalization of the movement. By the mid-1970s, two main nationalist movements emerged: the Zimbabwe African National Union and the Zimbabwe African People's Union. Using bases in nearby Zambia and Mozambique, the two groups began well-organized and effective guerrilla attacks against Rhodesia. Although Rhodesia received valuable financial and military support from white-ruled South Africa, within a few years it was clear that the white Rhodesians were losing the war. South Africa cut its support to Rhodesia and urged it to negotiate with Mugabe and the other nationalist leaders. Meanwhile, divisions within the liberation movements and the dreadful human costs of the conflict were creating pressure for a negotiated settlement. Britain skillfully took advantage of these dynamics to force the parties to the negotiating table.

In 1979 Britain hosted negotiations at Lancaster House in London that led to an agreement that paved the way for black rule in Rhodesia. The resultant political transition engendered considerable confidence among many within the country that Rhodesia—soon renamed Zimbabwe—would avoid the fate of so many other newly independent African states. These states had experienced a euphoric start only to have expectations dashed under the weight of growing political and economic decline. Unfortunately, Zimbabwe has not deviated from this African pattern.

CONSOLIDATING CONTROL

Elections in 1980 led to the election of Robert Mugabe as prime minister. His party, the Zimbabwe African National Union-Patriotic Front (ZANU-PF), won an overwhelming majority of parliamentary seats, defeating Mugabe's erstwhile Patriotic Front ally—the Zimbabwe African People's Union (ZAPU), headed by Joshua Nkomo. The new leader's domestic policies aimed to consolidate his political authority. One of his first steps was to bring many elements of civil society under its control, including trade unions and the university.

Efforts to consolidate political control initially failed when it came to Joshua Nkomo's defeated ZAPU. The party broke with the government over the slow pace of land reform and over ethnic tension between the majority Shona tribe of Mugabe and Nkomo's minority Ndebele tribe located in southwestern Zimbabwe. The resultant low-level civil war led to the deaths of thousands. By 1987 Nkomo and Mugabe had negotiated a truce, which led to the official merger in December 1989 of these two parties as the Zimbabwe African National Union-Patriotic Front (ZANU-PF). The agreement with Nkomo gave Mugabe a two-thirds majority in parliament. Mugabe thus was able to change the constitution to become Zimbabwe's first executive president, further enhancing his powers.

Voting irregularities and relatively low turnout due to voter intimidation allowed the socialist ZANU-PF to win the 1990 elections by a decisive margin. This victory afforded Mugabe the opportunity to try to amend the 1980 constitution that grew out of the Lancaster House agreement to create a de jure one-party state. Mugabe, in the face of strong domestic and international opposition, backpedaled, stating that Zimbabwe would be a de facto single-party state following Marxist-Leninist precepts.

To overcome growing disenchantment with his rule, Mugabe used the highly emotional issue of land reform to gain support. The president pushed through legislation in 1992 that cleared barriers to the compulsory purchase of land from white farmers. The populist measure helped shore up his support, although it alienated farmers, the business community, and international donors concerned over this property expropriation. ZANU-PF was able to win a fourth parliamentary election in 1995 and Mugabe—unopposed—won the presidency the next year. Mugabe was lucky to have poorly organized and funded opponents who failed to make common cause and withdrew before the elections.

The lack of an organized opposition movement did not, of course, translate into support for Mugabe. In 1997, in an attempt to appease popular protest against his rule, Mugabe announced a new policy of outright land seizure, approved by a now largely rubber-stamp parliament. White-owned farms were attacked, and black supporters of Mugabe moved onto the land. Farmers who resisted were killed, and the police refused to defend the farmers. In June 2001 Mugabe overcame the last remaining institutional opposition to him—the Supreme Court—by hinting that he could not guarantee the safety of Chief Justice Anthony Gubbay. The message was clear; the chief justice and several other senior judges resigned. Mugabe replaced them with his allies and took the extra precaution of expanding the court from six to eight judges. The government soon received a favorable ruling on land seizure, overturning an earlier court ruling. An independent judiciary had fallen.

OPPOSITION GROWS

In hindsight, the expenditure of the political capital needed to seize land may prove to be the beginning of the end for Mugabe. He achieved his goal but domestic and international opposition has grown louder and better organized as a result. After two decades of power, Mugabe's rule looks increasingly shaky after the results of the June 2000 elections. His ZANU-PF won the election, but barely avoided an upset by the sudden rise to power of a unified opposition party. The Movement for Democratic Change (MDC), led by Morgan Tsvangirai, won 57 of the 120 directly elected seats in parliament. Mugabe's ZANU-PF squeaked by with 62. Just a few months earlier the MDC had forced the first-ever defeat for Mugabe when Zimbabwean voters rejected a referendum on constitutional changes that would have benefited Mugabe.

Tsvangirai is a product of the trade unions. Born in 1952 in eastern Zimbabwe, he worked in the mines, and within a few years was heading the mineworkers' union. By the late 1980s he was leader of the Zimbabwe Congress of Trade Unions. From this position he was able to stage a series of strikes against Mugabe, forcing him to back down on plans to increase taxes. His natural constituency is younger urban workers in the two major cities of Harare and Bulawayo and the rural areas of the predominantly Ndebele ethnic area of southwestern Zimbabwe. MDC supporters are united largely by an "anyone who can beat Mugabe" consensus. The party has proved resilient and popular despite Mugabe's efforts to undermine its appeal. Party leaders have been arrested, beaten, and killed. MDC supporters have faced pro-Mugabe troops, paramilitary forces, and thuggish "veterans" from the war for independence.

Until 1994, international opposition to Mugabe's increasingly autocratic rule was muted by the need to show solidarity against apartheid in South Africa. Zimbabwe, as the most powerful state in Southern Africa outside the diplomatically and economically isolated South Africa, played a leading role in the subregion. Zimbabwe, for example, supported the Marxist government of neighboring Mozambique in its fight against South African-supported rebels. Absent this support, Mozambique might have fallen to anti-Marxist rebel forces. After the subsequent democratization of both South Africa and Mozambique in the early 1990s, the unfavorable contrast with Zimbabwe grew. But the new South African government, led by Nelson Mandela's African National Congress (ANC) and later Thabo Mbeki, was reluctant to criticize Mugabe because of his support for the ANC during its long struggle against the apartheid-era government.

The illegal land seizures and growing political violence did eventually galvanize South Africa and other African states. Another impetus was Zimbabwe's worsening economic climate due, in part, to the country's costly dispatching troops to Congo (the former Zaire) in 1998. Zimbabwe supported President Laurence Kabila against Rwanda- and Uganda-backed rebels seeking to topple him. Mugabe's supporters thus gained valuable mineral and timber concessions in Congo. In September 2001 a special British Commonwealth delegation of seven foreign ministers met in the Nigerian capital of Abuga. Three African states—South Africa, Kenya, and Nigeria—told Zimbabwe's foreign minister that the blame for Zimbabwe's troubles rested squarely on Mugabe's policies.

The international pressure on Mugabe intensified dramatically when later in September African leaders of the Southern African Development Community (SADC) called a special meeting in Zimbabwe's capital city, Harare. The summit leaders blamed Mugabe directly for many problems facing the country. The leaders also invited opposition leaders and farmers to speak. The public humiliation of Mugabe by his peers did not, however, stop the continued seizure of land and intimidation of opposition leaders and supporters.

To maintain pressure a committee was formed to continue scrutinizing events in the country. In January 2002, SADC leaders meeting in Malawi called for free and fair elections, which, if they were free and fair, would likely topple Mugabe. In addition, Nigerian President Olusegun Obasanjo spearheaded efforts to mediate the Zimbabwean crisis, which has undermined his and Thabo Mbeki's project to bolster Africa's economic growth. After the election, both Mbeki and Obasanjo met Tsvangirai and Mugabe, proposing a power-sharing arrangement and reconciliation.

The chorus of Western criticism against Mugabe has grown. The United States State Department, for example, repeatedly and publicly rebuked Mugabe. Congress has legislation pending that would levy travel and economic sanctions against political leaders responsible for political violence and intimidation. After the March 2002 elections the 54-state Commonwealth, at the strong urging of the former colonial power of Britain, made the highly unusual decision to suspend Zimbabwe from the group for a period of at least one year. And SADC election observers stated that the Mugabe government failed to create the conditions required for a free and fair election.

A CRUMBLING ECONOMY

Fueling the strength of the domestic and international political opposition to Mugabe's rule is the deterioration of Zimbabwe's economy. At independence Zimbabwe had one of the most highly developed economies in Africa. The war for independence, however, nearly bankrupted the country. The cost of rebuilding was clearly more than Zimbabwe could afford by itself. International aid was crucial in allowing the government to begin the task of reconstruction. Nevertheless, economic growth was far below the rosy independence-era estimates of 8 percent, due to a fall in commodity prices and a drought. (Average economic growth rates in the first decade after independence were 3.4 percent annually. As the economic crisis deepened, growth contracted by 6.1 percent in 2000, 4 percent in 2001, and is expected to fall another 5 percent in 2002.)

These external shocks were beyond the control of the new postcolonial leadership. The Mugabe government, however, began implementing economic policies that ultimately led to Zimbabwe's economic meltdown. To boost economic growth rates and reward supporters, Mugabe used government spending. This spending soared even as private investment shriveled. In a move that further soured the investment climate, Mugabe used import- and export-licensing systems that rewarded political supporters but led to severe distortions in the market. Although the system created a wealthy black class and provided money for government education and health programs, it led to corruption among politicians with connections and to rationing of goods due to firms' inability to obtain import licenses for needed production inputs.

The government also purchased companies not already state-owned. The board members of these parastatals were political supporters of the government, but lacked managerial expertise. Companies failed and had to be rescued by the government or sold at a loss. This further strained the government's budget as revenue fell and expenses grew. The government increasingly borrowed money to meet budget needs, and interest rates jumped due to government demand for money. This, in turn, increased the cost of borrowing to private firms, and businesses failed or stagnated. Unemployment therefore grew, and government tax revenue fell as consumers purchased less and paid fewer taxes. A vicious cycle had been created.

A decade after independence the Zimbabwean government was broke and the economy was in decline.

A decade after independence the Zimbabwean government was broke and the economy was in decline. Zimbabwe gained a short economic respite as conflicts wound down in Mozambique and South Africa, thus boosting trade. The Mugabe government obtained financial assistance from the World Bank and the International Monetary Fund, but these organizations demanded economic reform as a condition for the loans. The strings took the form of economic Structural Adjustment Programs (ESAPS), standard cure-alls for the ailing economies of heavily indebted developing countries. The ESAPS were designed to lessen the government's role in the economy and pro-

HIV/AIDS EPIDEMIC

IN ASSESSING a post-Mugabe Zimbabwe, the devastating impact of the HIV/AIDS epidemic must be considered. Clearly Mugabe is not responsible for the rapid spread of HIV in the country since the early 1990s. Indeed, the full impact of AIDS will not be felt until long after Mugabe departs from government service. For the next few decades Zimbabwe will be rebuilding the state in the midst of a plague that will be sapping its vigor.

According to the United Nations, approximately 25 percent of Zimbabwe's adult population has HIV. These rates are comparable to neighboring South Africa and Botswana, and are among the highest in Africa. In Zimbabwe, like the rest of Africa, heterosexual contact is the primary method of transmission. The rapid transmission of HIV is aided by low rates of condom use, high rates of other sexually transmitted diseases, and multiple sexual partners. The impact of HIV/AIDS is hardest on the most economically productive segment of society, those 15–49 years of age. The very young and old depend on this segment to survive. Approximately 40 percent of Zimbabwe's population is under 15, twice the rate in the United States. Adults are also responsible for caring for their elderly parents. Thus, the epidemic creates both orphans and destitute grandparents while reducing the resources available to those left behind.

The future impact on the country is already being taken into account in official statistics. Average life expectancies, which had steadily risen over the past few decades, are now being revised downward. A Zimbabwean's life expectancy with no AIDS epidemic had been projected to nearly 70 years of age by 2010. Factoring in AIDS reduces life expectancies to just under 40 years of age. Rates of natural increase, which averaged 1.7 percent annually during the 1990s, are falling due to the rise in death rates. Although AIDS may not lead to negative population growth rates, it will certainly slow the rate at which the country's population is increasing. It will also leave a much sicker population. Although future medical treatments and behavioral changes may eventually contain or end the epidemic, the next decade will see AIDS continue to take its toll on Zimbabwe.

R. L.

mote private enterprise. Zimbabwe was required to liberalize its economy by reducing subsidies to state-owned companies, abolishing import-export licensing requirements, and devaluing its currency.

The reforms were never fully implemented. Zimbabwe gained few of the advantages that reform would have brought in the long term but did experience the short-term costs of high inflation, the closure of flabby companies and the resultant unemployment, and a reduction in social services for the poor. By 1997 the cumulative impact of economic mismanagement caught up fully with Zimbabwe. High and growing unemployment, accelerating inflation, and increasing labor strikes led to rioting in the streets. Mugabe's decision to dispatch 11,000 troops to Congo led to enormous strains on government finances. The seizures of white-owned farmland led to major disruptions in the tobacco market, which is an important source of foreign exchange. The seizures also affected production of corn, which is a major food staple. Today the Zimbabwean economy continues to contract, with famine predicted due to policy-induced food shortages.

REBUILDING ZIMBABWE

The election results behind him, President Mugabe thus faces a daunting task. The economy is deteriorating, political stability is weakening, and a deadly AIDS epidemic is wasting the country. With the arguable exception of the AIDS epidemic, it did not have to be this way. Zimbabwe is in its current crisis due to decisions made during the past two decades. When Mugabe came to power, he used the resources of the state as his own to consolidate his political power. When these eventually proved inadequate, he turned to foreign sources for additional funds. When the economy finally crashed around him, he started plundering the last economic plum left in the state: the white-owned commercial farms.

Mugabe's continued rule delays for a few years the final day of settling accounts. A win by Morgan Tsvangirai would have provided international goodwill and a favorable reception to rescheduling Zimbabwe's debts. Zimbabwe, in the face of international political and economic isolation, no longer has this grace period. Mugabe will be forced to confront the consequences of his policies. His campaign promises before the election and statements after the election indicate that he believes he has received a mandate for the continuation of policies that led to Zimbabwe's current crisis. If more of the same is to be expected, then the debate shifts to preparing for a post-Mugabe Zimbabwe. Two questions arise. What steps will need to be taken to politically and economically rebuild Zimbabwe after Mugabe? And what steps need to be taken during this transitional period to facilitate a post-Mugabe future?

First, the legitimacy of the Zimbabwean government must be established. Without the support of Zimbabwe's citizens, a new government will be hard pressed to tackle the country's problems. The challenge for the new leader is to devise politically difficult policies that avoid the personalization of power that is a part of the political culture. In the long run, Zimbabwe may be little better off if a Morgan Tsvangirai becomes just another Mugabe in his approach to power.

The personalization of power, a pervasive aspect of African political culture, lies at the heart of Zimbabwe's present difficulties. The view that all the resources of the state—and indeed the state itself—are simply an extension of the leader is not uniquely African—it was Louis XIV's France that gave us "L'état c'est moi." Elsewhere in Africa, an extreme concatenation of state and personal power has caused the collapse of the state. In Zaire (now Congo), for example, the departure of President Mobutu Sese-seko in 1997 led to the disintegration of the Zairian state. Zimbabwe appears to have been on the path toward this end.

Reform of government institutions must be a priority for the new government. The form of the government will reflect both Zimbabwean political culture and short-term calculations. The government may be more authoritarian than democratic, but to avoid the trap of the personalization of power, the party and government must be separated and the parliament and judiciary must have some independence from the executive. There must be limits to the party in power and a political space for an opposition.

Economic liberalization is also a necessary condition for rebuilding the economy of the Zimbabwean state. The country is broke, the infrastructure needs maintenance, living standards are falling precipitously, and many of the most educated and skilled have long since departed for more opportune pastures. Two decades of living off its endowment has meant that the new government will find it difficult to obtain sufficient resources to provide government services. It is difficult to see how Zimbabwe can even begin to develop the resources needed for state building without major political and economic reform. Unfortunately, much of the population, having experienced the pain of structural reform but without seeing the benefits, will lack enthusiasm for a new round of reform. Attempts to raise revenue through tax increases will be skeptically viewed by those accustomed to a government that uses state resources for personal benefit.

A first step for a politically and economically renovated Zimbabwe is the establishment of the rule of law, particularly property rights. The stable, predictable, and legally transparent environment needed for investment must be established. The issue of land reform must be dealt with fairly and justly for all parties. This situation especially has festered for too long; Mugabe has been able to use it precisely because it is so tender. The new government must fashion a compromise that respects the legitimate property rights of white Zimbabweans while at the same time meeting some of the expectations of landless black farmers. It must be clear to all that a radical redistribution of land from commercial to subsistence farming would not improve Zimbabwe's economy.

Fortunately, the international environment necessary to support such an economic restructuring has never been better. The end of not only the cold war but also apartheid in South Africa and civil war in Mozambique (which provides Zimbabwe with its seaport) means the area is at peace. Regional organizations such as the Organization of African Unity and subregional organizations like the SADC are attempting to grapple more effectively with continentwide security and economic issues. These and other organizations provide an environment in which a new regime in Zimbabwe can seek ways to foster a better business climate and resolve disputes.

The winds of change are blowing against Robert Mugabe. Leaders of the African liberationist tradition of the 1960s and 1970s crushed dissent, distrusted free markets, and were often antagonistic toward the West. Beginning in the 1990s, the extreme personalization of power commonly seen in these early postindependence African governments began to yield to regimes that increasingly tolerated some degree of political and economic freedom. Given this broader trend, both African states and the international community need to begin planning for a post-Mugabe Zimbabwe.

ROBERT B. LLOYD *is a professor of international relations at Pepperdine University and director of international studies. His previous article for* Current History, *"Mozambique: The Terror of War, the Tensions of Peace," appeared in April 1995.*

From *Current History*, May 2002, pp. 219–224. © 2002 by Current History, Inc. Reprinted by permission.

Talking Peace, Waging War

*The present hostilities stem from the second
Palestinian intifada that erupted in September 2000,
sparked by Ariel Sharon's visit to the Temple Mount.*

CHRISTOPHER C. JOYNER

Over the past two years, events in the Middle East have produced the sensation of watching a slow-motion train wreck. Everyone sees the tragic outcome of the crash, but no one can stop it.

Though many good-faith efforts have been made to apply brakes to the conflict, none has succeeded in slowing it down. Many peace talks have been held, but the cascade of bloody violence, death, and destruction goes on, inevitably and unending.

The present hostilities stem from the second Palestinian intifada, the "people's uprising" that erupted in September 2000 after Ariel Sharon's visit to the Haram al-Sharif compound, home of the al-Aqsa Mosque on the Temple Mount.

The timing and motive behind Sharon's visit to Islam's third-holiest shrine seem clear enough: to provoke political unrest among Palestinians and gain the Likud Party's leadership over Benjamin Netanyahu. He succeeded profoundly on both counts.

Ironically, the tidal wave of intifada violence that followed contributed to catalyzing the most productive Middle East peace negotiations to date between Palestinians and Israelis. Under the Camp David plan, Yasser Arafat and then-Israeli Prime Minister Ehud Barak agreed that Palestinians would gain sovereignty over 95 percent of the West Bank, all of Gaza, Arab areas of Jerusalem, and Haram al-Sharif. The Israelis were to have control of the Western Wall at the base of the Temple Mount and Israeli neighborhoods in Jerusalem. Jerusalem was to be capital of both Israel and the Palestinian state, and the international community would provide economic assistance to implement the agreement.

The plan also called on the Palestinians to drastically scale back their demand for all refugees from the 1948–49 war and their descendants—about 4 million people—to have the right of return to their original homes. When Arafat refused to accept this provision, negotiations collapsed.

Though talks continued throughout January 2001, they eventually failed in February, just prior to a special Israeli election in which Barak was soundly defeated by the more hawkish Sharon. At that point, all proposals were taken off the table.

Lack of result, not effort

The Bush administration's policy toward the Palestinian-Israeli conflict has been distinguished more by lack of result than lack of effort. Its major initiatives—the Mitchell Plan, the Tenet Plan, and the June 24 Bush peace proposal—have made little progress in the absence of an essential ingredient for peace. That critical missing element has been the willingness of both Arafat and Sharon to make peace, even though each often looks to the United States as a broker to reduce his political risks at home.

The Bush administration began unveiling its Middle East policy in November 2001, when President Bush and Secretary of State Colin Powell announced that, for the first time, the United States supported a "viable Palestinian state." Powell then sent Gen. Anthony Zinni to open the door for securing agreement on the Mitchell and Tenet Plans.

In May 2001, the Mitchell Commission issued its conclusions calling for an immediate cease-fire, to be followed by confidence-building measures and ultimately renewed peace negotiations. It also asked Palestinians to jail terrorists and Israel to freeze all construction of Jewish settlements in the West Bank and Gaza.

The Palestinians accepted the findings, but Israel declared that it couldn't stop building because it must accommodate natural population growth in these settlements, which are home to more than 200,000 Israelis. The Mitchell Plan was thus rendered dormant.

Following renewed violence, CIA Director George Tenet visited the region and undertook negotiations that produced a detailed arrangement for ending the violence and resuming peace negotiations, given the consent of the two sides. The Tenet Plan contains measures that ask the Palestinians to crack down on militants and Israel to withdraw from territory seized during the intifada. This strategy sets out a mechanism for implementing a ceasefire, after which the Mitchell Plan—the blueprint for getting the peace process back on track—can be implemented. The Tenet Plan went into effect on June 13, 2001, but resumption of negotiations was conditioned on achieving a violence-free week. No such week has occurred. Instead, we have seen an epidemic of suicide bombers that has terrorized Israeli society.

Who Is to Blame?

- Yasser Arafat, not Ariel Sharon, has proved to be the greater obstacle to Middle East peace.

- The Palestinian leader has not made a sincere effort as demanded by the United States to prevent terrorism in Israel, especially suicide bomb attacks against Israeli civilians by militant Palestinian groups such as Islamic Jihad, Hamas, and the al-Aqsa Brigades.

- This has become all too evident as the number of Israeli dead and injured mounts.

- Escalation of suicide bombings culminated on March 27 in the Passover bombing of a Netanya hotel that left 27 Israelis dead and more than 50 injured.

- In retaliation, the Sharon government launched "Operation Defensive Shield" to attack Palestinian terrorists, their weapons factories, and their arms caches.

- Twelve days of fighting in Jenin took a severe human toll on both sides. Israel reported that 23 of its troops were killed. UN officials reported that 54 Palestinians died, and an additional 49 were missing.

Arafat, the stumbling block

Arafat, not Sharon, proved to be the major obstacle. The Palestinian leader made no sincere effort to comply with the U.S. demand to prevent violence and terrorism in Israel, as evidenced by the mounting use of suicide bombers against Israeli civilians by militant Palestinian groups such as Islamic Jihad, Hamas, and the al-Aqsa Brigades.

Suspicions about the Palestinian Authority's commitment to peace under Arafat deepened early in 2002. On January 3, Israel seized the *Karine A*, a Palestinian ship chartered by Arafat's closest aides that was carrying 50 tons of Iranian arms. Both the United States and Israel concluded that Arafat was planning to escalate violence while negotiating for a cease-fire.

This incident strongly suggested that, despite the impressive global consensus against terrorism after September 11, Arafat persisted in ratcheting up local violence, even to the verge of all-out war between Israel and the Palestinians. Politically, the *Karine A* episode severely undercut whatever diplomatic credibility Arafat had left.

Around the same time, the Bush administration embarked on an international campaign to mobilize support for intervening in Iraq, with the transparent goal of ousting Saddam Hussein. In early February, the White House announced that Vice President Cheney would tour the region, ostensibly to drum up Arab support for the next phase of the war on terrorism, which involved hunting down al Qaeda remnants throughout the region. Cheney's mission was to convey Bush's determination to remove Saddam, lest he acquire weapons of mass destruction.

Significantly, Cheney's impending visit prompted a surprising initiative from Saudi Crown Prince Abdullah, who proposed in February that Arab states accept "normal" relations with Israel, in exchange for Israel's withdrawal to pre-1967 borders, the transfer of Jerusalem, and the right of return. The Abdullah proposal, subsequently endorsed by the League of Arab States, is the first time that Arab governments formally expressed a willingness to accept Israel's political existence in the region and thus stands as a milestone.

The Cheney tour in March proved to be an education in regional politics for the administration. The plight of the Palestinians, not possible military threats from Iraq, was of concern to Arab leaders. Going after Hussein would be difficult; Iraq remained an integral part of the Arab world.

The situation thus became crystallized for U.S. policymakers. If Arab governments are to support a campaign against Iraq, the Palestinian-Israeli conflict must first be stabilized, with justice given to the Palestinians. Moreover, irrefutable links between Hussein and international terrorist groups would have to be demonstrated, in the Arab viewpoint.

'Operation Defensive Shield'

Late in March, Arafat's strategy of compelling international intervention bore political fruit. During the previous year, suicide bombers had killed more than 100

Israelis, culminating in the March 27 bombing of a Netanya hotel that left 27 Israelis dead and more than 50 injured. In retaliation, the Sharon government launched "Operation Defensive Shield" as a military offensive to attack Palestinian terrorists, their weapons factories, and their arms caches.

The swiftness and ferocity of this assault, especially the 10-day battle throughout the Jenin refugee camp, shocked the Palestinians, who did not believe the Israelis would venture into their cities with massive ground forces. Twelve days of fighting in Jenin took a severe toll on both sides. Israel reported that 23 of its troops were killed. UN officials reported that 54 Palestinians died and an additional 49 were missing.

The main purpose of Israel's attack on Jenin was to destroy or capture terrorist weapons. In Nablus, Israeli forces killed 78 Palestinians. Arafat's compound in Ramallah was destroyed; he was confined to a few rooms, along with suspects wanted for the murder of the Israeli minister of tourism the previous October.

In Bethlehem, 240 persons, including the 35 armed Palestinians who had seized the Church of the Nativity, were besieged for 39 days by the Israeli army. Though nearly all were released unharmed, the symbolism of Israeli power was unmistakable.

A week into the incursion, President Bush publicly demanded that Sharon withdraw Israeli forces without delay. He also called on Arafat to accept a cease-fire and suppress Palestinian terrorism and asked Arab governments to cease inciting Palestinian violence.

In addition, Secretary of State Powell was dispatched to help end hostilities. Powell's visit yielded few results. He brought no plan from Washington for a Palestinian state, to the disappointment of the Palestinian leadership. He secured no guarantee from Sharon of a full withdrawal from the West Bank.

Powell returned home with neither a truce nor any evidence that Israel would soon end its military offensive. His trip made it clear, however, that more than U.S. exhortation would be necessary to pull each side back from disaster. Both sides distrust each other so intensely that only outside mediators can create the conditions that supply sufficient confidence to allow them to step back from the brink. The United States must join with key Arab governments—Jordan, Saudi Arabia, and Egypt—to provide assurances for each side to obtain its main objectives.

Suicide bombers everywhere

Since September 2000, at least 75 suicide bombings have killed nearly 300 Israeli civilians and wounded some 5,000 others as they rode buses, shopped, sat in cafés, danced in clubs, or celebrated religious holidays. Given this brutality, UN Secretary-General Kofi Annan became involved in seeking ways to foster peace.

In June 2001, Annan embarked on a six-day tour of the Middle East, where he met with Syria's President Assad, Jordan's King Abdullah, various Lebanese leaders, Israeli government officials, and Arafat. He pushed the need to implement the Mitchell Commission's proposals and advocated an early resumption of peace talks, to give the Palestinian leadership some kind of incentive to abide by a cease-fire. The Israeli leadership rejected his plan, asserting that first there had to be six weeks free from violence. That demand sealed the fate of Annan's proposal.

President Bush wants two states living nonviolently side by side, in peace and security.

Annan also proved sympathetic to the Palestinian demand for an international protection force. In recent months, he asserted that the situation in the Middle East has become so serious that sending international peacekeepers to the occupied territories is imperative. Israel, not surprisingly, strongly resisted that idea. Similarly, UN efforts to investigate Israel's assault on the Jenin refugee camp in April 2002 were spurned by the Sharon government, sensing that this would amount to little more than a political vehicle to indict Israel for alleged war crimes.

The escalating carnage from suicide bombings pushed President Bush to announce a much-anticipated U.S. plan for resolving the conflict. Offered on June 24, his peace initiative produced a vision without a map, as it contained no policy suggestions for how either side might reach any of the end goals.

True, President Bush wants two states living nonviolently side by side, in peace and security. He approved a Palestinian state with provisional borders and sovereignty, pending a "final settlement." This, President Bush suggested, can be negotiated in three years, even though the most intractable problems remain, namely, the definition of borders, the right of return, and the future of Jerusalem. Further, Bush said that Israel must stop building settlements and withdraw from the occupied territories.

The most salient feature of the Bush plan, however, is his demand for new Palestinian leadership. New information linking Arafat to financial support of the suicide bombers prompted Bush to insist on negotiating with a new and different Palestinian leadership, one that was "not compromised by terror."

Acceptable democracy?

President Bush's call for the ouster of Arafat was greeted with cheers in Israel but jeers throughout the rest of the world, including the capitals of Europe. The rub comes with the high price of democracy. That is, the Palestinian people democratically elected Arafat in 1996 with 88 percent of the vote. If new elections, now slated for January 2003, should again endorse Arafat and his fellow Fatah leaders, no government has the right to undermine that outcome.

Still, there is no question that Arafat remains a considerable part of the problem. He presides over a corrupt administration in the Palestinian territories. The Israeli leadership loathes him. His own followers are disappointed at his inability to advance the peace process.

In Washington, Arafat comes across as the villain. He rejected the Clinton-Barak proposals, broke promises to control violence, and permitted more death and destruction in Israel. His strategy apparently seeks to foster mayhem through homicide bombings. At the same time, he seeks to attract international intervention to attain his avowed goals, which are to move Israel behind the June 1967 borders; obtain a Palestinian state in all of Gaza and the East Bank, with Jerusalem as its capital; and gain the right of return for all Palestinian refugees.

Now the American position has changed, putting more pressure on the Palestinians than the Israelis. The Palestinians might have to turn to a new guard of leadership: men such as Ahmed Qurei, speaker of the Palestinian Legislative Council; Mahmoud Abbas, Arafat's deputy head of the Palestinian Liberation Organization; Jibril Rajoub, security chief for the West Bank; and Marwan Barghouti, secretary-general of Fatah in the West Bank. The Bush position clearly marginalizes Arafat's participation in any future peace process and requires serious reform of the Palestinian Authority.

The Israelis see little advantage in renewing peace negotiations at present. Sharon maintains that suicide terrorism betrays the sincerity of Palestinians to accept peace, and, thus, Israel is again fighting for its very survival. His conclusion is that a sustained campaign of drastic self-defense must be undertaken. To that end, the Sharon government has yet again deployed troops to invade major West Bank towns; he announced in late June that Israeli forces would reoccupy territories previously ceded to the Palestinian Authority as long as Palestinian terrorism continues.

In addition, Israel unilaterally began constructing a massive border fence between the Jewish state and Palestinian-occupied areas. But if the past is prologue, Israeli military occupation cannot stop Palestinian resistance and may incite more of it. Retreat into "fortress Israel" furnishes no solution for a lasting peace in the region. Such repressive policies frustrate any hope of success for a new U.S.-backed Israeli-Palestinian peace process even before it is launched. Moreover, they play into the hands of Palestinian extremists.

It is apparent that the essence of the Middle East conflict no longer is Israel's rejection by the Arab world. Rather, it has become the struggle of the Palestinians for a viable state of their own against the stubborn determination of Sharon and the Israeli right wing to deny this ambition by preserving Jewish settlements in the occupied territories. The collapse of the Oslo peace process, nearly two years of Palestinian terrorism—punctuated by horrific homicide bombings that indiscriminately kill innocent men, women, and children—and brutal Israeli military reprisals have combined to obscure this reality.

If the Bush administration is serious about securing peace in the Middle East, which means security for Israelis and justice for the Palestinians, it must address all causes of Palestinian terrorism, including Sharon's rejection of a viable Palestinian state and his wholesale commitment to the settlements. Rhetorically, President Bush said as much in his June policy address, though he provided no clues as to how to attain those ambitions.

In the end, however, nothing can come of U.S. diplomacy unless both sides are willing and prepared to stop fighting. That is the preeminent challenge for U.S. policy in the region. The Bush administration must convince both Palestinians and Israelis that peace and justice, not violence, serve their mutual long-term interests. Until that happens, the bloody fighting will go on, and more Palestinians and Israelis will die or be maimed as revenge and retaliation drive the cycle of gruesome bloodshed on both sides.

Christopher C. Joyner is professor of international law in the government department at Georgetown University.

From *The World & I*, September 2002, pp. 20-25. © 2002 by The World & I, a publication of The Washington Times Corporation. Reprinted by permission.

ENDING THE DEATH DANCE

by RICHARD FALK

Few would deny that September 11 unleashed a fearsome sequence of reactions, and none so far worse than the anguishing fury of this latest cycle of Israeli-Palestinian violence. Surely the United States is not primarily responsible for this horrifying spectacle of bloodshed and suffering, but there is a gathering sense here and overseas that the US government has badly mishandled its crucial role for a long, long time, and especially since the World Trade Center attack. As the situation continues to deteriorate for both peoples, there is a rising chorus of criticism that paradoxically blames the United States both for doing too much on behalf of Israel and not enough to bring about a durable peace. Both lines of criticism seem justified.

There is little doubt that part of the recent escalation can be traced back to President Bush's overplaying of the antiterrorist card since Day One of the response to Al Qaeda. By overgeneralizing the terrorist threat posed by the September 11 attacks, Bush both greatly widened the scope of needed response and at the same time gave governments around the planet a green light to increase the level of violence directed at their longtime internal adversaries. Several important governments were glad to merge their struggle to stem movements of self-determination with the US war on global terror, and none more than Ariel Sharon's Israeli government. The Bush Administration has made several costly mistakes. By not limiting the response to the Al Qaeda threat, it has taken on a mission impossible that has no end in sight; even worse, the Administration embraces war in settings where it has no convincing relationship either to US or human security. Related to this broadening of the goal is the regressive narrowing of the concept of terrorism to apply only to violence by nonstate movements and organizations, thereby exempting state violence against civilians from the prohibition on terrorism. Indeed, this statist approach has been extended so far that it calls nonstate attacks on military targets such as soldiers or warships terrorism, while not regarding state violence as terrorism even when indiscriminately directed at civilian society, as

seemed the case at times during the Russian response to Chechnya's drive for independence and with respect to Israel's approach to occupation. Such a usage is ethically unacceptable, politically manipulative and decidedly unhistorical. It is important to recall that the usage of the word "terrorism" to describe political violence derives from the government excesses that spun out of control during the French Revolution.

The issue here is not one of political semantics but of analysis and prescription. By designating only Palestinian violence as terrorism, Israel's greater violence not only avoids stigma in the American context but has been officially validated by being treated as part of the struggle against terrorism. The point here is not in any way to excuse Palestinian suicide bombers and other violence against civilians, but to suggest that when a struggle over territory and statehood is being waged it can and should be resolved at the earliest possible point by negotiation and diplomacy, and that the violence on *both* sides tends toward the morally and legally impermissible. This contrasts with the challenge of Al Qaeda, a prime instance of visionary terrorism that can neither be neutralized by negotiation nor deterred, and must and can be disabled or destroyed in a manner that is respectful of moral and legal limits. To conflate these two distinct realities, as Bush has consistently done, is at the root of the US diplomatic failure to diminish to the extent possible the threats posed by the September 11 attacks and to offer the Palestinians and Israelis constructive guidance.

There is another feature of the situation that infects commentary from virtually every corner of the debate, also reflecting the mindlessness of a statist bias. Everyone from George Mitchell to George Bush seems entrapped in the mantra that it is of course to be expected that every sovereign state must react violently and punitively against any significant act of terror directed against it. Many of these commentaries also take note of the degree to which such counterterror gives rise to worse violence on the other side, revealing the bankruptcy of the approach. It is truly a vicious circle. At the same time, it never sees that the logic of such vengeful violence works reciprocally. If the dominant actor pursues such an approach, what of the weaker side? When the Palestinians strike, their actions are never understood here as reactive and understandable, always provocative. Never has this been truer than with respect to the horrifying Passover bombing at Netanya and the equally horrifying Israeli incursion with tanks and helicopters throughout occupied Palestine. If one is essentially acceptable, and the other condemned, it deforms our understanding.

The same dynamic applies to the endless discussion about Yasir Arafat's role. It is condemned, to varying degrees, while Sharon's bloody past is rarely mentioned. He is usually treated with respect or, at most, Palestinian intransigence is given as the reason Israelis chose such an extreme leader in a democratic election.

But the problems of US leadership cannot all be laid at the feet of the Bush presidency. Just as crucial was the insufficiency of the Oslo peace process, and the blame game that has been played ever since the outbreak of the second intifada in late September 2000. It has been endlessly repeated, without any demonstration, that the Israelis under Prime Minister Ehud Barak made a generous offer at Camp David in the summer of 2000. It is then alleged that Arafat rejected an offer he should have accepted, and resumed armed struggle. Further, it has been alleged that Arafat's rejection was tantamount to saying that the struggle was not about establishing a Palestinian state but about ending the existence of the Jewish state. It was this one-sided assessment, alongside others, that led to Sharon's election, which meant that Israel would henceforth be represented by a man with a long record of uncompromising brutality toward Palestinians and a disregard of their legitimate claims for self-determination.

But was Arafat to blame for the failure of the Oslo endgame? I think it was a most unfortunate failure of leadership by Arafat not to explain to the Palestinians, Israelis and the world why Barak's Camp David proposals were unacceptable. It should be remembered that Arafat at one point seemed on the verge of accepting them but backed away only when confronted by the unhappiness of a large proportion of his own people with the sort of Palestinian state that would result. It should also be remembered that the entire negotiation concerned 22 percent of the original British mandated territory of Palestine, about which the Palestinians were expected to strike compromises while leaving the 78 percent that was Israel out of account. Further, the future of the settlements in the occupied territories was to be addressed by Israeli annexation of half of them, including 80 percent or more of the settlers, despite the settlements' illegality and the degree to which their existence was a daily irritant to Palestinian sensibilities. And on refugees, there were evidently some signs of a compromise in the making at the supplemental negotia-

tions at Taba in January 2001, but nothing was written down, and it was far from clear that Barak could have delivered on what was offered even if re-elected, so strong were Israeli objections to any return by Palestinians to pre-1967 Israel. Beyond this, it was expected that the security of Israel was to be maintained in such a way as to put any emergent Palestine in a permanent position of subordination, thus denying the fundamental message of any genuine peace: insuring equivalence between the two states for the two peoples. The Palestinians would sooner or later challenge such a solution even if their leaders could be induced to sign on the dotted line. Many have forgotten that a widespread fear among Palestinians at the time of Camp David was that Arafat would sell his soul and that of his people (especially the more than 50 percent who were refugees) for the sake of a state, any state, as this was thought to be his sense of personal mission.

The Palestinian mainstream learned via Oslo that its cease-fire would not produce a fair solution in the form of sovereign and equal states.

Similarly, the widespread contention in American circles that Arafat opted for terrorism is also seriously misleading. Such thinking deforms perceptions of what is reasonable. Arafat was up against more militant forces in the Palestinian movement throughout this period, and was generally viewed as the most moderate voice among the Palestinian leadership, and had even shown an early willingness to incur the wrath of Hamas and Islamic Jihad militias by taking seriously his duty to prevent the territories under the administration of the Palestinian Authority from being used against Israel and Israelis. Beyond this, it was Sharon's own provocative visit to the Al Aqsa Mosque that started the second intifada. This visit proceeded despite fervent warnings about the explosion likely to happen, given privately to the Barak leadership by the most respected Palestinians, including the late Faisal Husseini, head of Orient House in Jerusalem.

The Palestinian demonstrations that followed were notably nonviolent at the outset. Israel countered from the beginning by using excessive force, killing and seriously wounding demonstrators in large numbers, and by its practice of extrajudicial assassination of a range of Palestinians living in the West Bank and Gaza. At this point the escalatory spiral was initiated, with Israel acting with ever more force at each stage, ratcheting up the stakes to such a level that the Palestinians were being attacked with among the most sophisticated weapons of warfare, including very modern tanks and helicopter gunships. It was in the course of this process that Palestinian resistance gradually ran out of military options, and suicide bombers appeared as the only means still available by which to inflict sufficient harm on Israel so that the struggle could go on. I was a member of a human rights inquiry appointed by the UN Human Rights Commission a year ago; our report fully supported this line of interpretation in its study of the second intifada, as did the overwhelming majority of the Security Council membership. The basic conclusion of these efforts at impartial understanding was that Israel was mainly responsible for the escalations, and that its tactics of response involved massive violations of international humanitarian law.

There is the closely related matter of continued Israeli effective occupation of the West Bank and Gaza, a reality that has been fully re-enacted in the past few weeks. It poses the question of what sort of right of resistance is enjoyed by an occupied people when the occupying power ignores international law and refuses to withdraw. Such a right of resistance does not permit unrestricted violence, but it certainly would seem to legitimize some armed activities. It puts in a different light the furor raised in January by the intercepted arms shipment that was evidently intended for Palestinian use. Should the opposition, in the context of the sort of struggle that has gone on for decades, have no right to gain the means of self-help while the occupying power can arm itself to the teeth, all the while denying international accountability and refusing UN authority?

Here is the essential point: The Palestinian mainstream learned via Oslo that its cease-fire would not produce a fair solution in the form of sovereign and equal states, and that its real interests had been sacrificed on the altar of geopolitics. In effect, negotiations would be bargains reflecting the realities of power and control rather than either a pathway to some mutually acceptable form of parallel states or what many Palestinians had expected—namely, resolution by reference to international law. It is important to appreciate that on virtually every issue in contention, the Palestinians have international law on their side, including the Israeli duty to withdraw from land taken during a war, the illegality of the settlements under Article 49(6) of the Fourth Geneva Convention, the right of refugees to a safe return to the country that wrongfully expelled them and the generalized support for a Jerusalem that belongs to everyone and no one. In other words, if fairness is understood by reference to international law, the outcome would look nothing like what was offered in the Barak/Clinton proposals. Such a result would come nowhere close to satisfying the right of self-determination as understood by almost all Palestinians, and as achieved long ago by the Israelis. The failure of the US government to uphold Palestinian rights and the inability of the UN to implement its authority was extremely disillusioning for moderate Palestinians, and this

tended to shift attention to the ouster of Israel from southern Lebanon through the use of force by Hezbollah.

What is worse, virtually all of the discussion about reviving the peace process, including that of the Palestinian leadership, is a matter of going back to a reconstituted Oslo—that is, negotiations between the parties after a cease-fire has been agreed upon. The Mitchell Commission report moves in this direction, as does the Tenet plan for putting a cease-fire into effective operation. Even these rather flawed initiatives have been stymied primarily by Sharon's hostility to the whole idea of peace negotiations under international auspices that would draw into question the settlements or address the grievances of the refugees and the sovereignty of Jerusalem in any way that would satisfy even the most moderate Palestinian expectations. The Palestinian Authority can also be faulted on the opposite basis, for too readily subscribing to the "honest broker" claim of the United States in relation to the peace process, despite abundant evidence over the years of the degree to which the US government pursues an unabashedly pro-Israeli foreign policy that is underpinned by massive annual foreign assistance, mostly for weapons purchases. At the very least, Palestinian leaders should point to the problem, and possibly seek more neutral auspices for these matters of life and death for their people. If real peace is the goal, we cannot get there from here!

It is this tragedy that continues to be played out in the most reprehensible ways. To say this is not to underestimate the difficulty of a good-faith peace process that meets the needs of both peoples. It would be a mistake to pretend that international law provides all the answers, although it does give guidance as to what is reasonable given the overall controversy. On refugees, for instance, implementing international law would surely doom any agreement, since almost all Israelis would regard an unrestricted Palestinian right of return as tantamount to the destruction of the Jewish state. My conversations with many Palestinians suggest that there would be a great willingness to find a formula that both sides could accept, possibly relying on an Israeli acknowledgment of the wrongfulness of the expulsions, especially in 1948, provisions for compensation for lost property and limited opportunities for return phased in over time. If the Israeli leadership were prepared to work for the establishment of a Palestinian state equal to their own, I would anticipate an outpouring of Palestinian efforts to reassure Israel of its own sovereign identity.

Oddly, despite its record of partiality, only the United States seems to have the current capacity to put the two states on such a genuine peace track, but it is not likely to do so until pushed hard from within and without. An American civic movement of solidarity with the well-being of both peoples is essential, as is a more active independent European and Arab involvement. Both latter possibilities are becoming more plausible with each new atrocity. The belated yet still welcome Saudi initiative, offering normalization of Arab diplomatic relations in exchange for Israeli withdrawal to 1967 borders, is an important contribution. And Europe seems ready to propose a more independent alternative to what Washington has been offering if the White House cannot do better. Bush's call for Israel to withdraw its military forces from Palestinian areas "without delay" was somewhat encouraging, although it was immediately neutralized by Sharon's insistence on "finishing the operation" and by the fact that Bush sent Secretary of State Colin Powell to exert pressure but allowed him to adopt the most nonurgent itinerary, including several intermediate stops in North Africa. Such a diplomatic pattern has been widely criticized as "incoherent" at best, but at least it is a modest improvement over backing Sharon's recent criminal assault on Palestinian cities and towns.

If the United States does do better, then these new forces of engagement could at last begin to draw the line between a process that puts the weaker side in the position of either accepting what is offered or getting blamed for not doing so, and a process that gives both sides what they need: security and sovereignty. Of course, it will be difficult to move forward with the present cast of leaders and mainstream assumptions. But we should at least be clear that Sharon is a much bigger obstacle to real peace than Arafat is or ever was.

Richard Falk is currently Visiting Distinguished Professor at the University of California, Santa Barbara. His most recent book is Religion and Humane Global Governance (*Palgrave*)

From *The Nation*, April 29, 2002, pp. 11-13. © 2002 by The Nation. Reprinted by permission.

UPDATE/COLUMBIA

COLOMBIA PEACE IN TATTERS

As we went to press, the Colombian peace process discussed here broke down again, this time seemingly irreparably. On February 20, after FARC guerrillas highjacked an airplane and kidnapped a Colombian senator, President Andrés Pastrana declared talks with the FARC over. He ordered the military to retake the demilitarized zone earlier ceded to the insurgents. Check our website www.nacla.org for further analysis of this volatile situation.

BY ADAM ISACSON

In the immediate aftermath of the September 11 attacks, it was not clear what would become of the United States' large and growing military aid program in Colombia. With higher-profile missions in Afghanistan and the "homeland," might Washington reverse its steady descent into Colombia's messy, complicated conflict? Or would Colombia, with three groups on the State Department's list of foreign terrorist organizations, become a new front in the "war on terrorism," breathing new life into the failed "Plan Colombia" drug-war strategy?

In early 2002, it seems as though the second scenario is being fulfilled. The "Bush Doctrine" is not yet calling for a headlong rush into Colombia's war. But its new aid proposals and statements have explicitly targeted the 18,000-strong Revolutionary Armed Forces of Colombia (FARC) guerrilla group, erasing the longstanding, though always artificial, distinction between counternarcotics and counterinsurgency.

After falling from the headlines after September 11, Colombia reappeared in Washington's consciousness in early January, when its long-suffering peace process entered its latest, and worst, crisis.[1] Since October, the FARC had refused to negotiate anything but a demand that the Colombian military lift controls, such as roadblocks and overflights, around the demilitarized zone where talks have taken place since January 1999. A frustrated Colombian government suddenly got up from the table on January 9, and President Andrés Pastrana threatened to start a 48-hour countdown for the military's re-entry into the zone.

With tens of thousands of government troops massing on the zone's perimeter, the FARC held two days of intensive talks with James Lemoyne, the UN Secretary General's representative to Colombia. Pastrana rejected the first FARC proposal to restart the talks, setting the 48-hour countdown in motion on January 12. On the 14th, hours before the troops were to move in, the FARC gave up its demands about military activities outside the zone, and the dialogues re-started. By January 20, both sides had agreed to a timetable for cease-fire talks. Had the timetable been followed, a cease-fire could have been in place by April 7.

The United States played a quiet but important role during the crisis. Ambassador Anne Patterson—who on January 8 presided over the delivery of 14 Blackhawk helicopters to Colombia's army—met several times with President Pastrana and Colombia's high command. Probably reflecting a lack of administration consensus on whether to continue supporting the talks, officials took pains to be noncommittal, their statements falling short of specifically endorsing the current peace process. "We support President Pastrana, his decisions regarding how to proceed," was the most that State Department spokesman Richard Boucher could say on January 14. A delegation of mostly Republican members of Congress, coincidentally in Bogotá for a four-day visit, was effusive in its praise for Pastrana's new hard line. Official U.S. opposition to the talks may soon become more explicit; a "confirmed intelligence report" leaked to the conservative *Washington Times* in mid-February contends that the FARC "is just buying time through unproductive peace talks while it mounts attacks and expands its highly lucrative cocaine operations."[2]

Despite these difficulties, there was reason for optimism about the future of the talks. UN representative LeMoyne, as well as the Colombian Episcopal Conference, a Catholic Church group, and ambassadors from the ten European and Latin American nations of the "group of friends" of the peace process, were given permanent seats at the table. It was hoped the third parties' "accompaniment" would speed the talks by holding both sides to an agenda and finding solutions to inevitable impasses. Another source of optimism was the mere fact that the FARC made a concession, perhaps its largest in three years of talks. Some observers hoped that this rare show of flexibility indicated that the guerrillas—who have too often resembled an armed band with no coherent political direction—were showing more concern for public opinion and the political impact of their actions.

Any goodwill the guerrillas' concession earned was squandered, however, with a large-scale FARC offensive in late January and early February. This round of attacks was more urban than most previous actions, marked by bombings of infrastructure (34 power pylons, as well as aqueducts and oil pipelines) and three car bombings.

While the offensive earned universal condemnation, analysts offered divergent interpretations of its meanings. Many saw it as evidence that the guerrillas are in fact uninterested in peace. Others, however, viewed the brutal rampage as the FARC's attempt to extract the best possible terms in a cease-fire agreement. As a truce would cut off the guerrillas' main sources of income (kidnapping and extortion) while having little effect on military capabilities, FARC leaders undoubtedly view a cease-fire as a period of relative weakening, and may be using the violence to press for maximum government concessions in exchange.

Whatever the reason, the offensive shattered most remaining public support within Colombia for the peace process and swelled the chorus of calls for a hard-line response. It has severely shaken the campaign for Colombia's May presidential elections. A poll taken in late January by several of the country's main media outlets found a third-party law-and-order candidate, Alvaro Uribe Vélez, ahead of his nearest challenger by a margin of 40% to 30%.[3] Uribe—who promised to send the military into the FARC zone as soon as he took office, if the guerrillas did not immediately cease attacks and kidnappings—had only 22% support in a November 2001 poll by the same organizations.

It was against this increasingly bellicose backdrop that the Bush administration, on February 4, made public its future plans for Colombia. As details about the 2003 foreign-aid budget request emerged, it became clear that the administration is ready to go where no U.S. administration has gone before, offering Colombia its first significant non-drug military aid since the Cold War.

In its foreign operations budget request, which includes all economic and most military aid, the State Department would give Colombia $538 million in 2003, $374 million of it (70%) for Colombia's military and police. Counting additional anti-drug aid through the defense budget (mostly construction, intelligence and training), the military/police aid figure approaches half a billion dollars, more than Colombia has ever received in a single year.[4] By comparison, at the January 2002 international donors' conference in Tokyo, the United States pledged just $290 million in economic aid to help rebuild Afghanistan.

Most of the foreign operations request for Colombia ($275 million in military aid and $164 million in economic aid) comes under the category of the "Andean Regional Initiative," a proposed $731 million outlay of funds by the State Department's drug control program for Colombia and six other countries. Begun in 2002 in an effort to regionalize the Clinton administration's "Plan Colombia" approach, this "initiative" has already nearly doubled military assistance to several of Colombia's neighbors (particularly Ecuador, Panama and Peru).

Much of the Colombia portion of the 2003 Andean request would maintain, with spare parts, fuel, follow-on training and other support, a 2,300-man Colombian Army Counternarcotics Brigade. This brigade, created with U.S. funds in 2000 and 2001, is based in Putumayo department in southern Colombia, near the Ecuadorian border. It is almost ready to begin full-scale operations, pending the mid-year delivery of thirty upgraded "Huey II" helicopters and the graduation of trained pilots for its expensive new Blackhawk helicopters.

Once operational, the brigade will embark upon the so-called "push into southern Colombia," an offensive against the armed groups that dominate Putumayo. The operation's stated goal is to "create the security conditions" for vastly expanded aerial herbicide fumigation of the zone's coca-growing peasants. The fumigation is carried out jointly by Colombia's police and employees of DynCorp, a Virginia-based contractor that supplies the State Department with dozens of spray-plane and helicopter pilots, mechanics, search-and-rescue personnel and other specialists. The contractors operate at some risk under conditions that would be off-limits for uniformed U.S. personnel; on January 18, three U.S. contractors on a spray mission were pinned down for hours by FARC fire after the guerrillas shot down their Huey helicopter in Caquetá department, just north of Putumayo. Five Colombian police died in the effort to rescue them.

The fumigation has brought a tidal wave of health and environmental complaints, whose scientific verification has been made difficult by security conditions.[5] The spraying was to go hand-in-hand with a multimillion-dollar campaign of development assistance, designed to provide legal alternatives to the peasants whose only viable economic option, coca, is to be eradicated. Delivery of this aid has been excruciatingly slow, however, with aid being delivered to only a small fraction of Putumayo residents who signed "social pacts" of aid in exchange for manual eradication. With trust in the U.S.-funded development program eroding quickly, the effort risks the outcome of past U.S. fumigations minus economic aid: discouraged peasants simply moving along to grow new coca, out of the spray planes' range.

It appears that the response to new coca-growing, foreseen in the 2003 Andean Initiative request, will be to create a second Colombian Army counternarcotics brigade to mount a second "push" into another part of the country. Though details are sketchy as of early February, this new unit may operate in the eastern departments of Guainía and Vichada, near the Brazilian border.

The most controversial proposal, however, is elsewhere in the aid request. The 2003 plan would give Colombia's army $98 million in Foreign Military Financing (FMF). FMF is the heir to the programs used for Central America's 1980s military buildup, and in recent years has mostly aided Israel and Egypt. The money would create (or re-train) yet another army brigade with a very different mission: to protect an oil pipeline.

The FMF-aided unit would be given a dozen UH-1 "Huey" helicopters and charged with protecting the Caño Limón-Coveñas pipeline, which runs from guerrilla-dominated Arauca department to paramilitary-dominated Sucre department in northern Colombia. Much oil in this pipeline, which was attacked 166 times by the FARC and ELN guerrillas in 2001, belongs to Los Angeles-based Occidental Petroleum.

Occidental, which many activists know as the company that pushed for oil exploration on land claimed by the U'Wa indigenous nation in Arauca, has spent years lobbying for additional military assistance to Colombia. The $98 million "Critical Infrastructure Brigade," as the Bush administration aid proposals call it, would be protecting a pipeline that, when operational, pumps about 35 million barrels per year. This adds up to nearly $3 per barrel in costs to U.S. taxpayers to protect a pipeline for which Occidental currently pays security costs of about 50 cents per barrel, according to the *Wall Street Journal*.[6]

The pipeline brigade represents a qualitative change in U.S. policy toward Colombia, one that would have been difficult to contemplate had the September 11 attacks not taken place. Before, U.S. military assistance was portrayed as limited to the counternarcotics mission. "We are not saying this is counter-drug—this is different," an unnamed U.S. official told reporters on February 5. "The proposition we are making to the government of Colombia and to our Congress is that we ought to take an additional step."

U.S. officials are now portraying Colombia's armed groups, particularly the FARC, as terrorist threats to U.S. security.

In fact, officials are now portraying Colombia's armed groups, particularly the FARC, as terrorist threats to U.S. security, and are indicating that the pipeline-defense unit is only a first step. "The terrorist threat also goes beyond Islamic extremists and the Muslim world," CIA Director George Tenet told the Senate Intelligence Committee on February 6. "The Revolutionary Armed Forces of Colombia poses a serious threat to U.S. interests in Latin America because it associates us with the government it is fighting against." According to a February 11 *Washington Times* report, "some Bush administration officials are advocating taking the fight—and U.S. involvement—even further. They want the [existing counternarcotics] brigades authorized to attack FARC units if intelligence shows they are about to attack a village or other target."

Administration officials have far less to say about Colombia's right-wing United Self-Defense Units of Colombia (AUC) paramilitary group, which is also on the State Department's terrorist list. Though the paramilitaries have tripled in size in the last four years (to 12,000 members), carry out the majority of non-combatant killings, and are also involved in the drug trade, the guerrillas are getting the lion's share of U.S. attention.

As of mid-February, the next step for Washington's military-aid plan in Colombia is the implementation of human rights conditions in the 2002 foreign aid law. The State Department must decide whether Colombia's Armed Forces are suspending officials suspected of violating human rights, cooperating with investigations and prosecutions, and actively cutting links with the paramilitaries. If the State Department cannot certify that this is happening, no aid can go to the Colombian military. On February 6, Human Rights Watch, Amnesty International, and the Washington Office on Latin America released a joint report with ample evidence that "Colombia's government has not, to date, satisfied these conditions."[7] But the State Department appears likely to certify anyway.

In Colombia, the next several months promise to be especially violent. Once the Putumayo counternarcotics brigade receives its dozens of helicopters, the launch of the "push into southern Colombia" and expanded aerial fumigation may cause a further spread of violence, drug cultivation and forced migration. This may particularly affect beleaguered Ecuador, which has already suffered some cross-border violence and the economic impacts of instability and fumigation in Putumayo.

In Washington, where the Bush administration has shed the counternarcotics fig leaf, we are likely to see an energetic debate about the long-term direction of U.S. policy in Colombia. If the administration does indeed seek to allow past counternarcotics aid (especially the dozens of helicopters given in the past few years) to be used for counterguerilla missions, Colombia will move to the very top of the U.S. foreign policy debate, right next to the so-called "axis of evil" countries.

It is still not clear how far U.S. foreign policy planners want to go with their new counterterror/counterinsurgency emphasis in Colombia. Hopefully, they will recall that the FARC, ELN and AUC are not small groups of shadowy cells, like Al Qaeda in Afghanistan or Abu Sayaaf in the Philippines, but large armies with long histories and control of territory. Confronting them militarily would require a large-scale counterinsurgency effort—something that is clearly under consideration, as indicted by recent conversations with government officials, some of whom cite the experience of El Salvador. A June 2001 study by the Rand Corporation, funded by the U.S. Air Force, in fact suggested that "the U.S. program of military assistance to El Salvador during the Reagan administration could be a relevant model."[8]

As the debate heats up, those who would repeat this model in Colombia might want to step back from the charged rhetoric of the war on terrorism, and recall that the United States spent $2 billion on El Salvador's military over a 12-year period, during which 70,000 people died and over a million were forced into exile. Colombia is fifty-three times larger than El Salvador.

Notes

1. For background on the participants in the peace process, Nazih Richani, "Columbia at the Crossroads: The Future of the Peace Accords," NACLA Report, XXXV No. 4, Jan–Feb 2002. Available on line at http://www.nacla.org/art_display.php?art=526

2. Rowan Scarborough, "FARC triggers concern in U.S.," Washington Times, February 11, 2002.

3. "El terremoto de enero," Cambio (Bogotá) February 3, 2002.

4. The "Plan Colombia" aid request covered two years (2000–2001). See Center for International Policy, "The 2003 aid request." http://www.ciponline.org/colombia/aid03.htm.

5. See Maria Alvarez, "Forests Under Fire" NACLA Report, XXXV No. 1 July/Aug 2001. Available online at http://www.nacla.org/art_display.php?art=452

6. Alexei Barrionuevo and Thaddeus Herrick, "Threat of Terror Abroad Isn't New For Oil Companies Like Occidental," the Wall Street Journal, February 5, 2002.

7. Human Rights Watch, Amnesty International, and the Washington Office on Latin America, "Colombia Human Rights Certification III," Washington, February 5, 2002. http://www.hrw.org/press/2002/02/colombia0205.htm

8. Angel Rabasa and Peter Chalk, "Colombian Labyrinth: The Synergy of Drugs and Insurgency and Its Implications for Regional Stability," Rand Corporation, June 2001. http://www.rand.org/publications/MR/MR1339/

Adam Isacson is a senior associate at the Center for International Policy where he coordinates a program that monitors and seeks limits on U.S. military assistance to the Western Hemisphere.

From *NACLA Report on the Americas*, March/April 2002, pp. 10-13. © 2002 by Adam Isacson.

UNIT 4

Political Change in the Developing World

Unit Selections

Key Points to Consider

- What are the current trends in democracy throughout the world?

- In what ways may the "democracy industry" undermine indigenous efforts to foster democracy?

- What are the challenges involved in rebuilding Afghanistan?

- What role did civil society play in the brief ouster of Venezuela's president Hugo Chavez?

- What shapes the prospects for democracy in Africa?

- What factors threaten to undermine democracy in Latin America?

 **Links: www.dushkin.com/online/**
These sites are annotated in the World Wide Web pages.

Greater Horn Information Exchange (GHIE)
   http://edcsnw3.cr.usgs.gov/ghai/ghai.html
Latin American Network Information Center—LANIC
   http://www.lanic.utexas.edu
ReliefWeb
   http://www.reliefweb.int/w/rwb.nsf
World Trade Organization (WTO)
   http://www.wto.org

Political change in the developing world has not necessarily produced democracy, in part due to the fact that developing countries lack a democratic past. Colonialism was authoritarian and the colonial powers failed to prepare their colonies adequately for democracy at independence. Even where there was an attempt to foster parliamentary government, the experiment frequently failed, largely due to the lack of a democratic tradition and political expediency. Independence-era leaders frequently resorted to centralization of power and authoritarianism, either to pursue ambitious development programs or often simply to retain power. In some cases, leaders experimented with socialist development schemes that emphasized ideology and the role of party elites. The promise of rapid, equitable development proved elusive, and the collapse of the Soviet Union discredited this strategy. Other countries had the misfortune to come under the rule of tyrannical leaders who were concerned with enriching themselves and who brutally repressed anyone with the temerity to challenge their rule. Although there are a few notable exceptions, the developing world's experiences with democracy since independence have been very limited.

Democracy's "third wave" brought redemocratization to Latin America during the 1980s, after a period of authoritarian rule. The trend toward democracy also spread to some Asian countries, such as the Philippines and South Korea, and by 1990 it also began to be felt in sub-Saharan Africa and, to a much lesser extent, in the Middle East. The results of this democratization trend have been mixed so far.

Although Latin America has been the developing world's most successful region in establishing democracy, its commitment to democracy has been shaken by widespread dissatisfaction due to corruption, inequitable distribution of wealth, and threats to civil rights. Threats to democracy were evident in the April 2002 coup that briefly ousted Venezuela's president Hugo Chavez from power. Argentina's economic collapse also threatens political instability as the middle class, deeply affected by the crisis, demonstrates little confidence in its political leadership.

Africa's experience with democracy has also been varied since the third wave of democratization swept over the continent beginning in 1990. Although early efforts resulted in the ouster of many leaders, some of whom had held power for decades, and international pressure forced several countries to hold multiparty elections, political systems in Africa range from consolidating democracies to states mired in conflict. South Africa's successful democratic consolidation in the face of major challenges stands in sharp contrast to the circumstances in such countries as the Democratic Republic of Congo, Cote d'Ivoire, and Burundi.

Efforts to develop a democratic political system in Afghanistan are complicated by ethnic differences. The assassination of the vice president and the attempt on the life of President Karzai demonstrate the fragility of the current political arrangement and the persistent challenge of regional warlords. In the Middle East, there has been little progress toward democratic reform. Much of the region remains under the control of monarchies or authoritarian rulers presiding over tightly constrained political systems. There are some interesting developments in Iran, where a popularly elected president is locked in a struggle with Islamic hardliners. There are also preliminary indications of a push for greater democracy within the Palestinian Authority. Should Iraq's Saddam Hussein be removed from power, democracy is unlikely to result in the near term. It is not clear that other countries in the region would welcome democracy in Iraq, in the unlikely event it were to occur, since it might represent a challenge to their own regimes.

In Asia, India's democracy continues to be tested by communal strife resulting from religious and caste differences. Pakistan's president, General Pervez Musharraf, who came to power in a 1999 coup, recently announced his intention to remain in power. Many of the former Soviet republics of central Asia appear to be turning away from democracy and toward an authoritarianism rooted in their cultural and political past. Indonesia's political situation remains uneasy, and although the military's power has been curbed, it is still lurking in the background.

While there has been significant progress toward democratic reform around the world, there is no guarantee that these efforts will be sustained. Although there has been an increase in the percentage of the world's population living under democracy, nondemocratic regimes are still common. Furthermore, some semidemocracies have elections but lack civil and political rights. International efforts to promote democracy have often tended to focus on elections rather than the long-term requirements of democratic consolidation.

POLITICS ABROAD

Democracies: Emerging or Submerging?

Anthony W. Pereira

THE SECOND half of the twentieth century was an age of democracy. The women's movement, anti-colonial struggles, and challenges to what W. E. B. DuBois called the "color line" won political inclusion for many people throughout the world. And starting in the mid-1970s, electoral democracies replaced authoritarian regimes in southern Europe, Latin America, and parts of Africa and Asia. This process was extended when the former Soviet Union broke up, and communist regimes collapsed there and in Eastern Europe. According to Freedom House, a nonprofit institution that issues an annual assessment of political and civil rights worldwide, there were only twenty-two democracies with 31 percent of the world population in 1950, out of a total of eighty sovereign states. Today, there are 120 electoral democracies representing 58 percent of the world population, out of 192 sovereign states. The percentage of the world's population living under some form of democracy nearly doubled in the last fifty years. This was a historic shift of epic proportions.

Western triumphalism about the "end of history," however, was never justified. Non-democratic regimes still exist. Morocco, Saudi Arabia, the Gulf States, and Swaziland are traditional monarchies. China, Cuba, North Korea, and Vietnam are nominally communist, one-party regimes whose leaders deeply distrust multiparty democracy. And military regimes of the kind so common in developing countries in the 1960s and 1970s survive. Burma (Myanmar), which the democratic activist Aung San Suu Kyi calls a "Fascist Disneyland," is one of them, as are Pakistan and, until 1999, Nigeria. Although they often declare their intent to supervise a democratic "transition," such regimes do not respect the democratic principles of universal suffrage and elected government.

In addition, some of the world's states have imploded, rendering the form of their regime irrelevant: Sierra Leone, the Congo, Colombia, and Sri Lanka today; Afghanistan, Angola, Burundi, Haiti, Liberia, Rwanda, and Somalia a few years ago. It is likely that there will be more, and that international intervention, when it occurs, will not be entirely successful in promoting peace and rebuilding the state.

Furthermore, many new democracies are illiberal, unable or unwilling to guarantee their citizens important political and civil rights, even if regular elections are held. Leaders with a dictatorial bent manipulate political systems to prevent genuine competition or to steal elections. Observers speak of these polities as "democracies without citizenship" or "delegative democracies"—political systems in which plebiscitary elections create mandates for powerful chief executives who rule virtually unchecked.

Finally, changes in patterns of economic and political organization have shifted authority upward to regional and global institutions (the European Union [EU], the North American Free Trade Agreement [NAFTA], the World Trade Organization [WTO], the International Monetary Fund [IMF], the World Bank, multinational corporations, and financial markets) and downward to local and provincial governments. The scope of the nation-state's power has diminished as its institutions have, more and more, been democratized.

The "Third Wave" and "Transitology"

Given all this, what can be said about the world's new democracies? A survey of the situation suggests that many of them are in trouble, and only sustained transnational pressure to democratize globally will create the political conditions in which they can flourish.

The creation of new democracies in the late twentieth century involved the demise of many different kinds of regimes. In

Portugal and Spain, long-lasting fascist dictatorships ended; in Latin America, brutal military dictatorships ran aground; in Africa and Asia, one-party machines lost their hegemony. Few would question the value of such events for people living in these countries. However, much of the current literature about the emerging democracies reflects assumptions that don't hold universally and agendas that embody the interests of powerful external actors.

U.S. political scientist Samuel Huntington christened the democratic transitions of the mid-1970s and after as the "third wave," a successor to the first wave of democratization in the nineteenth century and the second wave during and after World War II. Another political scientist, Philippe Schmitter, wryly refers to the work of Huntington and the legions of other scholars and practitioners who write about democratic transitions as "transitology." Despite its apparently neutral language, transitology reflects ideological assumptions about new democracies. For example, democracy tends to be defined minimally as competition between elites in competitive elections. It is seen primarily as the product of conscious crafting and deal-making and not, as an earlier generation of scholars believed, of long-term historical conflicts between rulers and ruled.

PERHAPS most important, certain kinds of democratic transition are taken to be inherently better than others. Portugal's 1974 revolution, for example, tends to be downplayed in the transitology literature, because it involved mass insurrection, the collapse of the old regime, and curbs on the prerogatives of business and private property in its early phase. In contrast, the more conservative and negotiated Spanish transition of 1975, even though it was actually more violent than its Portuguese counterpart, is the emblematic transition in the transitology literature.

Transitologists extract certain lessons from the Spanish model that they hope democratizers elsewhere will use: the "masses" threaten to destabilize transitions, and their mobilization should be feared, not encouraged; the prerogatives of the market, as defined by the most powerful actors in the global economy, should not be challenged; and negotiated transitions always produce better democracies than those that result from a regime collapse. Such a perspective is intrinsic to the U.S. government's "democracy promotion" bureaucracy—including the U.S. Agency for International Development (USAID) and the National Endowment for Democracy (NED)—which now spends roughly $700 million per year trying to promote a particular version of democracy overseas.

The third wave involved technology as well as ideas. If the generation of 1968 demanded a revolution, the generation of 1978 actually created one—but it was technological, not political. The information and communications revolution involving satellites, cellular telephones, digital technology, the Internet, fax machines, and the computer chip made data gathering, processing, and transmission cheaper and easier than ever before. It became markedly easier to organize elections in poor countries with limited transportation and communications infrastructures. An entire international regime of election specialists

mushroomed in the 1990s, and the United Nations monitored and sometimes ran elections in many places. New information technology was allied to the political transformations of the third wave to produce a sense among many analysts that the expansion of democracy was limitless, irreversible, and conducive to all sorts of other positive changes, such as economic development and international peace.

Actually Existing Democracy

In fact, transitology is of limited usefulness in understanding the politics of new democracies. Many of these political systems are marked by deep disjunctures between the sophistication of their elections, on the one hand, and the legitimacy of their governments and the actual enjoyment of rights by their citizens, on the other.

Although elected civilian governments are far more common in the world than they were thirty years ago, the continuing or revived popularity of former dictators is remarkable. In South Korea, for example, surveys reveal that the most popular former ruler of the country is the late Park Hung Chee, the autocratic leader of a military regime from 1961 to 1979. In a recent poll of Paraguayans, 70 percent said that they were better off under the dictatorship of Alfredo Stroessner (1954–89) than they are today. Polls of Russians reveal the same sentiments about the Brezhnev era. And in Brazil, only 47 percent of respondents in a recent poll agreed with the statement, "Democracy is preferable to any other form of government." Eighteen percent said that in some circumstances, authoritarianism was preferable, and 29 percent thought that it made no difference. Although many new democracies now exist, therefore, it is striking that large numbers of people in these countries prefer authoritarian leaders and are indifferent or hostile to current democratic governments.

Nor do regular elections necessarily prevent the rule of dictatorial leaders. For example, in Zimbabwe Robert Mugabe succeeded in terrorizing the rural population into at least partially endorsing the continued rule of his Zimbabwe African National Union–Patriotic Front (ZANU–PF) Party. He directly appointed twenty of the hundred and fifty members of parliament, while his loyal followers controlled the electoral system, and liberation war "veterans" (many too young to have actually fought in the war) replaced politically unreliable schoolteachers as polling station managers. Because of these manipulations and (it is widely suspected) ballot rigging, Mugabe was able to maintain his hold on power despite a strong showing by the opposition Movement for Democratic Change (MDC) in the parliamentary elections of June 2000.

Elections are also sometimes ignored when the results do not favor incumbents. In Burma in 1990, the opposition National League for Democracy (NLD) won an election by a large margin but then was prohibited from taking office by the military. Many NLD members, including Nobel Peace Prize winner Aung San Suu Kyi, remain under house arrest or in prison. Similarly, in Algeria in December 1991, a militant Islamic party won the country's first-ever parliamentary elections. In response, senior army officers forced then-president Chadli Ben-jedid to resign and suspended the electoral process, forming a regime that remains in power today

Political and civil rights are therefore nonexistent for large segments of the population in many new democracies. For this reason, Freedom House classifies thirty-five of the world's hundred and twenty democracies as "electoral" but not "liberal" democracies. The number of people in these countries is 20 percent of the world's population. Why is democracy in such political systems so tenuous?

THE QUESTION is complicated, but one way to answer it is to compare new democracies to what I've already called the emblematic case of third wave transition. Spain in 1975, the year of its democratization, had three advantages that are not shared by many other countries: a mature economy, a functioning state bureaucracy, and a society with a high degree of consensus. Each of these deserves some attention.

In the mid-1970s, Spain had a mature capitalist economy in which low-wage agriculture was no longer the leading sector. It had a relatively egalitarian distribution of income compared to that of most developing countries. Spain today has a more egalitarian distribution of income than the United States and a human development score (calculated by the United Nations Development Program) higher than that of Belgium and the United Kingdom. Furthermore, as Spain weathered its first crises as a new democracy, its economy continued to grow, and it faced the enticing prospect of joining the largest and most successful common market in the world, the European Union (EU)—which it did in fact join in 1986.

As a member of the EU, Spain received substantial funds due to policies aimed at reducing economic inequalities within the union. These funds improved infrastructure and boosted employment, especially in Spain's poorest regions. In the period between 1989 and 1993, Spain received about twelve billion ecus (about ten billion U.S. dollars; the ecu preceded the euro), or almost one-quarter of the EU's total structural-aid spending in that period. This was a significant transfer of resources that most countries in the developing world cannot hope to attain.

In addition, Spanish state institutions perform relatively well. Indeed, the judiciary performs too well from the point of view of former dictators such as Chile's Augusto Pinochet and Guatemala's Efraim Rios Mont, both of whom have been investigated and charged with crimes against humanity in Spanish courts. Once again, the European Union's influence on Spain was strong and significant. As democracy deepened, the EU encouraged Spain to modernize its legal system and incorporate Europe's human rights provisions.

Finally, Spanish society had a fairly high degree of consensus, at both the mass and elite level, at the time of its transition. The conflicts that had created the Franco regime lay in the distant past. The most violent period of repression had ended three-and-a-half decades earlier. There was a strong consensus on both the right and the left that the horrors of the civil war should not be repeated. With the exception of Basque terrorism, there was a relative lack of ethnic tension or separatist sentiment

in the country. And finally, a relatively moderate ruling class was willing to accept, without violence, the re-emergence of the unions and the left that the end of authoritarian rule engendered. What some analysts call the "civic micro foundations" of democracy—trust, tolerance, respect for opposition, and acceptance of equal citizenship—were firmly in place in Spanish society before its transition. All these were strengthened in the years after Franco's death as Spain entered the EU, continued to grow economically, and staged successive elections that resulted in the alternation of ruling parties.

BUT OTHER new democracies frequently lack one or more of the variables that contributed to the success of the Spanish transition. In some cases, a mature and growing economy, a well-functioning state, and civic micro foundations are all missing.

It is in economic matters that the differences are most striking. Certainly, the revolution in information technology has benefited some developing countries. South Korea manufactures silicon computer chips and India produces software. But for many countries, previously huge socio-economic inequalities with rich countries have widened, as have domestic inequalities.

In Latin America, for example, there has been a marked jump in informal-sector employment in recent years in a region that is already the most unequal in the world. In 1980, the International Labor Organization estimated that 40 percent of Latin America's nonagricultural workforce was in the informal sector. (In agriculture, where work tends to be seasonal, that figure would be far higher.) These workers have no pension benefits, no unemployment insurance, no paid leaves of any kind, and no employment contract—they can be fired at will. By 1998, informal-sector employment had climbed to 58 percent, a majority of the workforce. Among women in the nonagricultural labor market, the figure was even higher, 65 percent that same year. With so many people so vulnerable, elections in Latin America have a different character than those in rich countries. The poor seek patrons in the political system, and politicians give away material benefits to win votes. Patronage and clientelism rob elections of much of their deliberative content, and while they serve as a redistributive mechanism in societies without extensive welfare states, their ability to ensure democratic accountability is diminished. This is true in other parts of the world as well.

In addition, the globalized world economy is highly volatile. Short-term, speculative investments in currencies, stocks, and other liquid assets flow in and out of developing countries quickly and erratically, subject to the wild swings of the international investment community. The U.S. Federal Reserve Bank estimates that economic volatility, measured by the variation in the growth rate of the gross domestic product from one year to the next, is almost three times as high in Latin America as it is in the countries that belong to the Organization for Economic Cooperation and Development (OECD). When economies are volatile, so is politics. In short, the prevailing development model excludes large numbers of people who are not property-owning stakeholders, but who live instead with a high degree of economic insecurity and deprivation. This is not a foundation for successful democracy.

Much the same story could be told in the area of state institutions, which were seriously debilitated in many developing countries as governments cut spending to cope with the debt crisis of the 1980s. Courts are often neither legitimate nor competent. Widespread crime and violence vitiate the effective exercise of democratic rights for millions.

In many developing countries, but not in Spain, cold war conflicts were large-scale and recent. They make the McCarthy period in the United States look genteel by comparison. The killing fields of Cambodia are one horrific example of this kind of violence, but there are many others. Much of the violence had an ethnic dimension, making it particularly difficult to ameliorate after a democratic transition. Fear and suspicion remain entrenched, hampering the creation of the civic micro foundations of democracy. Violence continues, especially in rural areas, where landowners, state officials, and other powerful groups use armed forces of various kinds to enforce their rule. Elections certified as "free and fair" by international election observers may be marred by widespread intimidation prior to the vote that is not easily monitored by short-term visitors.

Finally, although new information technology is used by states to organize elections, ordinary citizens often lack access to it. Grassroots movements have on occasion used technology—the Zapatistas in Chiapas used the Internet effectively after 1994, for example, and the Grameenphone program has recently provided cellular phones to 100,000 subscribers in 250 villages in Bangladesh. Information technology has great potential to reduce the cost to grassroots organizations of spreading information and organizing. But all too often ordinary people in developing countries are still excluded from the new technology networks. Jeremy Rifkin estimates that 65 percent of the current world population has never made a telephone call. A *Wall Street Journal* survey estimates that 57 percent of those people who regularly surf the Web live in the United States. Only 1 percent live in the Middle East, 1 percent in Africa, and 5 percent in Latin America. In these regions, only affluent elites have access to the information that the Internet provides. The emergence of what political writer Andrew Sullivan calls "dot.communism," in which universal access to the Internet's plethora of free information, goods, and services creates genuine political equality, at least in cyberspace, is far from a reality.

The transitology literature, then, overestimates the transformative potential of third wave democratization. It exaggerates the degree to which deep historical divisions within countries can be crafted out of existence by the correct decisions of democratizing elites. And it uses as a model Spain's transition, which on closer inspection looks exceptional, with favorable conditions impossible to match in most other new democracies.

Global Economic Trends

In addition to these problems, the transitology literature fails to link democratic transitions with changes taking place in the global economy. While some economic changes facilitate openness, transnational networking, and similar developments supportive of democracy, others severely restrict the choices of governments in developing countries.

Beginning in the 1980s, the so-called "Washington Consensus," or neoliberal orthodoxy, came to dominate official thinking about economic development in the U.S. government, the IMF and World Bank, and private banks and foundations. The Washington Consensus requires sharp attacks on fiscal deficits (and therefore cuts in government spending), financial liberalization (letting the market determine interest rates), trade liberalization (lowering tariffs and other barriers), privatization of state-owned enterprises, deregulation, and more rigorous enforcement of property rights, including intellectual property rights. The supremacy of the market is aggressively asserted over all other values, and the supposedly "statist" policies of the past are characterized *tout court* as failures. In the words of political scientist Benjamin Barber, the triumph of market ideology means "no government intervention (however mild) is safe from criticism, and no market mechanism (however violent and unjust) is subject to rebuke."

Powerful international forces push developing countries to adopt a single prescription for their economic troubles, one that leaves them, especially the smaller ones, with little room for alternative policies. The preeminence of the market was reflected in terminology— developing countries became "emerging markets" in much of the literature, as if their only importance was as an open field for the exports and investments of rich countries. The fact that so many third wave democracies emerged precisely during the rise of the Washington Consensus meant, paradoxically, that their ability to manage their own economies was markedly curtailed. They became, in the words of Malawian political scientist Thandika Mkandawire, "choiceless democracies." They are so indebted to and dependent on international financial institutions that they are not really free to make basic democratic decisions about which of their goods and services are to be allocated through the price mechanism and which are not, how certain markets should be regulated, or how their economic reforms should be carried out.

The U.S. government, or at least parts of it, led the way in the forging of the Washington consensus. What the economist Jagdish Bhagwati calls the "Wall Street–Treasury complex" has been firmly in control of U.S. policy, insisting that any government's attempt to limit the mobility of capital is unacceptable, that developing countries should seek always to privatize their industries and lower tariffs, that the intellectual property rights of U.S. firms must take precedence over local needs, and that top priority must always be given to the wishes of the largest financial conglomerates and multinational corporations.

In its foreign economic policy, the United States has rarely deviated from the Wall Street–Treasury Department line. With a singlemindedness that would make the old *apparatchiks* of the Soviet Union proud, U.S. trade officials, State Department employees, and USAID technicians repeat the mantra in international conferences and in the capitals of developing countries: governments should be minimally involved in the economy; even surplus-generating state enterprises should be privatized;

social goals should not be taken into account when evaluating government performance. The only acceptable criterion of state reform is efficiency, measured by the bottom line of the business world; government should not seek to protect an overarching public interest distinct from the preferences aggregated by the market—indeed, there is no public interest distinct from the market; the U.S. government should aggressively represent the interests of U.S. exporters, investors, and other companies (who after all bankroll its elected representatives) rather than the whole of its population, and changes in developing countries should be evaluated solely in terms of the opportunities they provide to those interests. The relentless repetition of this economic formula does not amount to genuine global leadership. It is a politically unsophisticated form of salesmanship that looks hypocritical, since the U.S. government continues to maintain protectionist barriers against developing country exports in agriculture and other sectors.

The view of politics fostered by the Washington Consensus is a utilitarian one. Democracy is seen as useful because it can generate the marketizing and commercializing changes that are "necessary" for the world economy. But democratic politics is feared as well as desired, because it gives rise to views opposed to the consensus. For this reason, much of the writing about the global economy from Washington these days uses the term "governance" rather than "politics." The difference is telling. Governance refers to the administration of people and things. Politics signifies engagement in political affairs. The first assumes that the challenge facing the world economy is to make the management of the status quo more efficient; it takes the dominance of the current orthodoxy for granted. The second allows for conflict between fundamentally different visions and parties; it is open to the possibility of new political winners and meaningful change.

As THE LAST century began, socialist and other radical movements were attempting to democratize politics at the national level by applying the principles of parliamentary democracy to the management of the "commanding heights" of the economy. Today, the terminology is different, and no one is seriously talking about eliminating markets. But many social movements are engaged in a similar project at a global level, attempting to apply the principles of parliamentary democracy to the institutions that manage the international economy. This is the logical next step in the struggle for democracy, which alone can sustain the third wave of democratization.

Whenever new forms of political authority are consolidated, social movements can open new space for democratic contestation. With the rise of nation-states in Europe after the French Revolution, domestic movements with their tactics of strikes, petitions, demonstrations, and marches were born. When the political form of the nation-state was carried by European powers to Africa and Asia, those colonized people used the rhetoric of European nationalism to win their own political independence. And as globalization proceeds, new social movements are emerging, capable at the very least of stopping the most egregious corporate and governmental manipulations. Global institutions such as the WTO should not be opposed *in toto,* but reformed and used as structures within which democratization can take place. A retreat into fortified, nationalist, autarchic states would probably be the worst thing that could happen to democracy in the world today.

The goal of nineteenth-century radicals still lies visible on the horizon in the twenty-first century. The commanding heights of the global economy are still run by small oligopolies of "Davos men": people, corporations, and states rich enough to have an influence on the rules of the game. On the board of the International Monetary Fund, for example, two-thirds of the voting rights are in the hands of representatives of the United States, Japan, and Europe, who speak for less than one-eighth of the world's population. Global grassroots movements might be able to force this and other institutions to move closer to the principles of universal suffrage and democracy, and thus produce a more humane global economy. The managers of the global institutions of "governance, and the corporate and state interests that prop them up, will not accede to change without a fight. But the fight is well worth the effort.

The market is not sacrosanct. People can socialize it so that human values help to determine the conditions of production and the allocation of goods and services. But they can only do that by questioning and challenging the vision of economic and political change promoted by those who run the world's economy. Advocates of democracy need to transcend the narrow confines of the transitology approach and take account of the global economic environment in which democratic transitions do, or do not, take place. Democracy is increasingly about who makes the global rules, not just about who replaces authoritarian leaders in individual countries.

ANTHONY W. PEREIRA teaches political science at Tulane University in New Orleans. He is the author of *The End of the Peasantry,* which deals with the rural labor movement in northeast Brazil.

From *Dissent* magazine, Winter 2001, pp. 17-23. © 2001 by Dissent. Reprinted by permission.

Democracy Inc.

Hardly an election occurs outside the developed world today without
an international corps of observers flying in to certify the results.
But the outsiders sometimes do more harm than good.

by Eric Bjornlund

Friends and foes of the United States smirked last fall as the champion of the free world waded in embarrassment through Florida's electoral swamps. Even as U.S. government agencies and nonprofit groups were busily monitoring "troubled" elections in half a dozen foreign lands, from Haiti to Azerbaijan, America's presidential election was thrown into doubt by arthritic voting technology, sloppy voter registration, and partisan election officials—flaws that were supposed to afflict only "less developed" countries. One Brazilian pundit half-seriously called for international sanctions to force a new vote in Florida.

But American democracy has never been faultless, and—derisive comments in the international press notwithstanding—U.S. efforts to promote democracy abroad have never been predicated on its perfection at home. Indeed, the American groups that work to spread representative government overseas have drawn heavily on non-American models precisely because they recognize the shortcomings and idiosyncrasies of the U.S. system.

The real flaws in the global effort to foster democracy, meanwhile, have gone largely unnoticed—and they are flaws that threaten great harm to the democratic cause. The scattered and diffuse democracy movement of decades past has been transformed into a worldwide industry of sorts, led but not controlled by the United States. The industry has done much good. But it has also put a stamp of legitimacy on Potemkin-village democracies in Cambodia, Egypt, Armenia, and other countries. It has frustrated local democratic activists from Indonesia to Peru, and it has provided autocratic rulers with ammunition to dismiss courageous local democrats as lackeys of foreign powers. Worst of all, it has undermined efforts to apply uniform democratic standards around the world.

The democracy industry has its deepest roots in the United States. From the time of President Woodrow Wilson's crusade to "make the world safe for democracy" to the era of the Cold War, Americans of virtually all political persuasions shared an ideological commitment to advancing the democratic cause in the world. But only under the Reagan administration did the United States begin to focus and institutionalize its efforts. Washington now devotes some $700 million annually to democracy promotion. Much of it is channeled through the Agency for International Development—which parcels out the money to private consulting firms and more than a score of nongovernmental organizations, such as the Carter Center and the Asia Foundation—and a small but significant portion goes to the congressionally-chartered National Endowment for Democracy. It is a substantial commitment, equal to about 10 percent of the entire U.S. foreign aid budget.

The United States, however, is outspent by others. The European Union and developed countries such as Japan and Australia, along with multilateral organizations such as the United Nations, the World Bank, and the Organization for Security and Cooperation in Europe, also pour large amounts of monetary, human, and diplomatic capital into the global crusade. The stated purposes are the same: fighting corruption, establishing the rule of law, fostering civil society, developing democratic parliaments, and monitoring elections. But not all of the industry's "players" share the same commitment to democracy, and some are willing to sacrifice its pursuit to other foreign-policy goals.

The industry's rise has coincided with a revolutionary expansion of democracy around the world. What Harvard University political scientist Samuel Huntington has called the "third wave" of democratization began in the late 1970s with political transitions in Spain and Portugal, and spread in the 1980s to Latin America and Asia. Democracy swept through Eastern and Central Europe after the fall of the Berlin Wall in 1989 and continued after the breakup of the Soviet Union. The 1990s also saw dramatic political openings in Africa and Asia. Since 1988, a total of 50 countries have made the transition to democracy, from Poland and Brazil to Taiwan and Nigeria.

The democracy industry can't claim credit for the third wave, but it has reinforced the trend. Last year, under the weight of domestic and international pressure, repressive regimes in Yugoslavia and Peru fell after election monitors helped expose their attempts to manipulate national elections. Two decades ago, such a feat would have been almost unimaginable.

The industry has been fortunate: Its successes have been more sensational than its failures. But an examination of highly publicized elections, such as the recent ones in Cambodia and Indonesia, shows that its failures can be deleterious.

I worked in both countries as an official of the National Democratic Institute for International Affairs (NDI), one of the four main nongovernmental organizations supported by the National Endowment for Democracy. (Each of the two major political parties sponsors an organization, and organized labor and business sponsor the other two.) While my work involved several areas of democracy promotion, the monitoring of elections best illustrates the tensions caused by the involvement of foreign activists. I have seen outside monitors contribute to public confidence in the integrity of elections, provide invaluable moral support to democratic activists facing authoritarian regimes, and deter fraud. But I have also seen them stumble—and do great harm to many of the world's fragile democracies.

Cambodia suffered more violent turmoil than almost any other country in the 20th century. It endured intense American bombing during the last years of the Vietnam War, and three years (1975–78) of terror under the communist Khmer Rouge, which, according to some estimates, left nearly a quarter of the Asian country's eight million people dead. A Vietnamese invasion in 1978 was followed by more than a decade of civil war.

In 1991, a glimmer of hope appeared when Cambodia signed an internationally sponsored peace agreement calling for liberal democracy and genuine elections. The United Nations established the largest, most costly peacekeeping force in its history (15,000 troops and a budget of $2 billion) to organize the election of a new government and administer the country during its transition. But the 1993 election failed to bring either democracy or stability, and in 1997 First Prime Minister Prince Ranariddh was overthrown in a bloody coup by his putative coalition partner, former communist Hun Sen. The United States and other countries suspended aid, and the United Nations denied the new government a seat in the General Assembly.

Hun Sen eventually agreed to a new election. But the international community was far from united in its approach. Though democracy watchers around the world deplored Hun Sen's violent putsch, many diplomats and aid providers believed that Cambodia could not be governed effectively without him. To them, an election—even an imperfect one—that lent Hun Sen legitimacy but also preserved a niche for political opposition seemed to be the best Cambodia could hope for.

With decisions about the future of foreign aid and diplomatic relations hinging on judgments about whether the contest was "free and fair," the pressure was on to grant it a clean bill of health—giving Hun Sen a sense of how much he could get away with. Eager to end the political crisis, the European Union, the United Nations, Japan, and Australia offered money, equipment, and technical assistance for the administration of Hun Sen's far-from-perfect election.

The Americans were more squeamish about lending legitimacy to a dubious election. But the U.S. government tried to have it both ways: It declined to offer election aid—but watched from a distance, poised to resume aid and improve diplomatic relations if the process miraculously turned out well.

One Cambodian newspaper dubbed the Americans "idealists" and the Europeans "pragmatists." But the difference was rooted in more than attitude. The United States, with its long history of activism by independent human-rights and prodemocracy groups, has largely separated election monitoring from foreign policy. In its efforts to monitor elections abroad, the United States relies heavily on nongovernmental organizations such as NDI and the International Republican Institute (IRI). These groups have a single, clear mission: to further democracy. It's then up to the government to make decisions about whether and how to engage or aid foreign governments. Other players in the democracy industry assign diplomats and bureaucrats to monitor elections. Their judgments are inevitably colored by the fact that democracy is only one of the ends they seek.

Cambodia held its much anticipated election on July 26, 1998. Despite the atmosphere of intimidation created by Hun Sen and his followers, an astonishing 97 percent of Cambodian voters turned out to cast their ballots. Domestic monitoring groups described the process as relatively peaceful and well administered, as did the hundreds of assembled international observers. Speaking before a packed press conference at the plush Le Royale Hotel two days after the election, our own delegation's coleader, former representative Steven Solarz (D.-N.Y.), went so far as to speculate that the election might one day be seen as "the miracle on the Mekong."

Hun Sen's Cambodian People's Party (CPP) was declared the winner of approximately 42 percent of the ballots cast, which translated into a 64-seat majority of the 122-member assembly. Prince Ranariddh's constitutional monarchist party won 31 percent of the vote and 43 seats, and a second opposition party, led by activist Sam Rainsy, won 14 percent and 15 seats.

But the Cambodian election was no miracle. Politically motivated killings had been commonplace since the '97 coup, and they stopped only weeks before the election. Opposition members of Parliament, led by Ranariddh and Sam Rainsy, had fled the country in fear of their lives after the coup. Though opposition leaders were induced to return in early 1998, the violence hardly provided the backdrop for a "free and fair" democratic competition.

Violence was not all that marred the election: The CPP government denied opposition parties access to radio and television, threatened opposition supporters, and banned political demonstrations in the capital city of Phnom Penh during the campaign. Hun Sen's supporters freely exploited their control of the judiciary and security forces. Two weeks before election day, an NDI-IRI report concluded that the process up to that point was "fundamentally flawed."

Some foreign observers, however, failed to report these problems or blithely dismissed all signs of trouble. While the United States funded 25 long-term observers recruited through the Asia Foundation, none of their reports were made public or shared with other observer groups. The Joint International Observer Group (JIOG), a UN-organized umbrella organization of 34 delegations with some 500 members, didn't even wait for the initial ballot count or for its own observers to return from the field before it endorsed the process as "free and fair to an extent that enables it to reflect, in a credible way, the will of the Cambodian people."

Among the JIOG's grab bag of groups were delegations dispatched by the governments of Burma, China, and Vietnam—regimes hardly known for their democratic credentials. One JIOG delegation, which openly positioned itself as a Hun Sen apologist, urged even before balloting began that the election "not be discredited for reasons of international politics." Most troubling of all, however, was the tendency of the JIOG's *democratic* members—the "pragmatic" Europeans and Japanese—to gloss over the election's undemocratic features.

Notwithstanding Solarz's hyperbolic "miracle" remark, the NDI-IRI assessment as a whole was quite level-headed. It made clear our concern about "violence, extensive intimidation, unfair media access, and ruling party control of the administrative machinery." British politician Glenys Kinnock, speaking for the delegation from the European Union, rendered a terse and similarly restrained verdict—one that implicitly distanced the EU observers from both the "miracle" statement and the JIOG's unqualified endorsement. Indeed, Solarz himself had said that the election would prove a "miracle" *only* if the tranquility of election day prevailed, and if the subsequent grievance process and the formation of the government proceeded smoothly. But in most press accounts, little more than Solarz's sound bite survived.

The press, however, was not really to blame for the world's failure to come to terms with what happened in Cambodia. As they have in many other cases, international democracy groups erred by making election day the big media event. By bringing observers to Cambodia

only days before the polls opened, issuing much anticipated (and hastily composed) assessments of the polling, and hopping on the next plane home, monitoring groups encouraged journalists to zero in on "E-Day"—which constituted, after all, only 24 hours of a months-long process.

It didn't take long for things in Cambodia to fall apart, making the foreign observers' upbeat assessments of the election seem all the more disconnected from reality. ("Sometimes I wonder if we're in the same Cambodia," one exasperated local democrat said.) After struggling to complete the vote count, including a perfunctory attempt to conduct a recount in a few token locations, the CPP-dominated election commission and constitutional court summarily dismissed the numerous complaints filed by opposition parties. After election day, it was revealed that the election commission had secretly altered the formula for allocating seats, thus giving Hun Sen a majority in the National Assembly. There is some evidence that the commission was only responding to international advisers who wanted to correct their own technical mistake. No matter. The change was made in secret, depriving the election of whatever shred of legitimacy it might have claimed.

In Phnom Penh and other Cambodian cities, post-election protests turned violent. One man was killed. The formation of a new government stalled amid finger pointing and threats.

One month after the election, our group decried the violence and utter lack of an appeals process. But our warnings went unheeded. The army of observers and reporters was gone, and international attention had waned. Neither the United Nations, nor the European Union, nor the JIOG ever made a single additional public statement after their relatively positive assessments immediately following election day. That would have required them to confront uncomfortable facts.

Three years after the "miracle on the Mekong," Hun Sen presides over a corrupt and undemocratic regime. His security forces regularly harass opponents and commit rape, extortion, and extrajudicial killings with impunity. But, with American support, the Hun Sen regime has regained Cambodia's seat in the United Nations, and the flow of foreign aid, including American aid, has resumed.

A year after the dismal proceedings in Cambodia, Indonesia held a much happier and more legitimate election—its first genuinely democratic contest in 44 years. Many of the democracy-industry circuit riders who had been in Cambodia promptly turned their attention to the archipelago. But again, the global democracy industry made serious mistakes, perhaps missing a once-in-a-generation opportunity to shore up a fragile new democracy.

Indonesia's democratic opening came in May 1998, when public anger at the regime's corruption and economic mismanagement forced an aging President Suharto to step down after 32 years as the country's autocratic leader. Democratic activists in Indonesia quickly organized the most extensive domestic election-monitoring effort ever seen. The prospect of establishing democracy in the country with the world's largest Islamic population helped open foreign wallets. By early 1999, the United Nations Development Program and the interim government in Jakarta had launched an effort to raise $90 million in international contributions for election administration, voter education, and poll watching. Just over a third of the total was to come from the United States.

When the polls opened on June 7, 1999, more than half a million Indonesians and nearly 600 foreigners from 30 countries were on hand to monitor the proceedings. It can only be called a messy election—but the vote was undeniably democratic. In the subsequent indirect election of the president, moderate Islamic leader Ahdurrahman Wahid pulled out a surprising victory. Unfortunately, he has been ineffective, and the Indonesian national legislature now looks poised to remove him from office. Whether he stays or goes, Indonesia seems bound to endure a period of turmoil.

In Indonesia, the democracy industry inflicted a subtler form of damage than it did in Cambodia. In their drive to ensure fair procedures, the well-intentioned outsiders inadvertently disrupted the efforts of Indonesia's many democrats. They once again allowed too much attention to focus on election day. And they stole the spotlight from local groups that could have benefited from more media attention.

The sudden influx of foreign money—much of it dumped into the country only weeks, or even days, before the election—touched off a mad scramble among the Indonesian groups. At the very moment they should have been focusing on the logistics of election monitoring, they were pouring their time and energy into grant proposals and budgets. Huge sums encouraged infighting among the Indonesian organizations. Misguided donors often made things worse by favoring different groups or factions.

Foreign aid also encouraged the needless proliferation of new monitoring groups—organizations with little experience and even less commitment. Twelve months before the election, only one monitoring group existed in Indonesia. The next nine months witnessed the appearance of two more. But in the two months before the election, some 90 more groups elbowed their way to the table. New organizations sprang up like American dot-com companies in the heyday of high-tech opportunism—and many of them showed just as little resiliency.

Having created incentives for Indonesians to compete with one another, the donors then tried to compel them to join forces in ways that didn't always make sense. The monitoring groups, for example, were required to divide their responsibilities along geographical lines, which touched off new struggles as leaders haggled over their territories. The division could more effectively have been made along, say, functional lines, with some groups looking into such matters as pre-election complaints while others educated voters or monitored vote counts. In any event, it was a decision best left to local activists, not outsiders.

As the head of the NDI's 20-person professional team in Indonesia, I saw firsthand some of the ill effects of all this. Three weeks before election day, at a final planning meeting of the University Network for Free Elections held at the University of Indonesia in Jakarta, I was pained to see the group's leaders mired in arguments over money. For three days, student leaders from around the country complained about inadequate budgets, criticized the headquarters for hoarding money, and made apparently specious allegations of corruption. The urgent issues at hand—volunteer training, communications systems, vote count monitoring—went virtually undiscussed.

For the University Network's idealistic national leaders, such as human rights lawyer Todung Mulya Lubis and professor Smita Notosusanto, it was a profoundly dispiriting experience. After the election, they and their colleagues abandoned any ambitions of building a national grassroots prodemocracy network, instead creating a Jakarta-based advocacy organization called the Center for Electoral Reform.

The Indonesian experience is a reminder that elections are not an end in themselves; they are, rather, one step in the ongoing process of building democracy. Local organizations and networks created to monitor elections often go on to promote democracy in other ways, by fighting corruption, monitoring government performance, or engaging in civic education. They must be strengthened, with moral as well as material support, not treated like voting machines or ballot boxes to be stored away until the next election.

The democracy industry did a few things right in Indonesia. Not least, it helped ensure fair elections. And former president Jimmy Carter, the reigning celebrity in the international observer corps, offered a fine example of how foreign observers should behave. Carter was a careful student of the election, studying verification techniques, visiting polling stations, and listening to the reports of Indonesian monitors. On the day after the polls closed, he was enthusiastic. But hours before he was to address a press conference, he agreed to meet a small group of Indonesian democracy activists. They were worried about more talk of miracles. Carter listened, and he went before the television cameras with a very different message. He expressed optimism, but he also emphasized the need to pay attention in the days ahead as the votes were counted, the president was selected, and the new government took power. Carter focused attention where it belongs: on the long-term process of building democracy and the local groups that make it work.

With experience, attention, and care, many of the ills that beset the new global democracy industry can be overcome. Shifting attention from election day to the months before and after voters go to the polls is a matter of common sense. Such a shift would also underscore the broader point that genuine democratization takes time, and that those who are sincere in their efforts to help must commit for the long term. Democratization, says Cambodian opposition leader Rainsy, depends on political forces "who'll remain here, who'll fight here, who'll die here, and who are determined to fight for democracy—not just observers who come for a few days." There should be nothing controversial about helping local democratic activists become continuing players with a stake in their country's future. But because so many of the democracy industry's important actors regard representative government as just one goal, to be balanced against others, this will be difficult to achieve.

All elections must be judged honestly, by the same internationally recognized standards. We know what they are: In addition to fair balloting and counting, there must be opportunities for political parties to compete, reasonably equitable access to the news media, an impartial election administration, freedom from political intimidation, and prompt and just resolution of election-related grievances. But until international donors break the link between the promotion of democracy and other foreign-policy goals—something only the United States has attempted—diplomatic goals will inevitably dilute efforts to establish true democratic governance around the world.

The United States is often criticized for taking a retrograde stance on the environment, national missile defense, and other issues. But when it comes to promoting democracy, Americans are criticized for their crusading idealism. What the Cambodian and Indonesian elections show is that a little more idealism might not be so bad. American nongovernmental groups are motivated by an altruistic desire to help people establish democracy. Whatever the flaws of American assistance, the separation of activities such as election monitoring from the official role of government yields a special kind of commitment. Other countries would do well to emulate the U.S. approach.

In the last decades of the 20th century, democracy established itself as the world's dominant political ideal. Yet much of the world's population has yet to enjoy democratic rights, and the commitment of many ostensibly democratic countries remains questionable. If we are to deliver on the promise of global democracy, those who carry its banner must not compromise its simple principles.

ERIC BJORNLUND, *a Wilson Center fellow, is a former senior associate and director of Asia programs at the National Democratic Institute for International Affairs.* Copyright © 2001 by Eric Bjornlund.

From *The Wilson Quarterly,* Summer 2001, pp. 18-24. © 2001 by the Woodrow Wilson International Center for Scholars in Washington, DC.

Rebuilding Afghanistan

"In the past several decades, the international community has relied on three approaches to deal with countries that descend into chaos. It has supported strongmen capable of reimposing order by force; it has given up in despair, leaving the country to sort out its problems as best it can; and, most recently, it has embarked on ambitious projects to reconstruct the country in the image of a modern secular, multiethnic, and democratic state. None of these approaches should be used in Afghanistan."

MARINA OTTAWAY AND ANATOL LIEVEN

Afghanistan after the Taliban may easily turn into a quagmire for the international community, and the wrong kind of international strategies may easily worsen both its problems and America's. In particular, to begin with a grossly over-ambitious program of reconstruction risks acute disillusionment, international withdrawal, and a plunge into a new cycle of civil war and religious fanaticism.

Ambitious plans to turn this war-hardened, economically ravaged, deeply divided country into a modern democratic state are being proposed and have even been incorporated into the December 5, 2001 Bonn agreement among Afghan leaders. But nobody is proposing the full-fledged, long-term military occupation that would be required even to attempt such a transformation—one reason being that past occupations, whether British or Soviet, have ended in utter disaster. At most, the international community is speaking of a relatively lightly armed presence in Kabul and certain other centers.

The chances of successfully imposing effective modern democratic state structures on Afghanistan thus are negligible. Even with a massive Western military presence on the ground, the West has already run into serious problems in transforming tiny Bosnia. Afghanistan is a country 12 times the size of Bosnia with 26 million people; an extremely difficult terrain; an ethnically, tribally, and religiously segmented society; and a fearsome array of battle-hardened warlords who have no good reason to give up their power.

But the world cannot afford to turn its back on Afghanistan in frustration, as it has done in the past, lest the country again become a haven for terrorists and an international threat. Afghanistan needs a modest reconstruction program that does not require full-fledged military occupation and is tailored to the reality of the country.

A CENTURY OF TROUBLED STATE BUILDING

The Afghan state is a recent, partly colonial creation that has never commanded the full loyalty of its own citizens. Even today, many—perhaps most—Afghans give their primary allegiance to local leaders, ethnic groups, and tribes.

Afghanistan was only created at the end of the nineteenth century. All of its borders were determined by the British Empire, and reflected not an internal historical or ethnic logic, but an imperial one. Its northern border marked the furthest extent to which Britain was prepared to see the Russian empire advance. Its southern and eastern borders were the furthest limit to which the British Indian Empire felt it necessary and safe to extend itself. Within these borders an Afghan state with modern trappings was created by a confluence of British geopolitical interest and the ruthless government of King Abdur Rahman, the so-called Iron Amir, who reigned from 1880 to 1901. The king was a highly competent ruler who, by quite fiendish methods and with massive subsidies of money and weapons from the British, created the basis—albeit limited—for a centralized Afghan state.

Abdur Rahman's reign marked the start of the Afghan state-building process. In Europe, this process began in the early Middle Ages, stretched over several centuries with numerous catastrophic setbacks, and was attended by immense cruelty, resistance, and devastation. It therefore is hardly surprising that the very short Afghan state-building process met fierce resistance, had limited success, and ultimately collapsed—especially given the intensely warlike, independent, and anarchic traditions of many Afghan peoples, including the largest ethnos, the Pashtuns.

Abdur Rahman laid the foundations not only for the centralizing and modernizing Afghan state, but also for the alienation from that state of the religious, tribal, and ethnic groups that dominate Afghan society. This alienation helped bring about the failure of the Afghan constitutional monarchy in the 1960s and early 1970s and tore the country apart in the following decades.

Had the modern Afghan state succeeded in developing Afghanistan and bringing visible benefits to the mass of the population, hostility to the state would gradually have faded. But, as with state building in so much of the world, it failed to do so, and its one area of partial success helped seal its own fate. The modern education system, although limited to a small fraction of the population (and of course an even smaller proportion of women), created a mass of educated graduates and junior bureaucrats and military officers for whom no well-paying jobs could be found either in the impoverished private sector or state service. Their bitter frustration produced the communist revolution of 1978, which essentially was an attempt to relaunch the state's modernizing program in an ultraradical guise by returning to Abdur Rahman's savage methods.

The communists' program, like that of Abdur Rahman, depended critically on subsidies and weapons from an outside protector, in this case the Soviet Union. And as in the Iron Amir's time, this foreign support helped spark fierce resistance from a variety of religious, ethnic, and tribal groups. The resistance eventually triumphed, and between 1978 and 1992 it overthrew the communist regime and eventually the Afghan state itself, first in the mountains, then across most of the country, and finally in Kabul and the other main cities. Tragically, but not surprisingly, the resistance proved completely incapable of replacing this state with any unified authority of its own, except—after a period of violent chaos—in the pathological and temporary form of the Taliban.

The difficulty of creating an Afghan state based on anything but sheer coercion has been immensely complicated by the region's ethnic makeup. The original "state-forming" ethnic group, the Pashtuns, make up less than half the total population, with the rest divided among a wide range of different nationalities. Tajiks, Uzbeks, and Hazaras (Shias of Mongolian descent) are the largest groups and are mentioned most often, but several smaller ones play key roles in their own areas.

Equally important, the Pashtuns' own role in the history of the modern Afghan state has been profoundly ambiguous. Afghanistan is a Pashtun creation, achieved through a Pashtun dynasty, and to this day the Pashtuns constitute the core of the country. But Pashtun tribal society is highly segmented and thus radically unfit to serve as the basis for the formation of a unitary state. Pashtun and other tribal revolts against the state's modernizing policies, often led by local religious figures, plagued all Afghan rulers. They played a central part in the rebellion against communist rule, and in the general reaction against Western modernity and modern state institutions that followed.

THE CHOICES

In the past several decades, the international community has relied on three approaches to deal with countries that descend into chaos. It has supported strongmen capable of reimposing order by force; it has given up in despair, leaving the country to sort out its problem as best it can; and, most recently, it has embarked on ambitious projects to reconstruct the country in the image of a modern secular, multiethnic, and democratic state. none of these approaches should be used in Afghanistan, but something can be learned from each of them.

A compromise approach needs to be based on an awareness both of Afghanistan's past and its present conditions, not on an image of the modern state the West would like it to become. The international community must recognize that in the northern half of the country, the coherence of the Northern Alliance is unlikely to last for long without its raison d'être of resistance to the Taliban, whereas in the Pashtun areas confusion reigns. In short, it will be extremely difficult to create any unifying political structures.

Heavily armed tribal groups will not surrender their arms or their local power unless they are forced to do so by a national government with a powerful army of its own or by an overwhelming outside force. Because the international community is not prepared to produce an occupying force on the same scale as that deployed in Bosnia and Kosovo—thus, many times larger in absolute terms—the democratic-reconstruction model cannot be implemented. Indeed, it would almost certainly fail even if such a force were deployed. The strategy therefore needs to be less invasive.

The now-discredited strongman model is historically the favored method to stabilize a country in crisis; it was freely employed, for instance, by the United States during the cold war and by France as part of its neocolonial strategy in Africa. It is not ethically appealing, but it is cheap, can be effective for a time, and requires little effort on the part of international actors, who delegate the job of imposing order to local leaders. There is no conceivable strongman or strong organization for Afghanistan as a whole. There are, however, strongmen controlling different regions. They will remain part of the political scene, and the international community has no choice but to work with them as it has worked with other such leaders in the past.

Today's orthodox approach to restoring states is much more democratic, but also much more invasive and costly, yet not particularly successful. For the past 10 years, the explicit goal of the international community has been to transform countries in crisis into democratic states with a free market economy based on the argument that only such states benefit their citizens and safeguard the international need for stability in the long run. This Western-dominated sociopolitical engineering approach is becoming ever more complex and costly as experience reveals new areas where intervention is needed.

The components of the democratic-reconstruction model can be summarized as follows: the parties involved in the conflict must reach agreement on a new permanent political system. Elections must be held as soon as possible. The new state must be multiethnic, secular, and democratic—regardless of whether this has any basis in local tradition, or whether it is what the inhabitants of the country want. While the accord is being implemented, peace and order are guaranteed by an international force, as well as by the presence of a large number of UN administrators. The international financial institutions take on the restructuring of the country's economy. International nongovernmental organizations (NGOs) are funded to work in their specialized areas, ranging from humanitarian aid to election organizing.

Elements of the democratic-reconstruction model are already beginning to show up in the discussions of what to do in Afghanistan. The agreement reached by the Afghan factions in Bonn provides for the formation in six months of a broadly based interim government giving representation to all ethnic groups and to women, followed by elections two years later. Virtually all international organizations and NGOs demand strong action to promote women's rights. The World Bank's Afghanistan "Approach Paper" calls for helping the country to build a strong central bank and ministry of finance and for capacity building in all economic institutions. Other organizations target the strengthening of civil society. And this is only the beginning.

The international community must initially accept some version of ordered anarchy in Afghanistan and work to attenuate its worst shortcomings.

Not only is most of this impossible in Afghanistan today, but much of it fits only the wishes of a small minority of Westernized urban Afghans, many of whom have spent the past generation living in the West and are out of touch with their own society. They also, consciously or unconsciously, have a vested interest in Western strategies that would guarantee maximum employment and status for themselves. The model would need to be imposed on reluctant tribal leaders and warlords, on religious authorities, and probably on most ordinary Afghans, and would thus require a strong foreign military and civilian presence, projecting to the world the image of a Muslim country under foreign occupation. As in Somalia, the outcome would almost certainly be conflict between the international force and powerful local groups.

This conflict would most likely lead sooner or later to a swing in exactly the opposite direction, toward withdrawal and neglect, as happened in Somalia and in Afghanistan a decade ago. The reason was the same in both cases: the countries concerned did not appear sufficiently important to justify the effort to create order. The consequences of neglect were serious. Afghanistan became a haven for Al Qaeda. Somalia spawned not only harmless homegrown and clan-based Islamist groups but also al-Itihaad al-Islamiya, an organization aligned with Al Qaeda whose operatives were involved in the 1998 attacks on the United States embassies in Kenya and Tanzania.

In Somalia, however, neglect also had some positive consequences, and this lesson must be heeded in designing a strategy for Afghanistan. With no center to be held, and no pot of foreign aid to be fought over, fighting in Somalia was greatly reduced and mechanisms were developed to compensate for the absence of the state. This did not necessarily mean reverting to a completely primitive life within villages and clans. A new class of international traders emerged, for example, who are capable of financing complex transactions, making international payments, and developing markets.

The Somali experience has historical precedents. The "ordered anarchy" of medieval France, Germany, or Italy—characterized by multiple overlapping armed authorities—did not preclude the establishment of great and stable long-range trade routes and commercial and financial networks, major economic growth, and tremendous achievements in human culture. In the long run, these also laid the foundations for the growth of a modern judicial order, which in turn was an essential basis for the economic revolution and the modern state. The international community must initially accept some version of ordered anarchy in Afghanistan and work to attenuate its worst shortcomings.

THE RIGHT CHOICE

The international community's immediate aim for the Afghan government should therefore not be the impossible fantasy of a democratic government technocratically administering the country, but rather the formation of a loose national mediation committee functioning not just for the initial six months but indefinitely. This committee should seek not to create the entire apparatus of a modern state, but rather the minimal conditions for medieval civilization: the avoidance of major armed conflict, the security of main trade routes, and the safety and neutrality of the capital. These conditions should be secured not by an Afghan national army—another empty fantasy, given the present situation—but by an international force created by the United Nations and backed by the ultimate sanction of American airpower. An agreement on how to create such minimal conditions would be a greater accomplishment for the *loya jirga* called for by the Bonn agreement than would approval of a Western-style democratic constitution that could never be implemented.

Most Western aid therefore should not be directed through the Afghan government—even assuming that the appearance of a broadly based national government could be sustained—but should be provided directly to Afghanistan's regions. Aid should, moreover, be used in a quite clear-headed and tough way as an instrument of peacekeeping—as a way to give local warlords and armies an incentive not to go to war with each other. It would be a bribe of sorts, and might appear to perpetuate the power of warlords. But as Somalia and other African examples illustrate, greater risks would be involved in making the central government the chief channel for international aid, since this would make control of the government and the city of Kabul a vital goal for the country's various armed forces. Aid itself would become a source of future conflict.

Aid should also be provided directly at the local level, of course, to villages and local organizations. But the international community should have no illusion that it is possible to completely bypass warlords and tribal leaders in this fashion. In the end, as the experiences of aid agencies in many countries show, armed groups and powerful individuals always influence how aid is used in their areas.

The international strategy toward Afghanistan should therefore be based on these key principles:

- Discard the assessments of what help Afghanistan needs to become a modern democratic state and replace them with a sober evaluation of the minimal tasks a central administration needs to perform to allow a measure of normal life, economic activity, and, above all, trade.

- Work directly with regional leaders whose power is well established. Assign liaison officials to work with these leaders, monitor their behavior (especially their treatment of local ethnic minorities and their relations with other regions and ethnic groups), and make sure that they provide no shelter to terrorist groups.

- Instruct these liaison officials to work with international and domestic NGOs to ensure not only that they can work unhindered, but also that they do not become dangerously entangled in local politics.

- Create a corps of international civil servants to act as these liaison officials and otherwise assist Afghanistan. These officials should be paid generously in return for devoting a substantial term of service to this difficult and dangerous task and for investing in learning local languages, history, and customs; everything possible should be done to establish their position and prestige. A certain historical precedent here is provided by the British Empire's Indian Political Service, which managed—but, wisely, never tried to administer—the Pashtun tribal areas and handled relations with the Afghan monarchy.

- Give serious consideration to the standards that need to be met by local leaders in exchange for aid. Resist the temptation to impose unrealistic standards. Pick only a few battles to fight at one time. For example, make aid initially contingent on education for girls, but not on a comprehensive reform of legal or social codes governing the position of women in the family or major participation of women in administration. Incremental change is more likely to be sustainable.

- Accept that, even with checks and conditions, there will be corruption, and aid will help warlords consolidate their power and their client networks. Experience shows that corruption is inevitable whenever a country receives large amounts of aid, even if it is channeled through formal government institutions. Use aid quite consciously as a political tool to maintain peace.

- Establish certain basic national institutions in Kabul, but leave the question of a real national administration for Afghanistan for the distant future. Instead, treat the central government as a form of national mediation committee. Avoid

making Kabul and the central government prizes worth fighting over.

- Create a substantial United Nations—mandated international force to ensure the security and neutrality of the city of Kabul as a place where representatives from different areas can meet and negotiate, and where basic national institutions can be created. Be prepared to maintain this force for a period of several years, at least.

- Do not pursue democratic measures, such as organizing elections, that would increase competition at the center among different warlords or ethnoreligious groups: in present circumstances such elections could not possibly lead to stable democratic institutions.

WHAT IS NEEDED

The United States and the international community do not need Afghanistan to become a modern democratic state—even a united one—to protect their key interests. They require a cessation of serious armed conflict and sufficient access to all parts of the country to ensure that it will not again become a haven for international terrorist groups and a source of destabilization for its neighbors. Beyond this, America's interests and capabilities are highly limited.

If Afghanistan could be turned by fiat into a Scandinavian welfare state, well run and capable of delivering services to its population, its people surely would benefit greatly. But the international community cannot deliver such a state. At best, experience shows it can deliver institutions that conform to the appearance of the modern state, but that function inefficiently and corruptly and that generate new conflicts over control.

What the people of Afghanistan need most urgently, and the international community can help them obtain, is the cessation of war and the possibility of pursuing basic economic activities free from brutal oppression, ethnic harassment, and armed conflict. They need to be able to cultivate their fields, sell their products, go to market, send their children to school, receive basic medical care, and move freely around the country. In the long run, much more would be desirable, but the first step should simply be to reestablish a degree of normal life, even if it is not life in a modern state. Just to achieve this much will require many years of careful, concentrated effort by dedicated international workers on the ground. More ambitious state-building plans must be left for another generation, and to the Afghans themselves.

MARINA OTTAWAY *and* ANATOL LIEVEN *are senior associates at the Carnegie Endowment for International Peace. This essay is adapted with permission from "Rebuilding Afghanistan: Fantasy versus Reality" (Carnegie Endowment Policy Brief no. 12, January 2002).*

From *Current History*, March 2002, pp. 133-138. © 2002 by Carnegie Endowment for International Peace.

Venezuela's "Civil Society Coup"

Omar G. Encarnación

The events of this past April that led to the brief removal from power of Hugo Chávez Frías, Venezuela's mercurial president, caught many scholars and policymakers who had come to believe that coups were a thing of the past in Latin America by surprise. More significantly, the turmoil in Venezuela challenged another bit of conventional wisdom about contemporary politics (and indeed, a tenet of American foreign policy): that a strong and invigorated civil society is an unmitigated blessing for democracy.[1] This idea was put forward as early as the mid-1800s with the publication of *Democracy in America*, Alexis de Tocqueville's classic treatise on American political culture in the postcolonial period, and in the last decade it has enjoyed a robust renaissance in academic and policy circles.[2] "Tocqueville was right: democratic government is strengthened, not weakened, when it faces a vigorous civil society," writes Robert Putnam, a leading voice among the "new" Tocquevilleans.[3]

Putnam's views are shared by the international aid community, which in recent years has embraced the mission of fortifying civil society as a programmatic priority in nations that have recent inaugurated democratic governance.[4] The United States Agency for International Development (AID) and the National Endowment for Democracy (NED) have taken the lead in boosting the development of groups thought to comprise the heart of civil society: grass-roots social movements, unions, a free media, and a wide range of nongovernmental organizations involved in promoting such causes as human rights, governmental transparency, and protection of the environment. Presently, funding for "civil society assistance" exceeds that of any other initiative designed by AID to encourage democracy abroad. The agency's budget for 1999 designated $204 million for "civil society promotion," $147 million for "rule of law," $203 million for "governance," and $59 million for "elections and political processes."[5]

Few would dispute the importance of civil society to the creation and maintenance of a democratic public life, but civil society can only serve as an effective foundation for democracy where there are credible functioning state institutions and strong political parties with deep roots in society.[6] Under such conditions, the virtues of civil society—safeguarding society against abuse of power and socializing the citizenry to democratic practices—become

apparent. In their absence, however, civil society, especially an invigorated one, can become a source of instability, disorder, and even violence. The latter scenario, predicted by Harvard political scientist Samuel P. Huntington more than three decades ago in his classic work on political development, *Political Order in Changing Societies*, is currently being recreated in struggling democracies around the world.[7]

In particular, a mobilized and energized civil society in the midst of failing political institutions affords a highly auspicious environment for a "civil society coup." This is a shorthand term employed in this analysis to suggest the handling of governing crises by extraconstitutional, undemocratic means by such actors as the business community, organized labor, religious institutions, and the media. This distressing political phenomenon, an increasingly familiar feature of contemporary Latin American politics, found its latest and most dramatic manifestation in Venezuela.[8]

A civil society coup develops in three distinct phases, each with devastating consequences for democracy. The first is the institutional decay and eventual collapse of the political system (especially political parties), the result of corruption, incompetence, and neglect of the electorate's basic needs. The second is the rise of an antiparty, antiestablishment leader whose appeal to the masses is rooted in the failures of the political system and whose commitment to democracy is at best suspect. This development, in turn, makes civil society, rather than formally organized political forces, the principal opposition to the regime in power, and potentially the sole defender of democracy against encroaching state authoritarianism (itself a consequence of the lack of formal political opposition to the government).

The third phase is a confrontation between government and civil society, the result of the government's failure to deliver on its promises and its attacks on both civil society and the democratic system. In the absence of credible political institutions through which societal demands and dissatisfactions may be channeled, the streets—rather than the legislature, the courts, and the electoral system—become the principal setting for this confrontation. At this juncture, sectors of civil society are not only likely to become radicalized but are also vulnerable to being hijacked by antidemocratic forces.

A Decaying Political System

Hugo Chávez's brief and dramatic ouster from power on April 12 vividly brought to life the explosive nature of the convergence (or more accurately the clash) of an invigorated civil society and a failed political system. The roots of this predicament—and of the current political crisis—can be traced to the ossification and eventual collapse of the Venezuelan party system, a development that stands in striking contrast to the image of Venezuela as one of most successful democracies in postwar Latin America.

Prior to Chávez's rise to power, Venezuelan politics were the dullest, most predictable, and least typical in Latin America. This was owed to the organization of the country's political elites around a two-party system consisting of the Social Democratic Acción Democrática (AD) and the Christian Democratic Comité de Organización Política Electoral Independiente (COPEI). For several decades, these parties carefully maintained one of Latin America's most stable political regimes, making Venezuela the envy of every other country in the region. Indeed, from the 1960s through the 1980s, when many Latin American countries descended into military dictatorship, social unrest, civil war, and economic stagnation, Venezuela remained a sea of relative political tranquility and economic growth.[9]

Venezuela's postwar success made the country the darling of social scientists by giving credence to a new school of thought about democratization that favors the actions of political elites over the impact of socioeconomic, structural variables in the making of successful democracies.[10] Democracies are crafted, not born, was the primary lesson that scholars of Latin American politics took from Venezuela's democratic stability. Paradoxically, during the late 1980s, as scholars were touting the virtues of the Venezuelan model and countries such as Argentina, Brazil, and Chile were shedding their military regimes and looking to Venezuela as a beacon of democratic success, the country was coming apart at the seams.

In 1989, during the popular uprising commonly known as "El Caracazo," the country was rocked by a wave of food riots and violent protests that left hundreds of dead in its wake, after the government ordered the army to open fire on protestors. At the time, Venezuela's political elite vowed that such chaos would never occur again. But in 1992, the country once again fell into crisis, this time as a result of a failed military coup against President Carlos Andrés Pérez of the ruling AD party, led by none other than Hugo Chávez, who was then an army colonel. This attack on one of Latin America's oldest and most stable democracies did not surprise those intimately familiar with the deterioration of the Venezuelan political system, especially the institutional decay afflicting the leading parties. What was once Latin America's paradigmatic example of "party fortitude" had become by the mid-1990s an egregious case of "party deficit."[11]

The democratic stability projected by Venezuela through the 1980s was largely a mirage. It was built upon exclusionary and seemingly antidemocratic practices that in time gave way to a stifling political system and an alienated electorate. Beginning with the 1958 Punto Fijo Pact, which signaled both the end of the country's last military dictatorship under Gen. Marcos Pérez Jiménez and the founding of Venezuela's present democracy, the AD and COPEI conspired to keep the Communists and other left-wing organizations from attaining power.[12] From 1958 through the early 1990s, the AD and COPEI exchanged power on a regular and predictable basis and shared 90 percent of the national vote. In both substance and style, the "partyarchy" created by the AD and COPEI gave Venezuelan democracy the look and feel of the experiments with limited democracy typical of nineteenth-century Latin America.

Arguably more damaging to the country's democracy was the unrestrained corruption that characterized successive AD and COPEI administrations. The vast network of patronage and clientelism created and fed by corruption devoured the substantial earnings the country derived from the world's richest oil reserves outside of the Middle East and left the vast majority of Venezuelans economically deprived and resentful toward the political parties. Since the mid-1980s, 97 percent of the country's 23 million people have faced a steep decline in living standards, while the working class has seen its purchasing power cut to a third of what it was 20 years ago. At present, four out of five Venezuelans live below the poverty line. That these economic calamities have taken place in Venezuela, the land of plenty in postwar Latin America and home to a once-thriving middle class, makes the country's economic failures all the more glaring and disappointing.

Reenter Chávez

In 1998, Hugo Chávez, the quintessential antiparty, personalistic leader and former coup-maker, capitalized on the failures of the political establishment to gain control of Venezuela, this time through free and fair elections, and with 56 percent of the vote. This was the biggest margin of victory in Venezuela's electoral history.[13] The fact that the majority of Venezuelans were willing to overlook Chávez's previous attempt to seize power through a military takeover (for which he served two years in prison) spoke volumes about the bankruptcy of the political establishment. Also telling was Chávez's assertion that his attempt to unseat President Carlos Andrés Peréz in 1992 was a legitimate strike against corruption. This clearly resonated deeply with the general public during the 1998 elections, which came on the heels of Andrés Pérez's impeachment by the National Assembly on corruption charges.

It was the complete repudiation of the AD and COPEI in the 1998 elections, however, that told the real story of how discredited Venezuela's political establishment was. As one observer noted, "The pendulum of Venezuelan opinion seemed to swing to the opposite extreme, from placing all bets on parties to simply refusing to bet on parties at all."[14] For Venezuela's desperate and destitute masses, Chávez's electoral triumph was especially sweet since it was coupled with the evisceration of the two leading parties that had ruled the country for four decades. So deep was the internal crisis afflicting these parties in 1998 that neither bothered to put forward a presidential candidate. Instead, they banded together behind Henrique Salas Romer, an independent whose campaign was anchored on the sole claim that he was the "anti-Chávez" candidate. But the former ruling parties won only 9 percent of the vote. Adding insult to injury, a personality-driven party launched by a former Miss Universe bested both parties at the polls.

Chávez began upsetting the political status quo in Venezuela from the very inception of his government with a fiery populist rhetoric backed by a left-wing economic agenda. The key mechanism behind this enterprise was a new constitution enacted in December 1999 that made him the most powerful president in the country's 42 years of democracy. Approved by a clear majority of Venezuelans (70 percent) and a tightly controlled National Assembly that included Chávez's wife and brother, the 1999 constitution was aimed at reconstructing the country's politics in both substance and style.[15] In an important symbolic change, the country was now to be known as the "Bolivarian Republic of Venezuela", in honor of Chávez's hero, Simón Bolivar, South America's liberator from the Spaniards, and a symbol of regional unity and independence.

The new constitution also increased the role of the state in the economy, significantly reinforced presidential power at the expense of weakening political representation by abolishing the Senate, loosened civilian control over the military, and potentially secured the presidency in Chávez's hands until 2012. In asking the country to approve these dramatic changes, Chávez argued that they were needed to root out corruption.

Following approval of the new constitution, Chávez moved to implement a "Bolivarian" revolution by passing 49 "revolutionary" laws aimed at reconfiguring the nation's economic system. At their heart were provisions for implementing a sweeping land reform program that would take land from private hands without compensation and give it to the poor, and for securing tight control of the state-owned oil company, Petróleos de Venezuela (PDVSA). Control of the latter is critical to Chávez's anti-poverty programs since oil sales are responsible for about 80 percent of the country's foreign earnings and about 50 percent of government revenues.

After 1999, Chávez also began to organize his followers (drawn mainly from the poorest sectors of Venezuelan society) in so-called Bolivarian circles. This expansive network of grass-roots associations was designed by the Chávez government to meet several intertwined purposes, mostly of a political nature. The first was to channel the delivery of social services to the masses—including healthcare, job training, and short-term credit—through a paternalistic and clientelistic model of state-society relations that would promote Chávez's image as the defender of the poor and destitute. In the absence of a strong governing political party and alliances with established groups from the popular sector, such as organized labor, the Bolivarian circles were meant to provide Chávez with both symbolic and real protection against his increasingly long list of enemies. During the April coup, the members of the Bolivarian circles (many of them rumored to have been armed by the government) took to the streets to demand the return of the man they perceive to be nothing short of a messiah.

In an effort to chart an independent course in foreign policy, Chávez became a thorn in the side of the United States. He made alliances with Cuba (to which he began selling oil at bargain basement prices). He also paid high-profile visits to Iraq and Libya. He irritated American officials by refusing to declare his support for the U.S.-backed Colombian government in its struggles against its Marxist insurgency. Instead, he declared Venezuela neutral in this conflict, despite the fact that Colombia's civil war is rapidly spilling over the Colombian-Venezuelan border. In 1999, Chávez also refused American assistance in rebuilding a vital coastal highway destroyed by mudslides, even though U.S. military ships were already on their way. Most recently, Chávez failed to condemn the September 11 terrorist attacks on New York and Washington, and was quick to criticize the U.S. bombing of Afghanistan.

Within Venezuela, Chávez dealt harshly, at least rhetorically, with anyone who opposed him.[16] Members of the Venezuelan media, many of whom criticized Chávez's autocratic governing style and refused to serve as mouthpieces for his speeches and decrees, were demonized as "traitors" and "counterrevolutionaries," and lived under constant threat of having their radio and television licenses revoked by the government. Chávez has described the Venezuelan landed gentry as "rancid oligarchs and squealing pigs," and he tagged members of the business community who sought to derail his economic program as "subversives who should move to Miami." Even the Catholic clergy managed to incur Chávez's verbal wrath. After they had expressed disapproval of his penchant for class warfare and Marxist rhetoric, Chávez called church leaders "devils in vestments."

Chávez's tirades against his enemies, together with his generally erratic leadership and failure to deliver upon his promises to the poor, further polarized Venezuelan society by exposing raw and angry divisions between rich and poor, and set the stage for his ouster from power. During Chávez's tenure in office, Venezuela entered its worst economic crisis in 30 years. The national economy has contracted by 10 percent, the result of capital flight and declining production at the state-owned oil company, and a quarter of the working population is unemployed, a key factor in propelling the masses into the streets to protest declining living standards. Graffiti proclaiming "We Are Hungry" is frequently sighted in the streets of Caracas.

A Civil Society Coup

With the long-established political parties in disarray, civic groups, unions, and business organizations emerged as the main opposition to the Chávez government and defender of the democratic process against Chávez's increasingly authoritarian tendencies. It is hardly surprising, therefore, that a heterogeneous coalition of civil society spanning business, labor, the Catholic Church, and the media moved to address the country's mounting problems. Protests were its main weapon, and there were massive, cross-sectoral, multi-class demonstrations throughout 2001. These protests succeeded at various points in getting Chávez to back down, but more often than not he continued to ignore the demands of this loose coalition.[17]

The straw that broke the camel's back was Chávez's firing on April 6 of seven high-level PDVSA executives. The oil company had enjoyed a long history of autonomy, and Chávez's attempt to replace its leadership with left-wing cronies elicited a strong reaction.[18] In a rare sign of cross-class solidarity, Carlos Ortega, the leader of the Confederación de Trabajadores de Venezuela (CTV), the nation's leading trade union, and Pedro Carmona Estanga, the leader of the Federación de Cámaras de Comercio y Producción (FEDECAMARAS), the nation's leading business association, with the support of many other civil society organizations, called for a general strike on April 9. The unions and the employers were reacting to the erosion of self-management within PVDSA as well as to Chávez's attempt to undermine traditional labor representation in the oil sector. By April 10, CTV and FEDECAMARAS had decided to continue the general strike indefinitely while raising the strikers' demands from the reinstatement of those fired from the state oil company to Chávez's immediate resignation.

Civil society's victory over the Chávez government came suddenly, but not in the manner expected by the demonstrators. On April 11, the third day of the general strike, National Guard troops and pro-Chávez gunmen opened fire on the 500,000 anti-Chávez protestors gathered in front of the Miraflores palace (the official residence of the Venezuelan president), killing 17 people. This event unleashed a wave of violence throughout the country that left as many as 50 dead and hundreds injured, and prompted the military to throw its support behind the new, temporary government led by Pedro Carmona. The military held Chávez responsible for the killings, took him into custody, and forced him to resign.

However, Chávez's removal from power was in the making long before the theater of popular protest began to play in the streets of Caracas. For weeks, if not months, prior to the protests, civil society leaders were busy plotting behind closed doors which military leaders who were also disenchanted with Chávez's rule. And they apparently enlisted U.S. help. It has been widely reported that for at least a year before the coup, business and media leaders from Venezuela held talks with officials from the State Department and the White House on how to deal with the Chávez problem.[19]

More worrisome still were the actions of the sectors of civil society fronting the coup during the 48 hours that the Chávez government was out of power. Indeed, the reign of civil society in Venezuela was anything but civil and stands as an example of autocratic rule with few parallels in recent Latin American history. Nonetheless, Carmona employed the term "civil society" as a rhetorical tool to consolidate his newly acquired power. "I want to tell the country that all the actions I took as representative of civil society were never done with the goal of reaching this position," were Carmona's chosen words to introduce himself to the Venezuelan public.[20] He also declared that the transitional government was being formed with "the consensus of civil society and also from the military."[21] Its aim, he said, was to "create a pluralistic vision of a democratic society in keeping with the implementation of the law." The actions of the new government belied these lofty intentions, however. For a start, representatives of civil society itself were conspicuously absent from Carmona's short-lived government.

In what can be best described as a coup within a coup, Carmona jettisoned the plan for a broad civilian-military coalition government in favor of an interim government staffed almost exclusively with ultraconservatives, much to the dismay of the popular groups that had supported the coup, organized labor in particular. "The CTV was not consulted, and we feel betrayed by this violation of the rights and freedom of the workers' movement," remarked Carlos Ortega to the Spanish daily *El País*.

Once in control of the country, the new government moved quickly to dismantle Chávez's legacy. After nullifying the "revolutionary" laws passed by the Chávez administration and reversing the country's name to the Republic of Venezuela, Carmona announced that elections would be held within a year. He dissolved the National Assembly, fired the Supreme Court, and abolished the constitution that the majority of Venezuelans had approved in 1999. The new government also moved rapidly to reverse the course of Chávez's foreign policy. Foreign minister-designate José Rodríguez announced the new government's desire for "tight relations with Washington." He also announced that Cuba would not be getting any more oil and declared Colombia's "narco-guerrillas" to be "double enemies of humanity."

Other Latin American governments immediately condemned the takeover. The leaders of Mexico and Argentina, who were attending a heads of state meeting in San José, Costa Rica, said they would not recognize the new government, citing a flagrant "alteration of constitutional order." In their condemnation of the coup, Latin American leaders referred to the Inter-American Democratic Charter, signed in Lima on September 11, 2001 (the very day of the terrorist attacks in New York and Washington), by the members of the Organization of American States (OAS). The charter, which, ironically, was approved with a dissenting vote from Venezuela, requires the immediate condemnation of the overthrow of any democratically elected government of a member state.

Trampling on constitutional procedures, however, did not appear to concern business, union, and church groups and media organizations that actively supported the coup, at least at its inception. Especially troublesome was the role of the media. The Venezuelan media was not only an opponent of the Chávez administration (the result of the president's hostile treatment of the press and his contempt for democracy) but turned itself into an accomplice of the opposition movement that supported his ouster. Venezuelan newspapers and television stations provided extensive (some would say excessive), wall-to-wall coverage of the protests leading up to the coup, including the general strike called by the unions and the employers that drew some 150,000 people to the streets on its first day. But they fell silent on the unconstitutionality of Chávez's removal from power. Nor did they report the problems that immediately besieged the new government.

On April 13, as the interim government was falling apart and the Chávez government was being restored, television stations were showing old Hollywood movies and cooking shows, while the country's independent newspapers essentially shut down. This silence created, in the words of a columnist for the local daily *El Nacional*, "a diabolical blackout that left most of Venezuela misinformed about what was happening to the country." The few newspapers that chose to go to press the day following the coup sought to create a sense of inevitability about Chávez's departure from office. "It's over," gleefully declared the headline in *El Universal*, a Caracas newspaper, announcing the presumed end of Chávez's tenure in office.

Ironically, it was the military, hardly a friend of democracy in Latin America, that emerged as the defender of the democratic constitutional order in Venezuela. A section of the military opposed Carmona's dissolution of the National Assembly and the courts, actions it criticized for "undermining democracy." More surprising, perhaps, military officers objected to Carmona's appointments as being too representative of the business community and unrepresentative of the trade unions. On April 14, once key sectors of the military withdrew their support, Carmona's interim government collapsed and Chávez was reinstated as president.

Having demonstrated a disregard for democracy as callous as that exhibited by Chávez throughout much of his tumultuous time in office, Carmona now sits under house arrest awaiting trial and facing 20 years in prison. His farewell to the nation vividly highlighted the manner

in which "civil society," both as a political construct and as a political slogan, is being used and understood to represent opposition to the existing government in many struggling democracies. As pro-Chávez demonstrators were celebrating the announcement that "the tyrant has been deposed," Carmona insisted: "I will not go away from that civil society. I have been an important person in civil society and when this passes I will be there again."[22] The civil society groups that once offered support, however, are now distancing themselves from him and his short-lived dictatorship, and promising a referendum seeking Chávez's recall within two years.

The Role of U.S. Civil Society Assistance

The attack on a democratically elected government perpetrated in the name of civil society in Venezuela did not draw any overt criticism from the Bush administration. Indeed, for a brief but highly symbolic moment, the events of April 12 in Venezuela took U.S.-Latin American relations back to a time when American officials did not particularly care whether the region's political leaders were democrats or despots so long as they did not threaten U.S. interests. "A victory for democracy" was the White House's premature response to Chávez's ouster.

White House officials also denied that a coup had taken place in Venezuela and referred to the situation as a crisis that Chávez had brought upon himself. But once it had become clear to everyone but the White House that a coup had in fact derailed democracy in Venezuela, and Secretary of State Colin Powell had acknowledged as much, the administration began to scramble to distance itself from any direct involvement. The spotlight was turned on Powell's staff, especially the assistant secretary for Western Hemisphere affairs Otto Reich, already a controversial figure in Latin American circles given his role as adviser to the Contras in Nicaragua during the 1980s. Reich denied rumors that he had encouraged the takeover in conversations with members of Chávez's opposition in the months leading up to the coup.

Despite Powell's corrective statement that "defending democracy by undemocratic means destroys democracy," the fact that American officials all but applauded Carmona's insurgency set a troubling precedent for the future of democracy in Latin America. The American endorsement of the coup, clearly taken without prior consultation with Latin American heads of state, dramatically reversed years of steady American support for democracy in the region. During the 1990s, the Clinton administration had criticized the authoritarian tendencies of such democratically elected leaders as Alberto K. Fujimori in Peru and Jorge Serrano in Guatemala, and it had joined forces with other Latin American leaders to derail military takeovers in Paraguay and Ecuador.

Just as troubling as any direct involvement the Bush administration might have had in supporting a coup against Chávez are the revelations that American aid granted under the rubric of "civil society assistance" may have been used by groups in Venezuela to subvert a democratically elected government. It has been reported that as conditions in Venezuela deteriorated and Chávez clashed with various business, labor, and media groups, the National Endowment for Democracy stepped up its civil society assistance, quadrupling its budget for Venezuela to more than $877,000.[23] Whether any of that money was put into direct efforts to oust Chávez is the subject of much conjecture. The State Department has put on hold a $1 million grant to the NED pending an investigation into whether any recipients of the agency's funds went to groups that actively plotted against Chávez.

However, there is no doubt that U.S. civil society assistance has gone to organizations that actively supported the coup once it was underway. Among the recipient of such aid was the Venezuelan trade union CTV, which worked closely with Pedro Carmona to oust Chávez. Moreover, some of the American organizations through which the NED channels its aid endorsed the coup. "Last night led by every sector of civil society, the Venezuelan people rose up to defend democracy in their country," said George A. Folsom, head of the International Republican Institute, an organization affiliated with the Republican Party that is active in Venezuela and a recipient of NED grants.

The NED's leaders have distanced themselves from Folsom's statement and deny any involvement with those who staged the coup. They add that their objective in Venezuela was "to create political space for opponents to Mr. Chávez, not to contribute to his ouster." According to the NED's senior project officer for Latin America, the agency's funds went to specific projects designed "to bolster the democratic opposition in Venezuela—including training in civics, journalism and conflict resolution—and did not contribute to the attempted ouster of Mr. Chávez."[24]

These denials only serve to highlight the highly ambiguous and unpredictable role civil society assistance plays in fledgling democracies. The fact that this aid is targeted exclusively at advocacy or watchdog organizations suggests an impoverished understanding of what civil society actually is and what role it plays in the maintenance of a democratic public life. For "neo-Tocquevillean" scholars such as Robert Putnam, civil society is made up of all kinds of voluntary groupings (not just pro-democracy advocates), especially those far removed from the political sphere. It is bowling leagues, choral societies, and bird-watching clubs, rather than politically minded NGOs, that Putnam credits with promoting the social trust and solidarity that greases the wheels of democracy. To be sure, civil society assistance programs have been very useful in helping pro-democracy groups survive the weakening of civil society that comes with any successful transition to democracy; but whether such aid is employed for democratic or antidemocratic purposes remains beyond the control of its donors.

Lessons from Venezuela

The most obvious lesson to be gained from the coup in Venezuela is that civil society, widely hailed as the cornerstone of stable and healthy democracies, can also be a source of social unrest, political instability, and even antidemocratic behavior. To be sure, the point being stressed here is not that the actions of Venezuelan civil society in promoting and supporting Chávez's ouster warrant a collective indictment of "antidemocratic." Since Chávez came to power, civil society has been the only actor in Venezuelan society consistently opposing his increasingly authoritarian rule. Indeed, a legitimate argument can be made that the events of April 11–14, despite their extra-constitutional nature, were an act of democratic self-defense provoked by Chávez's tirades against civil society. It is also important to note that once some segments of civil society (such as the business community) revealed their antidemocratic leanings, other sectors withdrew their support from the coup, thereby preventing the consolidation of Carmona's short-lived dictatorship. The point is that the potential for instability, and even violence, exhibited by civil society in Venezuela exposes the myth of civil society as an infallible democratic miracle worker. Civil society can both aid and harm democracy and the key variable determining these outcomes is the health of the political system.

More importantly, the coup in Venezuela revealed the limitations of civil society as a protector of democracy, especially in the absence of credible political institutions, political parties in particular.[25] Under such conditions, we cannot expect civil society to prevent the rise of a Hugo Chávez, nor to marshal societal resources to counter an authoritarian leader's pernicious impact upon the polity. The problem is not the inability of civil society to rid itself of an unpopular government (it certainly can do so, as we have seen in Venezuela and elsewhere), but rather its inability to manage popular discontent and coordinate the activities of the citizenry into effective political solutions. These functions are the province of political institutions, especially political parties, which serve as conduits for citizens' demands and are a means of making sure that all players abide by the rules. Only a strong, competitive party system with deep roots in society can keep the democratic game in good health.

These lessons about the primacy of political institutions in keeping civil society and democratic practices in good standing appear to have gotten lost in the enthusiasm for civil society generated in recent years by the Left and the Right. The rhetoric of many of the social movements that animates the so-called New Left (environmentalists, feminists, human rights activists, and others) has led to widespread skepticism about the importance of political institutions (parties in particular) in contemporary democratic politics. Ardent civil society advocates from the right regard political parties as colonizing pariahs whose penetration of civil society leads to the demise of democracy.

The expectation of both camps is that an invigorated civil society will either fix whatever ails political institutions, including the state and the party system, or else render them obsolete. As Thomas Carothers, a critic of civil society assistance programs, writes, "The rise of civil society induces some to see a nearly state-free future in which tentative, minimalistic states hang back while powerful non-governmental groups impose a new, virtuous order."[26] But this logic, as suggested by the recent Venezuelan experience, is deeply flawed and even dangerous. Only under stable and effective political institutions can democracy thrive and survive.

Notes

1. Although there is no consensus among scholars and policymakers on the precise meaning of civil society, this term is generally thought to represent the realm of organized social life created outside of the state, the family, and the marketplace. More concretely, civil society incorporates a complex welter of organizations including civic, religious, and recreational groups. For a broader discussion of the meaning of civil society, see Alan B. Seligman, *The Idea of Civil Society* (Princeton, N.J.: Princeton University Press, 1991).

2. Alexis de Tocqueville, *Democracy in America* (Garden City, N.Y.: Anchor, 1968).

3. Robert D. Putnam, *Making Democracy Work: Civic Traditions in Modern Italy* (Princeton, N.J.: Princeton University Press, 1993), p. 182.

4. See, Alison Van Rooy, ed., *Civil Society and the Aid Industry* (London: Earthscan, 1998) and Omar G. Encarnación, "Tocqueville's Missionaries: Civil Society and the Promotion of Democracy," *World Policy Journal*, vol. 17 (spring 2000).

5. Thomas Carothers, *Aiding Democracy Abroad: The Learning Curve* (Washington, D.C.: Carnegie Endowment for International Peace, 1999), p. 50.

6. This study is not the first to cover this theoretical terrain. Indeed, neglect of the political sphere in affecting the behavior of civil society is a dominant theme of some of the better-known critiques of the civil society revival. See, for example, Sheri Berman, "Civil Society and the Collapse of the Weimar Republic," *World Politics*, vol. 49 (April 1997).

7. Samuel P. Huntington, *Political Order in Changing Societies* (New Haven: Yale University Press, 1968).

8. Recently, clashes between a mobilized civil society and a discredited government have occasioned the removal from power of the presidents of Ecuador, Peru, and Argentina.

9. On this period in Venezuelan politics, see John Martz, *Acción Democratica: Evolution of a Modern Political Party in Venezuela* (Princeton, N.J.: Princeton University Press, 1966).

10. See Juan J. Linz and Alfred Stepan, *The Breakdown of Democratic Regimes* (Baltimore: Johns Hopkins University Press, 1979); and Guillermo O'Donnell and Philippe Schmitter, *Transitions from Authoritarian Rule: Tentative Conclusions About Uncertain Democracies* (Baltimore: Johns Hopkins University Press, 1986).

11. Javier Corrales, "Strong Societies and Weak Parties: Regime Change in Cuba and Venezuela in the 1950s and Today," *Latin American Politics and Society*, vol. 43 (summer 2001).

12. The pact was signed by the AD, COPEI, and the Unión Republicana Democrática (URD), a non-communist left-wing party. The agreement committed the signatories to observe the democratic rules of the game and to a power-sharing formula on a proportional basis among the participating parties.

13. For a broader view of these events, see Jennifer McCoy, "Chavez and the End of Partyarchy in Venezuela," *Journal of Democracy*, vol. 10 (July 1999).

14. Corrales, "Strong Societies and Weak Parties," p. 103.

15. The apparent strong support that the Venezuelan people gave to their new constitution merits discussion. First, there was an abstention rate of around 54 percent due to torrential rains that coincided with the referendum and left at least 37 people dead and 10,000 homeless. Second, the level of opposition to the new constitution (29 percent) was the highest among all Latin American countries that voted on new constitutions during the 1990s.

16. Chávez's zealous foes have resorted to name-calling as well. "Fascist," "murderer," and "Communist thug" were among the epithets hurled at Chávez and his supporters during the demonstrations leading to the coup.

17. During late 2000 and throughout 2001, Chávez ruled virtually by decree and showed little interest in engaging the opposition in negotiations. Moreover, during this period, he antagonized many sectors of Venezuelan society with his ill-fated attempt to take over the CTV (the labor syndicate) and replace the rector of the Central University and the head of PDVSA.

18. This summary of events leading to Chávez's ouster draws from the reporting that appeared in *El País* (Madrid), April 6–14, 2002.

19. Christopher Marquis, "Pentagon to Investigate Its Role in Venezuela," *New York Times*, April 23, 2002.

20. Juan Forero, "Manager and Conciliator—Pedro Carmona Estanga," *New York Times*, April 13, 2002.

21. Juan Forero, "Venezuela's Chief Forced to Resign; Civilian Installed," *New York Times*, April 13, 2002.

22. Juan Forero, "Venezuela's Leader for a Day Denies There Was a Rebellion," *New York Times*, April 19, 2002.

23. Christopher Marquis, "U.S. Bankrolling Is Under Scrutiny for Ties to Chávez Ouster," *New York Times*, April 25, 2002.

24. Ibid.

25. I am indebted to Javier Corrales for this important insight.

26. Thomas Carothers, "Civil Society," *Foreign Policy*, no. 117 (winter 1999–2000), p. 26.

Omar G. Encarnación is assistant professor of political studies at Bard College. He is the author of Civil Society in the Age of Democratization: Myths, Realities and Lessons, *St. Martin's Press, forthcoming.*

From *World Policy Journal*, Summer 2002, pp. 38-47. © 2002 by the World Policy Journal. Reprinted by permission.

The Many Faces of Africa
Democracy Across a Varied Continent

JOEL D. BARKAN

A decade ago, seasoned observers of African politics including Larry Diamond and Richard Joseph argued that the continent was on the cusp of its "second liberation." Rising popular demand for political reform across Africa, multiparty elections, transitions of power in several countries, and negotiations toward a new political framework in South Africa led these experts to conclude that the prospects for democratization were good. Today, these same observers are not so sure. They describe Africa's current experience with democratization in terms of "electoral democracy," "virtual democracy," or "illiberal democracy," and are far more cautious about predicting what is to come. What is the true state of African democracy? And what is its future?

Governance Before the 1990s

Africa's first liberation was precipitated by the transition from colonial to independent rule that swept much of the continent, except the south, between 1957 and 1964. The West hoped that the transition would be to democratic rule, and more than 40 new states with democratic constitutions emerged following multiparty elections that brought new African-led governments to power. The regimes established by this process, however, soon collapsed or reverted to authoritarian rule—what Samuel Huntington has termed a "reverse wave" of democratization. By the mid-1960s, roughly half of all African countries had seen their elected governments toppled by military coups.

In the other half, elected regimes degenerated into one-party rule. In what was to become a familiar scenario, nationalist political parties formed the first governments. The leaders of these parties then destroyed or marginalized the opposition through a combination of carrot-and-stick policies. The result was a series of clientelist regimes that served as instruments for neo-patrimonial or personal rule by the likes of Mobutu Sese Seko in Zaire or Daniel Arap Moi in Kenya—regimes built around a political boss, rather than founded in a strong party apparatus and the realization of a coherent program or ideology.

This pattern, and its military variant (as with Sani Abacha in Nigeria), became the modal type of African governance from the mid-1960s until the early 1990s. These regimes depended on a continuous and increasing flow of patronage and slush money for survival; there was little else binding them together. Inflationary patronage led to unprecedented levels of corruption, unsustainable macroeconomic policies that caused persistent budget and current account deficits, and state decay, including the decline of the civil service. Most African governments still struggle with this structural and normative legacy, which has obstructed the process of building democracy.

Decade of Democratization?

Africa's second liberation began with the historic 1991 multiparty election in Benin that resulted in the defeat of the incumbent president, an outcome that was replicated in Malawi and Zambia in the same year. The results of these elections raised expectations and created hopes for the restoration of democracy and improved governance across the continent. By the end of 2000, multiparty elections had been held in all but five of Africa's 47 states—Comoros, the Democratic Republic of Congo, Equatorial Guinea, Rwanda, and Somalia.

Along with the new states of the former Soviet Union, Africa was the last region to be swept by the so-called "third wave" of democratization, and as with many of the successor states of the former Soviet Union, the record since has been mixed. In stark contrast to the democratic transitions that occurred in the 1970s and 1980s in Southern and Eastern Europe and Latin America (excluding Mexico), most African transitions have not been marked by a breakthrough election that definitively ended an authoritarian regime by bringing a group of political reformers to power. While this type of transition has occurred in a small number of states, most notably Benin and South Africa, the more typical pattern has been a process of protracted transition: a mix of electoral democracy and political liberalization combined with elements of authoritarian rule and, more fundamentally, the perpetuation of clientelist rule. In this context, politics is a three-cornered struggle between authoritarians, patronage-seekers, and reformers. Authoritarians attempt to retain power by permitting greater liberalization and elections while selectively allocating patronage to those who remain loyal. Meanwhile, patronage-seekers attempt to obtain the spoils of office via electoral means, as reformers pursue the establishment of democratic rule. The boundaries between the first and second of these groups, and sometimes between the second and third, can be blurred because political alignments are very fluid. Liberal democracy is unlikely to be consolidated until reformers ascend to power.

Politics is a three-cornered struggle between authoritarians, patronage-seekers, and reformers.

The result is what Thomas Carothers has termed a "gray zone" of politics, describing countries where continued progress toward democracy beyond elections is limited and where the consolidation of democracy, if it does occur, will unfold over a long period, perhaps decades. This characterization does not necessarily mean that the third wave of democratization is over in Africa. Rather, we should expect Africa's democratic transitions to be similar to those of India or Mexico. In the former, the party that led the country to independence did not lose an election for three decades, and periodic alternation of power between parties did not occur until after 40 Years. In the latter, the end of one-party rule and its replacement by an opposition committed to democratic principles played out over five elections spanning 13 years rather than a single founding election. Such appears to be the pattern in Africa, where two-thirds of founding and second elections have returned incumbent authoritarians to power, but where each iteration of the electoral process has usually resulted in a significant incremental advance in the development of civil society, electoral fairness, and the overall political process.

That many African polities fall into the gray zone is confirmed by the most recent annual *Freedom in the World* survey conducted by Freedom House. Of the 47 states that comprise sub-Saharan Africa, 23 were classified by the survey as "partly free" based on the extent of their political freedoms and civil liberties. Only eight (Benin, Botswana, Cape Verde, Ghana, Mali, Mauritius, Namibia, and South Africa)

were classified as "free" while 16, including eight war-torn societies (Angola, Burundi, the Democratic Republic of Congo, Ethiopia, Eritrea, Liberia, Rwanda, and Sudan) were deemed "not free."

The overall picture revealed by these numbers is sobering. Less than one-fifth of all African countries were classified as free, and of these, only two or three (Botswana, Mauritius, and perhaps South Africa) can be termed consolidated democracies. On the other hand, if one excludes states in the midst of civil war, one-fifth of Africa's countries are free, one-fifth not free, and three-fifths fall in-between. That is to say, four-fifths of those not enmeshed in civil war are partly free or free, a significant advance over the continent's condition a decade ago. Only a handful are consolidated democracies, but few are harsh dictatorships of the type that dominated Africa from the mid-1960s to the beginning of the 1990s. As noted by Ghana's E. Gyimah-Boadi, "Illiberalism has persisted, but is not on the rise. Authoritarianism is alive in Africa today, but is not well."

Optimists and Realists

The current status of democracy in Africa varies greatly from one country to the next, and one should resist generalizations that apply to all 47 of the continent's states; one size does not fit all. Notwithstanding this reality, those who track events in Africa have divided themselves into two distinct camps: optimists and realists. Those in the United States who take an optimistic view—mainly government officials involved in efforts to promote democratization abroad, former members of President Bill Clinton's administration responsible for Africa, members of the Congressional Black Caucus, and the staff of some Africa-oriented nongovernmental organizations—trumpet the fact that multiparty elections have been held in nearly 90 percent of all African states. They note that most African countries have now held competitive elections twice and that some, including Benin, Ghana, and Senegal, have held genuine elections three times, at least one of which has resulted in a change of government. The optimists further note that the quality of these elections has improved in some countries, both in terms of efficiency and of fairness. Electoral commissions seem to have been more independent, even-handed, and professional in recent elections than in the early 1990s. Opposition candidates and parties have

greater freedom to campaign and have faced less harassment from incumbent governments. The presence of election observers, both foreign and domestic, is now widely accepted as part of the process. Perhaps most significant, citizen participation in elections has been fairly high, averaging just under two-thirds of all registered voters.

Recognizing that elections are a necessary but insufficient condition for the consolidation of democracy, the optimists also point to advances in several areas, listed below in their approximate order of accomplishment. First, there has been a re-emergence and proliferation of civil society organizations after their systematic suppression during the era of single-party and military rule. Second, an independent and free press has also re-emerged, spurred on by the privatization of broadcast media in several countries. Third, members of the legislature have increasingly asserted themselves in policymaking and overseeing the executive branch. Fourth, the judiciary and the rule of law have been strengthened in countries such as Tanzania, and human rights abuses have also declined. Fifth, there have been new experiments with federalism—the delegation or devolution of authority from the central government to local authorities—to enhance governmental accountability to the public and defuse the potential for ethnic conflict, most notably in Nigeria but also in Ethiopia, Ghana, Tanzania, Uganda, and South Africa. One or more of these trends, especially the first two and perhaps the third, can be found in most African countries that are not trapped in civil war.

Optimists also point to less exclusive membership in the governing elite, which has expanded into the upper-middle sector of society far more than during the era of authoritarian rule. In country after country, repeated multiparty elections have resulted in significant turnover in the national legislatures and local government bodies, sometimes as high as 40 percent per cycle. While the quality of elected officials at the local level remains poor, members of national legislatures are younger, better educated, and more independent in their political approach than the older generation they have displaced. Although further research is needed to confirm any major change in the composition of these bodies, new politicians and legislators also appear more likely than their predecessors to be democrats and to focus on issues of public policy and less likely to be patronage seekers.

Finally, public opinion across Africa appears to prefer democracy over any au-

THE PEOPLE'S PERSPECTIVE

How Democratic is Your Nation?

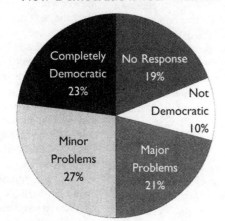

Completely Democratic 23%

No Response 19%

Not Democratic 10%

Minor Problems 27%

Major Problems 21%

How Preferable is Democracy?

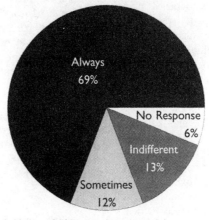

Always 69%

No Response 6%

Indifferent 13%

Sometimes 12%

These two figures, which represent the attitudes of citizens of African nations toward democracy, suggests a strong preference for democratic rule but uncertainty as to its prevalence.

Afrobarometer at www.afrobarometer.org

thoritarian alternatives. Surveys undertaken for the Afrobarometer project in 12 African countries between 1999 and 2001 found that a mean of 69 percent of all respondents regarded democracy as "preferable to any other kind of government," while only 12 percent agreed with the proposition that "in certain situations, a non-democratic government can be preferable." Moreover, 58 percent of all respondents stated that they were "fairly satisfied" or "very satisfied" with the "way democracy works" in their country.

Realists—who criticize what they contend was a moralistic approach to US foreign policy by the Clinton administration and disparage the use of democratization as a foreign policy goal—take a far more cautious view of what is occurring on the continent. Considering the same six developments that optimists cite as examples of democratic advances, realists note that all six are present in fewer than six countries. They also see much less progress than the optimists when nations are considered one by one. First, regular multiparty elections across the continent have resulted in an alternation of government in only one-third of the countries that have held votes. Moreover, only about one-half of these elections have been regarded as free and fair, with results accepted by those who have lost. It is also debatable in most of these countries whether recent elections have been of higher quality than those held in the early 1990s.

Second, although the re-emergence of civil society and the free press is a significant advance from the era of authoritarian rule when both were barely tolerated or systematically suppressed, civil society remains very weak in Africa compared to other regions and is concentrated in urban areas. Political parties are especially weak and rarely differentiate themselves from one another on the basis of policy. Apart from the church, farmers' organizations, or community self-help groups in a smattering of countries (such as Kenya, Cote d'Ivoire, and Nigeria), civil society barely exists in rural areas where most of the population resides. The press, especially the print media, is similarly concentrated in urban areas and thus reaches a relatively small proportion of the entire population. Only the broadcast media penetrates the countryside, but it is largely state-owned. Although private broadcasting has grown in recent years, especially in television and FM radio, stations cater almost exclusively to urban audiences. With a few exceptions, AM and short- and medium-wave radio—the chief sources of information for the rural population—remain state monopolies.

Third, while the legislature holds out the promise of becoming an institution of countervailing power in some countries, it remains weak and has rarely managed to effectively check executive power. Fourth, the judicial system in most countries is ineffectual, either because its members are corrupt or because it has too few magistrates and too poor an infrastructure to keep pace with the number of cases. Human rights abuses also continue, though less frequently and with less intensity than a decade ago. Fifth, Africa's experiments with federalism, though apparently successful, are confined to six states. Finally, the extent to which Africans have internalized democratic values is hard to judge. Although the Afrobarometer surveys indicate broad support for democracy, the results also suggest that such support is "a mile wide and an inch deep." An average of only 23 percent of respondents in each country described their country as "completely democratic."

Both optimists and realists are correct in their assessments of what is occurring in Africa. But how can both views be valid? The answer is that each presents only one side of the story. On a continent where the record of democratization is one of partial advance in over one-half of the cases, those assessing progress toward democratization, or lack of it, tend to dwell either on what has been accomplished or on what has yet to be achieved. These divergent assessments are proverbial examples of those who view the glass as either half-full or half-empty. Optimists and realists also draw their conclusions from slightly different samples. Whereas optimists focus mainly on states that are partly free or free, realists concentrate on states that are partly free or not free.

Optimists and realists are also both right because there are several Africas rather than one. In fact, at least five Africas cut across the three broad categories of the Freedom House survey. First are the consolidated and semi-consolidated democracies—a much smaller group of counties than those classified as free. This category presently consists of only two or three cases, such as Botswana, Mauritius, and perhaps South Africa. The second group consists of approximately 15 aspiring democracies, including the remaining five classified by Freedom House as free but not yet consolidated democracies plus roughly 10 classified as partly free where the transition to democracy has not stalled. All these states have exhibited slow but continuous progress toward a more liberal and institutionalized form of democratic politics. In this group are Benin, Ghana, Madagascar, Mali, Senegal, and possibly Kenya, Malawi, Tanzania, and Zambia. Third are semi-authoritarian states, countries classified as partly free where the transition to democracy has stalled. This category consists of approximately 13 countries including Uganda, the Central African Republic, and possibly Zimbabwe. Fourth are countries that are not free, with little or no prospect for a democratic transition in the near future. About 10 countries make up this group, including Cameroon, Chad, Eritrea, Ethiopia, Rwanda, and Togo. Finally, there are the states mired in civil war, such as Angola, Congo, Liberia, and Sudan. Each of these five Africas presents a different context for the pursuit of democracy.

Inhibiting Democracy

Several conditions peculiar to the continent make Africa a difficult place to sustain democratic practice. They explain why Africa lags behind other regions in its extent of democratic advance, why political party organizations are weak, and why the ties between leaders and followers are usually based on clientelist relationships. These conditions in turn create pressures for more and more patronage, a situation that undermines electoral accountability and leads to corruption.

Africa is the poorest of the world's principal regions: per capita income averages US$490 per year. This condition does not affect the emergence of democracy but does impact its sustainability. On average, democracies with per capita incomes of less than US$1,000 last 8.5 years while those with per capita incomes of over US$6,000 endure for 99. The reasons for this are straightforward. Relatively wealthy countries are better able to allocate their resources to most or all groups making claims on the state, while poor countries are not. The result is that politics in a poor country is likely to be a zero-sum game, a reality that does not foster bargaining and compromise between competing

interests or a willingness to play by democratic rules.

Almost all African countries remain agrarian societies. With few exceptions like South Africa, Gabon, and Nigeria, 65 percent to 90 percent of the national populations reside in rural areas where most people are peasant farmers. Consequently, most Africans maintain strong attachments to their places of residence and to fellow citizens within their communities. Norms of reciprocity also shape social relations to a much greater degree than in urban industrial societies. In this context, Africans usually define their political interests—that is to say, their interests as citizens visa-vis the state—in terms of where they live and their affective ties to neighbors, rather than on the basis of occupation or socioeconomic class.

Several conditions peculiar to the continent make Africa a difficult place to sustain democratic practice.

With the exceptions of Botswana and Somalia, all African countries are multiethnic societies where each group inhabits a distinct territorial homeland. Africans' tendency to define their political interests in terms of where they live is thus accentuated by the fact that residents of different areas are often members of different ethnic groups or sub-groups.

Finally, African states provide much larger proportions of wage employment, particularly middle-class employment, than states in other regions do. African states have also historically been large mobilizers of capital, though to a lesser extent recently. Few countries have given rise to a middle class that does not depend on the state for its own employment and reproduction. In this context, people seek political office for the resources it confers, for their clients' benefit, and for the chance to enhance their own status. In the words of a well known Nigerian party slogan, "I chop, you chop," literally, "I eat, you eat."

Likely Scenarios

Given these realities, what is the future of democratization on the continent? The answer varies greatly by type of polity. Over the next decade, the small handful of states that are currently classified as consolidated or semi-consolidated democracies is likely to gain perhaps six additional members. This category, however, will remain the smallest of the five Africas because of the limited number of polities that can be realistically considered as candidates. These include countries classified as free in the Freedom House surveys but that have yet to experience a turnover, much less a double turnover, of elected government. Even Botswana and South Africa have yet to pass this test.

Within the current category of aspiring democracies, we can reasonably expect a process of further political liberalization and growth of civil society, including the emergence of interest group politics that will challenge continuing clientelist arrangements. Civil society will also become better organized in the countryside. The legislature and judiciary are likely to become stronger and more independent. Greater electoral turnover of government will occur simply because some countries, such as Kenya, Tanzania, and Zambia, have adopted term limits at the presidential level. Some members of this category will be promoted to the ranks of consolidated democracies. All or nearly all should see their performance improve.

Among semi-authoritarian states where democratization has stalled, the prospects for progress are much worse. This Africa consists of approximately 13 polities, all of which are in the lower range of the partly free category. As a result, there is a limited measure of political space open for the growth of civil society and the press on the one hand, and competitive electoral politics on the other. One should not write off this category, however. Rather, one should acknowledge that democratization in these states will be an especially protracted process.

The fourth and fifth Africas—those countries classified as not free and those mired in civil war—are obviously the least promising. The immediate challenges in countries classified as not free is halting human rights abuses and beginning a process of political liberalization, the first steps toward democratization. The first challenge for countries mired in civil war is to stop the fighting and reconstitute the state. State decay is probably the greatest problem for these two groups and must be addressed before any meaningful process of democratization can begin.

In conclusion, one can be cautiously optimistic about the prospects for further democratization in between one-third and one-half of the states on the African continent, but one must be realistic about the prospects in the remainder, especially the bottom one-third of countries. Future progress is likely to be incremental, played out over a long period of at least a decade or more. Moreover, progress will not occur on a linear basis through a series of well-defined stages but unevenly and haltingly. Further progress, especially in the third Africa of semi-authoritarian states and to a limited extent in the fourth category of states ranked as not free, will depend on the outcome of a continuous three-cornered struggle between authoritarians, patronage-seekers, and reformers. One must also expect some erosion or reversals among states currently regarded as semi-consolidated democracies and aspiring ones.

Finally, it is important that US policymakers and those of other established democracies seeking to support the process appreciate these constraints and are honest about the varied nature of the challenge. The most difficult task will be to nurture the process in states falling into the third and fourth categories. Indeed, there is already a tendency to over-celebrate progress in the first and second Africas while retreating from the challenges in the third and fourth groups. Such an approach will undermine both the accomplishments to date and the prospects for further democratic advance. Critics of a foreign policy that stresses democratization should also reflect on the fact that the development of democratic systems in the West evolved over a period of more than a century. While one can appreciate the sense of impatience among many realists, this is no time to walk away from the Kenyas, Ugandas, and Zimbabwes, no matter how halting the democratic process seems.

JOEL D. BARKAN is Professor of Political Science at the University of Iowa and Fellow at the Woodrow Wilson International Center for Scholars.

Past Successes, Present Challenges

Latin American Politics at the Crossroads

Gerardo Munck

As it enters the 21st century, Latin America is marked by the co-existence of democratic regimes alongside a range of deeply disturbing political and socio-economic factors, from rampant corruption and widespread violence to wrenching inequalities. This configuration is quite unique. Latin America is a world of stark contrasts, historic achievements, and colossal challenges. On the one hand, modern Latin America bears the distinction of being the most consistently democratic region outside the advanced industrial world. On the other hand, Latin America is characterized by negative features that contrast sharply not only with the wealthy Western European and English-speaking democracies but also, in some aspects, with other regions in the developing world. For example, one long-standing feature of Latin America is its infamous distinction as the most unequal region in the world.

This characterization has significant implications for ongoing discussions about the question of democratic governance. Specifically, it offers an essential baseline to the discussion of future scenarios facing Latin America, and it breaks with the all-too-common teleological view that sees Latin American countries as traveling along the path blazed by Western European democracies. By emphasizing democratization at the expense of other desirable changes toward good governance, peace, and inclusive prosperity, this characterization suggests that efforts to think about democratic governance in Latin America fail to note the distinctiveness of Latin American realities, and thus have limited use. Even more problematic, these approaches can lead to misguided analyses and policy suggestions.

Confronting the current reality of Latin America and leaving aside empty hopes of replicating the path followed by successful European cases is a sobering exercise, providing little ground for optimism. Unfortunately, the odds are tilted toward a continuation or worsening of the situation in Latin America. However, this exercise does point to two important suggestions. One is that the unfolding of an optimistic scenario hinges on the active construction of an alternative but realistic vision in addition to a social force capable of overriding powerful actors who have a stake in perpetuating current arrangements. A second is that the path to such an end runs through the institutions of democratic regimes and will only open inasmuch as Latin Americans engage in a profound re-evaluation of politics.

Historic Achievements

Starting in the late 1970s, well before the collapse of communism in the East, Latin America underwent a vast and positive political transformation. Closed and repressive military political regimes in South America came to an end with the re-establishment of democratically elected regimes. The war-torn societies of Central America moved toward peaceful conflict resolution and began what was in most instances their first real experience with democracy. The gradual political opening in Mexico culminated in 2000, when there was a changeover in the party holding the presidency for the first time in over 70 years. Thus, as the 21st century began, Latin Americans could claim several remarkable political achievements. A region that in the late 1970s was overwhelmingly under the control of authoritarian rulers—at that time, only Costa Rica, Colombia, and Venezuela bucked the trend— had become almost entirely democratic.

Exceptions to this generalization exist and should be noted. Most obviously, Cuba is still undemocratic, and the democracies installed in the Andean region in the 1980s were severely threatened and even collapsed during the 1990s. But the broad trend is still strong. Never before have so many Latin American countries been democratic, and never before have Latin American democracies proven so durable. Moreover, the consensus that democracy is the preferable form of regime has never been so widespread and so deep. This consensus has been evident in the seriousness with which Latin American countries have responded to each threat to the stability of democracy, starting with the 1991 coup in Haiti and the 1992 *auto-golpe* or "self-coup" by President Alberto Fujimori in Peru. The signing of the Inter-American Democratic Charter by all members of the Organization of American States in September 2001 further underlines this trend. The Charter, which declares in its first article that "the peoples of the Americas have a right to democracy," should not be considered anything more than a document expressing intent. Nonetheless, it is hard to envision governments of any region outside Western Europe that could muster the political will to take such a step. The democratic feats achieved by Latin Americans are of truly historical proportions.

Colossal Challenges

These impressive democratic achievements, however, have not led to unqualified popular approval of the current state of affairs in the region. Dissatisfaction among the population at large is clearly registered in the annual survey by Latinobarometer, an organization dedicated to assessing the sentiments of the populations of various Latin American countries. Moreover, those members of the international community who have turned their attention to issues of democratic governance are expressing growing worries. The problem at the root of these expressions of discontent and concern does not pertain directly to the democratic nature of the political regimes. Rather, the problem is that the democracies of Latin America have coexisted with some extremely troubling factors.

One such factor is the prevalence of corruption in Latin America. This fact is driven home by the 2001 Corruption Perceptions Index compiled by Transparency International. With the clear exception of Chile and the partial exceptions of Uruguay and Costa Rica, the other Latin American countries are ranked in the lower half of the countries in this index. The prevalence of corruption has even affected business practices to such an extent that it has been criticized by US corporations that resent falling behind due to their inability under US law to offer bribes as corporations of other national origins do. This is telling not just in that it offers some sense of the magnitude of the problem but, more significantly, because it reminds us that it takes two to tango: corruption is a matter of politicians and businesspeople. Such alliances of complicity are often ignored in discussions of corruption that offer little more than justified but essentially moralistic cri-

tiques of corrupt politicians. Nevertheless, these alliances are essential to an understanding of the nature and impact of corruption.

At a broad level, these alliances of complicity show that, for all the talk about neoliberalism since the 1980s, the form of capitalism that has developed in Latin America is actually political capitalism, in which many businesses adopt strategies oriented more toward securing rents from state action than toward gaining profits through legitimate competition in the market. Thus, the problem of corruption is directly linked with economic productivity. But the problem goes even deeper because the prevalence of corruption also does much to protect businesses and elites from contributing their fair share to national development. The failure of Latin American states to tax wealthy businesses fairly is an important factor behind both the fiscal crisis of the state and the International Monetary Fund-sanctioned remedy that budget deficits should not be reduced by raising taxes, something seen as practically unfeasible, but rather by slashing social programs. As egregious as the surface manifestation of corruption is, the roots of corruption run deeper, involving well established, powerful interests, and the effects reach more widely, affecting the prospects of economic development.

Latin America's failure to achieve the felicitous combination of democracy, good governance, peace, and inclusive prosperity that characterized post-World War II Western Europe is a fact that can no longer be ignored.

A second disturbing aspect of Latin American democracies is the widespread violation of civil rights. The problem is not one of state terror, which was practiced by authoritarian regimes in the 1960s and 1970s. Nor is it the result of warring parties bent on asserting their power, a problem that afflicts many developing countries. After all, within Latin America, only the unfortunate case of Colombia continues to fit the description of a war-torn society. Rather, the civil rights issues that are currently of concern to Latin Americans take two new forms. First, the problem involves acts of repression by agents of the state, such as the police, within the context of democratic regimes. Second, the problem is more a failure of the state—whether due to a lack of political will or to a lack of capacity—to ensure the physical security of its citizens.

Beyond noting the changing forms of violence in Latin America, it is imperative to highlight the astounding scale

POWER BROKERS

Who Rules?

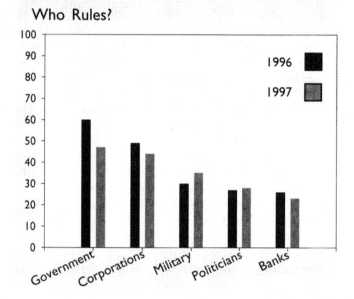

This figure represents Latin American citizens' assessment of the relative power of major social institutions as a percentage of respondents. It suggests a shift in power away from government and toward the military.

Latinobarometer at www.latinobarometer.org

of civil rights violations. A sad comment on global awareness of modern Latin America is that, the important efforts of human rights organizations notwithstanding, this darker side of Latin America is seriously underreported now that violence is not linked with overt ideological agendas or the struggle for democracy and few victims belong to the middle class. The facts, however, could hardly be more glaring. From a global perspective, only South Africa ranks as a more deadly society in terms of murder rate than El Salvador, Guatemala, and Colombia, with Brazil following close behind. And in sheer quantitative terms, the tragic reality is that in some countries more people die now as a result of these various forms of violence than did during the military dictatorships and civil wars of the 1970s and 1980s. It is hard to find more conclusive evidence that something is sorely wrong with current Latin American democracies.

Yet another negative factor must be noted: the obscene levels of inequality (and also of poverty) that are quite standard in the region. In this regard, the key fact is that Latin America's distinction as the most unequal region in the world continues to be matched only by the depressing lack of any sign that the situation is likely to improve. The extreme cases of Brazil and some Central American countries seem absolutely resistant to change. Where change has occurred, it has taken a negative direction. Thus, equality has markedly decreased in Argentina, which for a long time was a moderately egalitarian society, and in Colombia, which for many decades was a model of economic development that successfully generated economic growth and also made headway on the country's

debt. Even Chile, the Latin American economic success story since the mid-1980s, has made no noticeable progress toward reducing a grossly distorted pattern of income and wealth distribution.

In sum, Latin America is a region of stark contrasts. While the establishment of democratic regimes constitutes a historic landmark, these democracies coexist with many deeply troubling trends that define a set of colossal challenges remaining to be tackled. Unfortunately, it is all too obvious that achieving democracy has not offered a direct solution to other critical aspects of the broad problem of fostering democratic governance.

Rethinking Democracy

Latin American thinkers increasingly recognize that the replacement of authoritarian regimes with democratic ones, while an important step forward, has left some of the most critical determinants of the human condition unresolved. After an initial period when democracy was not only considered valuable in itself but also efficacious in achieving other normatively valued ends, the stubborn and persistent character of the unsavory correlates of Latin American democracy is becoming increasingly clear. Current attempts to rethink the question of democratic governance are decidedly pessimistic in tone. Latin America's failure to achieve the felicitous combination of democracy, good governance, peace, and inclusive prosperity that characterized post-World War II Western Europe is a fact that can no longer be ignored. Furthermore, the scenarios facing Latin America offer very little ground for optimism.

One scenario suggests that the most likely future of Latin America is a continuation of current trends. This scenario is based on the idea that the combination of democratic regimes and corruption, violence, and inequality, though morally perverse, can be understood as a system in equilibrium rather than a set of elements that are not compatible. As counterintuitive as this view may seem, it is based on the key insight that the problematic features of Latin America are not a mere matter of superficial coincidence but are actually firmly founded on the relations of power in Latin American societies. This scenario draws attention to a fundamental point that any discussion of possible scenarios must take into consideration: a defining feature of the current situation in Latin America is that those actors who have an interest in change do not have the power to bring about change, while those actors who have the power to introduce change do not have an interest in doing so.

Beyond this quite pessimistic scenario, it is useful to think of two others, which envision change but, assigning a different value to politics, have radically contrasting implications. One scenario of change can be projected on the basis of the growing anti-political sentiment among large segments of the population. These sentiments are, it should be stressed, rooted in legitimate grievances, especially concerning the corrupt practices of politicians. Nonetheless, anti-political attitudes also open the door to political manipulation by authoritarian actors and raise the prospect of an eventual breakdown of democratic regimes. There are signs in Latin America that the failure of democratic leaders to ensure good governance and deliver upon the demand for basic civil and social rights could very well lead to significant losses in terms of popular faith in democracy, the one aspect of democratic governance where the region has made advances.

The only optimistic scenario that is open to Latin America—and it is necessary to be blunt about this—hinges directly upon a re-evaluation of politics. This is not an argument against foreign actors and institutions but it does imply the need to develop national responses, based on a dialogue among national actors and ratified through democratic channels. At its core, this scenario rests on the simple but radical proposition that if Latin America is going to make any headway in confronting the colossal challenges it faces, the solution will have to be based on creative political action rather than on new attempts to replicate imported models. Furthermore, it will also have to be formulated and implemented through democratic institutions.

The odds that Latin Americans will be able to develop a distinctively Latin American solution to the problem of democratic governance are, however, low. The power of traditions and vested interests work in favor of the first scenario, which is a continuation of the status quo. The sentiment of the people increasingly seems to tilt toward the second scenario, a worsening of the current situation. But Latin Americans have, on occasion, constructed their history through democratic collective action when the odds were stacked against them. This is one of the lessons of the democratic struggles of the 1970s and 1980s. However, it is still an open question whether Latin Americans will be able to respond to the set of challenges they face by formulating an alternative that constitutes a real, viable solution and by constructing a social force powerful enough to democratically implement such an alternative.

The author would like to recognize his indebtedness to Guillermo O'Donnell and Sebastian Mazzucca, whose works have influenced many of the ideas discussed in this article.

GERARDO MUNCK is Associate Professor of Political Science at the University of Illinois at Urbana-Champaign.

UNIT 5

Population, Development, Environment, and Health

Unit Selections

Key Points to Consider

- How do current population trends represent a departure from the past?

- Why do some regard environmentalism as a new form of imperialism?

- How does environmental degradation affect developing countries?

- How might conservation create environmental refugees in the developing world?

- What are the consequences of the hazardous waste trade?

- What are the development consequences of the AIDS epidemic?

- What are the constraints on developing new drug therapies for diseases in poor countries?

 **Links: www.dushkin.com/online/**
These sites are annotated in the World Wide Web pages.

Earth Pledge Foundation
http://www.earthpledge.org

EnviroLink
http://envirolink.org

Greenpeace
http://www.greenpeace.org

Linkages on Environmental Issues and Development
http://www.iisd.ca/linkages/

Population Action International
http://www.populationaction.org

The Worldwatch Institute
http://www.worldwatch.org

The developing world's population continues to increase at an annual rate that exceeds the world average. The average fertility rate (the number of children a woman will have during her life) for developing countries is 3.1, while in sub-Saharan Africa it is as high as 5.8. Although growth has slowed somewhat, world population is currently growing at the rate of approximately 80 million per year, with most of this increase taking place in the developing world. Increasing population complicates development efforts, puts added stress on ecosystems, threatens food security, and affects population migration patterns.

World population surpassed 6 billion toward the end of 1999 and, if current trends continue, could reach 9.8 billion by 2050. Even if, by some miracle, population growth were immediately reduced to the level found in industrialized countries, the developing world's population would continue to grow for decades. Approximately one-third of the population in the developing world is under the age of 15, with that proportion jumping to 43% in the least developed countries. The population momentum created by this age distribution means that it will be some time before the developing world's population growth slows substantially. Some developing countries have achieved progress in reducing fertility rates through family planning programs, but much remains to be done. Funding for family planning programs has not met targets set at the 1994 International Conference on Population and Development, in part due to a lack of funding related to the controversy over abortion in the United States.

Over a billion people live in absolute poverty, as measured by a combination of economic and social indicators. As population increases, it becomes more difficult to meet the basic human needs of the developing world's citizens. Indeed, food scarcity looms as a major problem among the poor as population increases, production fails to keep pace, demand for water increases, per capita cropland shrinks, and prices rise. Larger populations of poor people also place greater strains on scarce resources and fragile ecosystems. Deforestation for agriculture and fuel has reduced forested areas and contributed to erosion, desertification, and global warming. Intensified agriculture, particularly for cash crops, has depleted soils. This necessitates increased fertilization, which is costly and also produces runoff that contributes to water pollution.

Economic development, regarded by many as a panacea, has not only failed to eliminate poverty but has exacerbated it in some ways. Ill-conceived economic development plans have diverted resources from more productive uses. There has also been a tendency to favor large-scale industrial plants that may be unsuitable to local conditions. Where economic growth has occurred, the benefits are often distributed inequitably, widening the gap between rich and poor. If developing countries try to follow Western consumption patterns, sustainable development will be impossible. Furthermore, economic growth without effective environmental policies can lead to the need for more expensive clean-up efforts.

Divisions between North and South on environmental issues became more pronounced at the 1992 Rio Conference on Environment and Development. The conference highlighted the fundamental differences between the industrialized world and

developing countries over causes of and solutions to global environmental problems. Developing countries pointed to consumption levels in the North as the main cause of environmental problems and called on the industrialized countries to pay most of the costs of environmental programs. Industrialized countries sought to convince developing countries to conserve their resources in their drive to modernize and develop. Divisions have also emerged on the issues of climate and greenhouse gas emissions. The Johannesburg summit on sustainable development, a follow-up to the Rio conference, grappled with many of these issues, achieving some modest success in focusing attention on water and sanitation needs.

Rural-to-urban migration has caused an enormous influx of people to the cities, lured there by the illusion of opportunity and the attraction of urban life. In reality, opportunity is limited. Nevertheless, most choose marginal lives in the cities rather than a return to the countryside. As a result, urban areas in the developing world increasingly lack infrastructure to support this increased population and also have rising rates of pollution, crime, and disease. Additional resources are diverted to the urban areas in an attempt to meet increased demands, often further impoverishing rural areas. Meanwhile, food production may be affected, with those remaining in rural areas having to choose either to farm for subsistence because of low prices or to raise cash crops for export.

Poverty and urbanization also contribute to the spread of disease. Environmental factors account for about one-fifth of all diseases in developing countries and also make citizens more vulnerable to natural disasters. Hazardous waste also represents a serious health and environmental risk. The HIV/AIDS epidemic has focused attention on public health issues, especially in Africa. Africans account for 70% of the 40 million AIDS cases worldwide. Aside from the human tragedy that this epidemic creates, the development implications are enormous. The loss of skilled and educated workers, the increase in the number of orphans, and the economic disruption that the disease causes will have a profound impact in the future. Meanwhile, the development and availability of drugs to treat AIDS and other common diseases in the developing world is constrained by patent and profitability concerns.

The Population
IMPLOSION

Be careful what you wish for. After decades of struggling to contain the global population explosion that emerged from the healthcare revolution of the 20th century, the world confronts an unfamiliar crisis: rapidly decreasing birthrates and declining life spans that might set back the progress of human development.

By Nicholas Eberstadt

It may not be the first way we think of ourselves, but almost all of us alive today happen to be children of the "world population explosion"—the momentous demographic surge that overtook the planet during the course of the 20th century. Thanks to sweeping mortality declines, human numbers nearly quadrupled in just 100 years, leaping from about 1.6 or 1.7 billion in 1900 to about 6 billion in 2000.

This unprecedented demographic expansion came to be regarded as a "population problem," and in our modern era problems demand solutions. By century's end, a worldwide administrative apparatus—comprised of Western foundations and aid agencies, multilateral institutions, and Third World "population" ministries—had been erected for the express purpose of "stabilizing" world population and was vigorously pursuing an international anti-natal policy; focusing on low-income areas where fertility levels remained relatively high.

To some of us, the wisdom of this crusade to depress birthrates around the world (and especially among the world's poorest) has always been elusive. But entirely apart from its arguable merit, the continuing preoccupation with high fertility and rapid population growth has left the international population policy community poorly prepared to comprehend (much less respond to) the demographic trends emerging around the world today—trends that are likely to transform the global population profile significantly over the coming generation. Simply put, the era of the worldwide "population explosion," the only demographic era within living memory, is coming to a close.

Continued global population growth, to be sure, is in the offing as far as the demographer's eye can see. It would take a cataclysm of biblical proportions to prevent an increase in human numbers between now and the year 2025. Yet global population growth can no longer be accurately described as

"unprecedented." Despite the imprecision of up-to-the-minute estimates, both the pace and absolute magnitude of increases in human numbers are markedly lower today than they were just a few years ago. Even more substantial decelerations of global population growth all but surely await us in the decades immediately ahead.

In place of the population explosion, a new set of demographic trends—each historically unprecedented in its own right—is poised to reshape, and recast, the world's population profile over the coming quarter century. Three of these emerging tendencies deserve special mention. The first is the spread of "subreplacement" fertility regimens, that is, patterns of childbearing that would eventually result, all else being equal, in indefinite population decline. The second is the aging of the world's population, a process that will be both rapid and extreme for many societies over the coming quarter century. The final tendency, perhaps the least appreciated of the three, is the eruption of intense and prolonged mortality crises, including brutal peacetime reversals in health conditions for countries that have already achieved relatively high levels of life expectancy.

For all the anxiety that the population explosion has engendered, it is hardly clear that humanity will be better served by the dominant demographic forces of the post-population-explosion era. Nobody in the world will be untouched by these trends, which will have a profound impact on employment rates, social safety nets, migration patterns, language, and education policies. In particular, the impact of acute and extended mortality setbacks is ominous. Universal and progressive peacetime improvements in health conditions were all but taken for granted in the demographic era that is now concluding; they no longer can be today, or in the era that lies ahead.

Where Have All the Children Gone?

2000 Fertility Rates

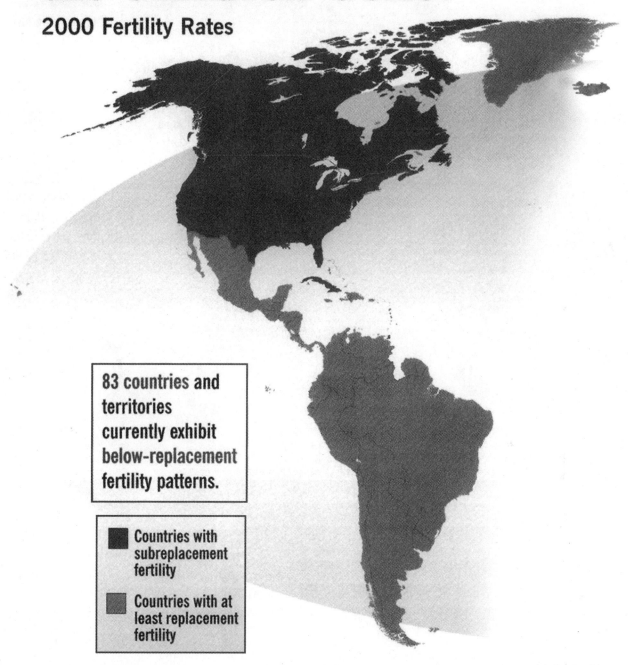

83 countries and territories currently exhibit below-replacement fertility patterns.

■ Countries with subreplacement fertility

▨ Countries with at least replacement fertility

(continued on next page)
Source: U.S. Bureau of the Census, International Data Base

THE GLOBAL BABY BUST

In arithmetic terms, the 20th-century population explosion was the result of improvements in health and the expansion of life expectancy. Human life expectancy at birth is estimated to have doubled or more between 1900 and 2000, shooting up from approximately 30 years to nearly 65 years. Population growth rates accelerated radically thanks to the concomitant plunge in death rates. Despite tremendous population growth, rough calculations suggest that the world's population would be over 50 percent larger today in the absence of any other demographic changes.

The world's population currently totals about 6 billion, rather than 9 billion or more, because fertility patterns also

Where Have All the Children Gone? 2000 Fertility Rates *Continued*

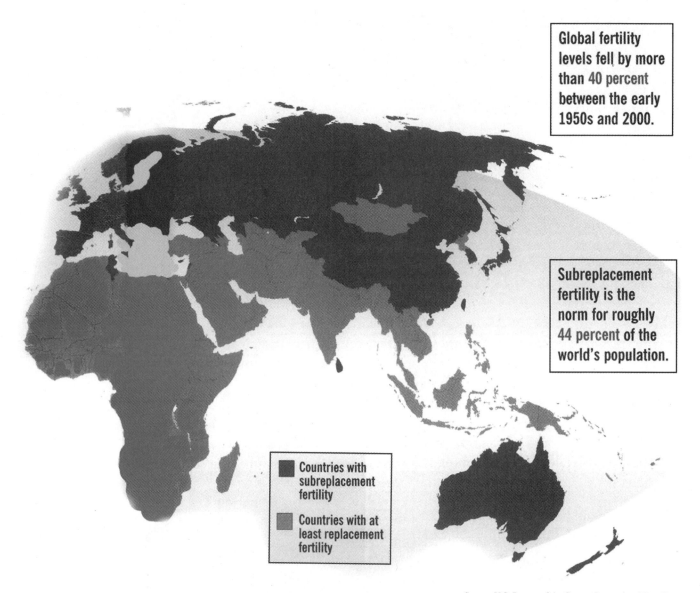

Global fertility levels fell by more than 40 percent between the early 1950s and 2000.

Subreplacement fertility is the norm for roughly 44 percent of the world's population.

■ Countries with subreplacement fertility

■ Countries with at least replacement fertility

Source: U.S. Bureau of the Census, International Data Base

changed over the course of the 20th century. And of all those diverse changes, without question the most significant was secular fertility decline: sustained and progressive reductions in family size due to deliberate birth control practices by prospective parents.

Within the full sweep of the human experience, secular fertility decline is very, very new. It apparently had not occurred in any human society until about two centuries ago in France. Since that beginning, secular fertility decline has spread steadily, if unevenly, embracing an ever rising fraction of the global population. In the final decades of the 20th century, subreplacement fertility made especially commanding advances: According to estimates and projections by the U.S. Census Bureau and the United Nations Population Division, fertility levels for the world as a whole fell by more than 40 percent between

the early 1950s and the end of the century—a drop equivalent to over two births per woman per lifetime.

Indeed, subreplacement fertility has suddenly come amazingly close to describing the norm for childbearing the world over. In all, 83 countries and territories are thought to exhibit below-replacement fertility patterns today [see map on this page]. The total number of persons inhabiting those countries is estimated at nearly 2.7 billion, roughly 44 percent of the world's total population.

Secular fertility decline originated in Europe, and virtually every population in the world that can be described as of European origin today reports fertility rates below the replacement level. But these countries and territories today currently account for only about a billion of the over 2.5 billion people living in "subreplacement regions." Below-replacement fertility is thus

The Population Explosion Fizzles
Fertility Rates For Populous Low-Income Countries: 1975, 2000, 2025

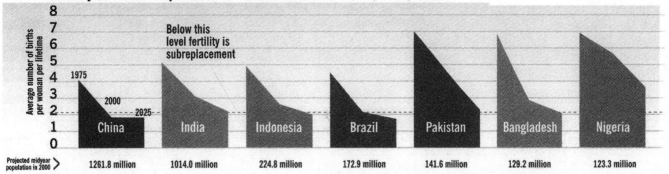

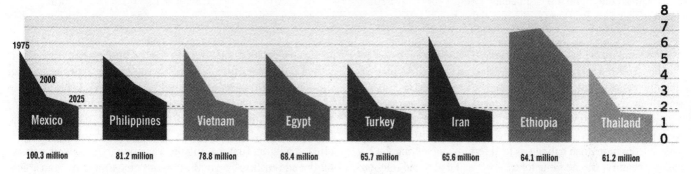

CHARTS BY TRAVIS C. DAUB

Source: U.S. Bureau of the Census, International Data Base, United Nations Population Division, World Population Prospects (New York: United Nations, 1998)
Note: 1975 rates interpolated from estimated 1970/75 levels. 2000 and 2025 rates are projected.

no longer an exclusively—nor even a predominantly—European phenomenon. In the Western Hemisphere, Barbados, Cuba, and Guadeloupe are among the Caribbean locales with fertility rates thought to be lower than that of the United States. Tunisia, Lebanon, and Sri Lanka have likewise joined the ranks of subreplacement fertility societies.

The largest concentration of subreplacement populations, however, is in East Asia. The first non-European society to report subreplacement fertility during times of peace and order was Japan, whose fertility rate fell below replacement in the late 1950s and has remained there almost continuously for the last four decades. In addition to Japan, all four East Asian tigers—Hong Kong, the Republic of Korea, Singapore, and Taiwan—have reported subreplacement fertility levels since at least the early 1980s. By far the largest subreplacement population is in China, where the government's stringent antinatal population control campaign is entering its third decade.

The singularity of the Chinese experience, however, should not divert attention from the breadth and scale of fertility declines that have been taking place in other low-income settings. A large portion of humanity today lives in countries where fertility rates are still above the net replacement level, but where secular fertility decline is proceeding at a remarkably rapid pace.

A glance at the 15 most populous developing countries illustrates the magnitude of fertility change over the last quarter century [see graph "The Population Explosion Fizzles"]. These countries account for about three quarters of the current population of the "less developed regions," and three fifths of

the total world population. In addition to China, Thailand is believed to be below the replacement level. Three other countries (Brazil, Iran, and Turkey) are thought to be just barely above the replacement level. Another four (Bangladesh, Indonesia, Mexico, and Vietnam) are slightly higher. Today, in other words, nine of the 15 largest developing countries are believed to register fertility levels lower than those that characterized the United States as recently as 1965. And over the last quarter century, fertility decline in this set of countries has been pronounced: In eight of those 15, fertility dropped by over half.

The regions where fertility levels remain highest, and where fertility declines to date have been most modest, are sub-Saharan Africa and the Islamic expanse to its north and east—more specifically, the Middle East. Those areas encompassed a total population of about 900 million in 2000, less than a fifth of the estimated total for less developed regions, and a bit under a seventh of the world total. Even for this grouping, however, the image of uniformly high "traditional" fertility patterns is already badly outdated. A revolution in family formation patterns has begun to pass through these regions. In 2000, in fact, the overall fertility level for North Africa—the territory stretching from Western Sahara to Egypt—was lower than the U.S. level of the early 1960s. Perhaps even more surprisingly, secular fertility decline appears to be unambiguously in progress in a number of countries in sub-Saharan Africa. For instance, Kenya's total fertility rate is believed to have dropped by almost four births per woman over the past 20 years.

Go Forth and Multiply

Population Projections from 2000

Region	Projected Midyear 2000 Population (in millions)	Projected 2025 Population (in millions)	Absolute Change 2000-2025 (in millions)	Percent Change 2000-2025
World	6,080	7,841	1,761	29
More Developed Countries	1,186	1,239	53	4
Less Developed Countries	4,895	6,602	1,707	35
Sub-Saharan Africa	661	1,071	410	62
North Africa	145	203	58	40
Middle East	171	280	109	64
Asia (excluding Middle East)	3,444	4,387	943	27
Latin America and the Caribbean	520	671	151	29

Source: U.S. Bureau of the Census, International Data Base

The remarkable particulars of today's global march toward smaller family size fly in the face of many prevailing assumptions about when rapid fertility decline can, and cannot, occur. Poverty and illiteracy (especially female illiteracy) are widely regarded as impediments to fertility decline. Yet, very low income levels and very high incidences of female illiteracy have not prevented Bangladesh from more than halving its total fertility rate during the last quarter century. By the same token, strict and traditional religious attitudes are commonly regarded as a barrier against the transition from high to low fertility. Yet over the past two decades, Iran, under the tight rule of a militantly Islamic clerisy, has slashed its fertility level by fully two-thirds and now apparently stands on the verge of subreplacement.

For many population policymakers, it has been practically an article of faith that a national population program is instrumental, if not utterly indispensable, to fertility decline in a low-income setting. Iran, for instance, achieved its radical reductions under the auspices of a national family planning program. (In 1989, after vigorous doctrinal gymnastics, the mullahs in Tehran determined that a state birth control policy would indeed be consistent with the Prophet's teachings.) But other countries have proven notable exceptions. Brazil has never adopted a national family planning program, yet its fertility levels have declined by well over 50 percent in just the last 25 years.

What accounts for the worldwide plunge in fertility now underway? The honest and entirely unsatisfying answer is that nobody really knows—at least, with any degree of confidence and precision. The roster of contemporary countries caught up in rapid fertility decline is striking for the absence of broad, obvious, and identifiable socioeconomic thresholds or common preconditions. (Reviewing the evidence from the last half century, the strongest single predictor for any given low-income country's fertility level is the calendar year: The later the year, the lower that level is likely to be.) If you can find the shared, underlying determinants of fertility decline in such disparate countries as the United States, Brazil, Sri Lanka, Thailand, and Tunisia, then your Nobel Prize is in the mail.

Two points, however; can be made with certainty. First, the worldwide drop in childbearing reflects, and is driven by, dramatic changes in desired family size. (Although even this observation only raises the question of why personal attitudes about these major life decisions should be changing so commonly in so many disparate and diverse locales around he world today.) Second, it is time to discard the common assumption, long championed by demographers, that no country has been modernized without first making the transition to low levels of mortality and fertility. The definition of "modernization" must now be sufficiently elastic to stretch around cases like Bangladesh and Iran, where very low levels of income, high incidences of extreme poverty, mass illiteracy, and other ostensibly "nonmodern" social or cultural features are the local norm, and where massive voluntary reductions in fertility have nevertheless taken place.

SEND YOUR HUDDLED MASSES ASAP

Barring catastrophe, the world's total population can be expected to grow substantially over the coming quarter century: U.S. Census Bureau projections for 2025 would place global population at over 7.8 billion, almost 30 percent larger than today. Yet, due to declining fertility, population growth is poised to decelerate markedly over the coming generation. The projected annual rate of world population growth in 2025 is just under 0.8 percent, considerably slower than the current projected rate of 1.3 percent, and far below the estimated 2.0 percent annual growth rate of the late 1960s. The great global birth wave will have crested and begun ebbing by 2025. In fact, by those projections, slightly fewer babies will be born worldwide in the year 2025 than in any year over the previous four decades.

The prospective pace of population growth for the different regions of the world is highly uneven over the coming generation [see table above]. The most dramatic increases will occur in sub-Saharan Africa, followed by countries in North Africa and the Middle East. By 2025 more people may be living in Africa than in all of today's "more developed countries" taken together.

This Old World

Percent of Population 65 and Above (Projected)

Country	2000	2025
Algeria	4.0	7.1
Brazil	5.3	11.3
China	7.0	13.5
Ethiopia	2.8	2.7
India	4.6	7.8
Iraq	3.1	4.3
Saudi Arabia	2.6	5.6
South Africa	4.8	9.6
Germany	16.2	23.1
Japan	17.0	27.6
Russia	12.6	18.5
Spain	16.9	23.5
United States	12.6	18.5

Source: U.S. Bureau of the Census, International Data Base

The natural growth of population in the more developed countries has essentially ceased. The overall increase in population for 2000 in these nations is estimated at 3.3 million people, or less than 0.3 percent. Two thirds of that increase, however, is due to immigration; the total "natural increase" amounts to just over 1 million. Over the coming quarter century, in the U.S. Census Bureau's projections, natural increase adds only about 7 million people to the total population of the more developed countries. And after the year 2017, deaths exceed births more or less indefinitely. Once that happens, only immigration on a scale larger than any in the recent past can forestall population decline. (The specter of population decline in more developed countries looms even larger if the United States, with its relatively high fertility level and relatively robust inflows of immigrants, is taken out of the picture. Excluding the United States, total deaths already exceed total births by almost half a million a year.)

For Europe as a whole (including Russia), the calculated long-term volume of immigration required to avert overall population decline is nearly double the recent annual level—an average of 1.8 million net newcomers a year, versus the roughly one million net entrants a year in the late 1990s. To prevent an eventual decline in the size of the 15 to 64 grouping (often termed the "working-age" population), Europe's net migration will have to nearly quadruple to a long-term average of about 3.6 million a year. Migration of this magnitude would change the face of Europe: By 2050, under these two scenarios, the descendants of present-day non-Europeans will account for approximately 20 to 25 percent of Europe's inhabitants.

Even more dramatic are the prospects for Japan, where current net migration levels are close to zero. To maintain total population size, Japan would have to accept a long-term average of almost 350,000 newcomers a year for the next 50 years,

and it would need nearly twice that number to keep its working-age population from shrinking. Under the first contingency, over a sixth of Japan's 2050 population would be descendants of present-day *gaijin* (foreigners); under the second contingency, that group would account for nearly a third of Japan's total population.

Europe and Japan will not lack immigration candidates in the years ahead. If Europe's needed immigration flows continue to come largely from North Africa, the Middle East, sub-Saharan Africa, and South Asia, those migrants will account for only about 3 to 7 percent of the population growth in their home countries. By the same token if Japan, for reasons of history and affinity, relies upon China and Southeast Asia for all its new national recruits, it will require just 2 to 4 percent of those countries' total envisioned population increase over the next 25 years. And as long as a huge income gap separates these more developed and less developed locales, there will be a compelling motive for such migration.

The issue clearly will not be supply, but rather demand. Will Western countries facing population decline opt to let in enough outsiders to stabilize their domestic population levels? Major and sustained immigration flows will entail correspondingly consequential long-term changes in a country's ethnic composition, with accompanying social alterations and adjustments. Such inflows will also require a capability to assimilate newcomers, so that erstwhile foreigners (and their descendants) can become true members of their new and chosen society.

The current outlook for "replacement migration" varies dramatically within the more developed regions. Throughout Europe, vocal (but still marginal) antiforeign political movements have taken the stage in recent years, while more tolerant sectors of the public have worried about the impact of immigration on their welfare states. Yet the continent, populated as it has been by successive historical flows of peoples, possesses traditions and capacities of assimilation that are not always fully appreciated.

The situation looks very different for Japan, where no major influxes of newcomers have been recorded over the past thousand years, and where the delicate distinctness of the Japanese *minzoku* (race) is a matter of intense, if not always enunciated, public consciousness. Despite reforms in Japanese immigration laws, a community of ethnic Koreans in Japan—many of them fourth-generation residents of the country—still does not enjoy Japanese citizenship. Indeed, Japan naturalizes fewer foreigners each year than tiny Switzerland.

It is extraordinarily difficult to imagine any circumstances under which the Japanese public might acquiesce in "replacement migration." Socially and politically, long-term demographic decline seems likely to be a much more acceptable alternative. But these are the only two choices, and over the coming decades all the more developed countries must decide between them. For all societies with long-term fertility rates significantly below the replacement level, the only alternative to an eventual decline of the total population—or of key age groups within that total population—is steady and massively enhanced immigration.

A GRAY WORLD

The world's population is set to age markedly over the coming generation: The longevity revolution of the 20th century has foreordained as much. The tempo of social aging, however, has been accelerated in many countries by extremely low levels of fertility. In 2025, there will likely remain a few pockets of the world in which populations will remain as youthful as those from earlier historical epochs. For example, the median age in sub-Saharan Africa in 2025 will be just 20 years, that is, as many people would be under 20 as over 20. (Such a profile probably characterized humanity from the Neolithic era up until the Industrial Revolution.) Throughout the rest of the world, however, the phenomenon of aging will transform the structure of national populations, often acutely.

Population aging will be most pronounced in today's more developed countries. By the U.S. Census Bureau's estimates, the median age for this group of countries today is about 37 years. In 2025, the projected median age will be 43.

Due to its relatively high levels of fertility and immigration (immigrants tend to be young), the population of the United States is slated to age more slowly than the rest of the developed world. By 2025, median age in the United States will remain under 39 years. For the rest of the developed world, minus the United States, median age will be approximately 45 years. And for a number of countries, the aging process will be even further advanced [see table "This Old World"].

Current developed countries grew rich before they grew old; many of today's low-income countries, by contrast, look likely to become old first.

In Germany, for example, the projected median age in the year 2025 is 46. Greece and Bulgaria are both ascribed median ages in excess of 47. Japan would have a median age of over 49. In this future Japan, more than a fifth of the citizenry would be over 70 years of age, and nearly one person in six would be 75 or older. In fact, persons 75 and older would outnumber children under 15 years of age.

Population aging, of course, will also occur in today's less developed regions. Current developed countries grew rich before they grew old; many of today's low-income countries, by contrast, look likely to become old first. One of the most arresting cases of population aging in the developing world is set to unfold in China, where relatively high levels of life expectancy, together with fertility levels suppressed by the government's resolute and radical population control policies, are transforming the country's population structure. Between 2000 and 2025, China's median age is projected to jump by almost 9 years. This future China would have one-sixth fewer children than contemporary China, and the 65-plus population would surge by over 120 percent, to almost 200 million. These senior citizens would account for nearly a seventh of China's total population.

Caring for the elderly will inexorably become a more pressing issue for China under such circumstances, but nothing remotely resembling a national pension system is yet in place in that country. Even with rapid growth over the next quarter century, China will still be a poor country in 2025. Coping with its impending aging problem promises to be an immense social and economic issue for this rising power.

DEATH MAKES A COMEBACK

Given the extraordinary impact of the 20th century's global health revolution, well-informed citizens around the world have come to expect steady and progressive improvement in life expectancies and health conditions during times of peace.

Unfortunately, troubling new trends challenge these happy presumptions. A growing fraction of the world's population is coming under the grip of peacetime retrogressions in health conditions and mortality levels. Long-term stagnation or even decline in life expectancy is now a real possibility for urbanized, educated countries not at war. Severe and prolonged collapses of local health conditions during peacetime, furthermore, is no longer a purely theoretical eventuality. As we look toward 2025, we must consider the unpleasant likelihood that a large and growing fraction of humanity may be separated from the planetary march toward better health and subjected instead to brutal mortality crises of indeterminate duration.

In the early post-World War II era, the upsurge in life expectancy was a worldwide phenomenon. By the reckoning of the U.N. Population Division, in fact, not a single spot on the globe had a lower life expectancy in the early 1970s than in the early 1950s. And in the late 1970s only two places on earth—Khmer Rouge-ravaged Cambodia and brutally occupied East Timor—had lower levels of life expectancy than 20 years earlier. In subsequent years, however, a number of countries unaffected by domestic disturbance and upheaval began to report lower levels of life expectancy than they had known two decades earlier. Today that list is long and growing. U.S. Census Bureau projections list 39 countries in which life expectancy at birth is anticipated to be at least slightly lower in 2010 than it was in 1990. With populations today totaling three quarters of a billion people and accounting for one eighth of the world's population, these countries are strikingly diverse in terms of location, history, and material attainment.

This grouping includes the South American countries of Brazil and Guyana; the Caribbean islands of Grenada and the Bahamas; the Micronesian state of Nauru; 10 of the 15 republics of the former Soviet Union; and 23 sub-Saharan African nations. As might be surmised from the heterogeneity of these societies, health decline and mortality shocks in the contemporary world are not explained by a single set of factors, but instead by several syndromes working simultaneously in different parts of the world to subvert health progress.

Russia has experienced a prolonged stagnation and even decline in life expectancy, and its condition illuminates the problems facing some of the other former Soviet republics [see graph "The Days of Our Lives"]. After recording rapid postwar reductions in mortality in the 1950s, Russian mortality levels stopped falling in the 1960s and began rising for broad groups of the population. By 1990, overall life expectancy at birth in Russia was barely as high as it had been 25 years earlier. With the end of communist rule in 1991, Russia suffered sudden and severe declines in mortality, from which it has not yet fully recovered. By 1999, overall life expectancy at birth in Russia had regressed to the point where it had been four decades earlier.

Although many aspects of Russia's continuing health crisis remain puzzling, it appears that lifestyle and behavioral risks—including heavy smoking and extremely heavy drinking—figure centrally in the shortening of Russian lives. A weak and rudderless public health system, combined with apparent indifference in Moscow to the nation's ongoing mortality crisis, also compromises health progress. Although Russia is an industrialized society with an educated population and a large indigenous scientific-technical cadre, such characteristics do not automatically protect a country from the sorts of health woes that have befallen the Russian Federation.

In sub-Saharan Africa, a different dynamic drives mortality crises: the explosive spread of the HIV/AIDS epidemic. In its most recent report, the Joint United Nations Programme on HIV/AIDS (UNAIDS) estimated that 2.8 million died of AIDS in 1999, 2.2 million in sub-Saharan Africa alone. UNAIDS also reported that almost 9 percent of the region's adult population is already infected with the disease. By all indications, the epidemic is still spreading in sub-Saharan Africa. As of 2000, UNAIDS projected that in several sub-Saharan countries, a 15-year-old boy today faces a greater than 50 percent chance of ultimately dying from AIDS—even if the risk of becoming infected were reduced to half of current levels.

HIV/AIDS may not be the only plague capable of wrenching down national levels of life expectancy over the coming quarter century.

Given sub-Saharan Africa's disappointing developmental performance and conspicuously poor record of governance over the post-independence period, the pervasive failure in this low-income area to contain a deadly but preventable contagion may seem tragic but unsurprising. Yet it is worth noting that the AIDS epidemic appears to have been especially devastating in one of Africa's most highly developed and best-governed countries: Botswana.

Unlike most of the region, Botswana is predominantly urbanized; its rate of adult illiteracy is among the subcontinent's very lowest; and over a generation in which sub-Saharan economic growth rates were typically negative, Botswana's was consistently positive. Yet despite such promising statistics, Botswana's population has been decimated by HIV/AIDS over

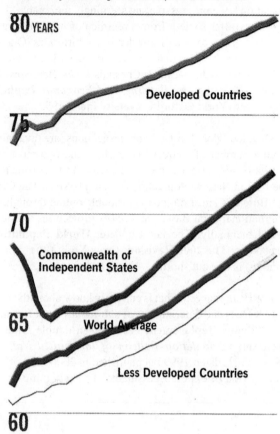

The Days of Our Lives
Life Expectancy at Birth, 1990-2025

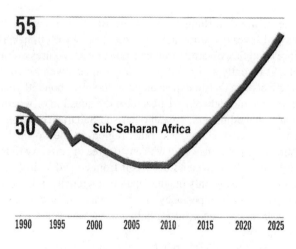

Source: Latest projections, U.S. Bureau of the Census, International Data Base

the last decade. Between 1990 and 2000, life expectancy in Botswana plummeted from about 64 years to about 39 years,

Want to Know More?

Over the last two centuries, world population trends have been dominated by what is known as "the demographic transition": the progressive and seemingly inexorable shift in society after society from a regimen of high birthrates and death rates to one of low death and birthrates. One of the best introductions to this powerful, but still mysterious, phenomenon is Jean-Claude Chesnais's *The Demographic Transition: Stages, Patterns, and Economic Implications* (New York: Oxford University Press, 1992).

Detailed global population projections are available from a number of sources, but perhaps the two most authoritative of these are the U.S. Bureau of the Census and the United Nations Population Division (UNPD). The Census Bureau's latest outlook is available online through its International Data Base. The UNPD's projections are published biennially. The latest update, **World Population Prospects: The 2000 Revisions** (New York: UNPD, forthcoming) is due out shortly.

For a summary of what is currently know about the global contours and determinants of sub-replacement fertility, read **"Below Replacement Fertility,"** a double issue of the UNPD's *Population Bulletin of the United Nations* (Nos. 40/41, dated 1999 but published in 2000) that offers a collection of expert assessments. The economic and

policy implications of social aging are examined in a series of publications and papers by the Orginisation for Economic Cooperation and Development (OECD), including **"Maintaining Prosperity in an Ageing Society"** (Paris: OECD, 1998) and **"Reforms for an Ageing Society"** (Paris: OECD, 2000). The UNPD report *Replacement Migration: Is It a Solution to Declining and Ageing Populations?* (New York: UNPD, March 2000), available online, examines the potential of migration to forestall depopulation and social aging. Demetrios G. Papademetriou and Kimberly A. Hamilton argue that Japan must end its restrictive immigration policies if it is to ensure its place as a global leader in *Reinventing Japan: Immigration's Role in Shaping Japan's Future* (Washington: Carnegie Endowment for International Peace, 2000).

On health setbacks in Russia and other post-communist countries, see Nicholas Eberstadt's *Prosperous Paupers and Other Population Problems* (New Brunswick: Transaction Publishers, 2000) and **"Russia: Too Sick To Matter?"** (*Policy Review*, June/July 1999).

•For links to relevant Web sites, as well as a comprehensive index of related FOREIGN POLICY articles, access **www.foreignpolicy.com.**

---

that is to say, by almost a quarter century. Recent projections for 2025 envision a life expectancy of a mere 33 years. If this projection proves accurate, Botswana will have a much lower life expectancy 25 years from now than it had nearly half a century ago.

One of the disturbing facets of the Botswanan case is the speed and severity with which life expectancy projections have been revised downward. Assuming most recent figures are accurate, as recently as 1994 expert demographers were overestimating Botswana's life expectancy for 2000 by about 30 years. Such abrupt and radical revisions raise the question of whether similar brutal adjustments await other sub-Saharan countries— or, for that matter, countries in other regions of the world. This question cannot be answered with any degree of certainty today, but we would be unwise to dismiss it from consideration. HIV/AIDS may not be the only plague capable of wrenching down national levels of life expectancy over the coming quarter century. Twenty-five years ago, HIV/AIDS had not even been identified and diagnosed.

Surprisingly, sub-Saharan Africa's AIDS catastrophe is not projected to alter the region's population totals dramatically. That speaks to the extraordinary power of high fertility levels. Given the region's current and prospective patterns of childbearing, the subcontinent's population totals in 2025 may prove

to be unexpectedly insensitive to the scope or scale of the disasters looming ahead. Yet it is the mortality patterns that will do much to define the quality of life for those human numbers— and to circumscribe their economic and social potential.

THE SHAPE OF THINGS TO COME

Looking toward 2025, we must remember that many 20th-century population forecasts and demographic assessments proved famously wrong. Depression-era demographers, for example, incorrectly predicted depopulation for Europe by the 1960s and completely missed the "baby boom." The 1960s and 1970s saw dire warnings that the "population explosion" would result in worldwide famine and immiseration, whereas today we live in the most prosperous era humanity has ever known. In any assessment of future world population trends and consequences, a measure of humility is clearly in order.

Given today's historically low death rates and birthrates, however, the arithmetic fact is that the great majority of people who will inhabit the world in 2025 are already alive. Only an apocalyptic disaster can change that. Consequently, this reality provides considerable insight into the shape of things to come.

By these indications, indeed, we must now adapt our collective mind-set to face new demographic challenges.

A host of contradictory demographic trends and pressures will likely reshape the world during the next quarter century. Lower fertility levels, for example, will simultaneously alter the logic of international migration flows and accelerate the aging of the global population. Social aging sets in motion an array of profound changes and challenges and demands far-reaching adjustments if those challenges are to be met successfully. But social aging is primarily a consequence of the longer lives that modern populations enjoy. And the longevity revolution, with its attendant enhancements of health conditions and individual capabilities, constitutes an unambiguous improvement in the human condition. Pronounced and prolonged mortality setbacks portend just the opposite: a diminution of human well-being, capabilities, and choices.

It is unlikely that our understanding of the determinants of fertility, or of the long-range prospects for fertility, will advance palpably in the decades immediately ahead. But if we wish to inhabit a world 25 years from now that is distinctly more humane than the one we know today, we would be well advised to marshal our attention to understanding, arresting, and overcoming the forces that are all too successfully pressing for higher levels of human mortality today.

Nicholas Eberstadt holds the Henry Wendth Chair in Political Economy at the American Enterprise Institute in Washington, D.C.

Saving the planet:
Imperialism in a green garb?

Developing countries feel that protecting the world's resources is just another way for rich nations to retain the upper hand in the international trade game

SHIRAZ SIDHVA *Unesco Courier* JOURNALIST

For nearly a decade, international efforts to address global environmental concerns have been frustrated by a deep rift in perceptions between rich and poor countries. Economists and environmentalists in developing nations argue that the agenda for environmental negotiations is almost exclusively driven by the North. Under the pretext of saving the planet, they say, the industrialized world is wielding a new brand of dominance, "ecoimperialism."

Developing countries like India and China continue to resist global environmental protocols, like the 1989 Montreal Accord to cut the production of CFC gases (used for example in refrigerators) by 50 percent, or the Clean Development Mechanism (CDM), part of the climate change negotiations initiated under the 1997 Kyoto Protocol. "These are viewed as instruments to make the Third World pay for damages caused primarily by the North," says Indian environmentalist Vandana Shiva. India and China account for two percent of CFC consumption, while the United States consumes 29 percent. "This 'ecoimperialism' undermines national sovereignty, while generating new costs for those once marginalized by colonialism," claims Shiva.

The spectre of imperialism is likely to vitiate the next round of climate change talks in Bonn (Germany) this July, when policymakers finalize the terms on which the CDM will be implemented. Negotiated by industrialized countries to gain some flexibility in meeting the emission reduction targets pledged in Kyoto, few issues in recent environmental diplomacy are proving as contentious.

Critics say the mechanism is the latest in a string of attempts to dominate poor countries, which are being virtually "bribed" so that rich nations can continue business as usual. By financing forestry schemes and other energy-efficient projects, industrialized countries could exploit the mechanism to avoid reducing their own greenhouse gases. Environmentalists fear this could turn the Amazon and other primeval forests into "carbon sinks" to absorb pollution, but with side effects which disregard the developmental needs of southern countries.

Lopsided negotiations

"The Northern bias continues to dominate discussion of the global climatic crisis," explains Shiva. "The threat to the atmospheric commons has been building over centuries, mainly because of industrial activity in the North. Yet discussions seem to focus more on developing countries: the North refuses to assume extra responsibility for cleaning up the atmosphere. No wonder the Third World cries foul when it is asked to share the costs."

"The whole effort to bring about ecological change is very one-sided," says Chow Kee, who represents Malaysia at the climate negotiations. "The developed countries don't want to give up their extravagant lifestyles, but plan to curtail our development."

Beyond negotiations on climate, efforts to link environmental concerns to trade are sparking more allegations of imperialism. "There is an attempt by rich countries to stunt the growth of developing nations like India, and we are fighting it tooth and nail," says Pramod Mahajan, India's minister for information technology. "They are practising protectionism under the garb of environmental protection." Economists argue that sanctions could spell further economic marginalization for developing countries, which often lack the means to set up expensive quality-control systems.

The 1989 Basel Convention, for instance, imposed restrictions on trade in scrap metals and recyclable materials, claiming they were hazardous to the environment. Economists say it prohibits poor countries from competing in the lucrative world market for computer parts, scrap metals, and recyclable products.

Other examples of trade restrictions are cited. In the early 1990s, Malaysia and Indonesia fought to overturn an ecolabelling law introduced by Austria ostensibly to safeguard the Asian rain forests. Austria refused to import timber that was not from sustainably managed forests, but no such curbs existed for wood from temperate areas. The protectionist flavour of the measure was overt, and Austria eventually revoked it.

In other trade-environment disputes over the last decade, the United States has been accused of protectionism in banning the import of Mexican tuna because dolphins were getting ensnared and killed in nets meant for the fish. Shrimp from India, Pakistan, Thailand and Malaysia, which paid no heed to sea-turtle protection, were similarly banned in 1996. The sanctions may have been motivated by a desire to protect dolphins and turtles, but the poorer countries claimed that they were a pretext for suppressing competition in the global fish market.

A green agenda to stop growth?

Deepak Lal, professor of international development studies at the University of California, Los Angeles, cites these examples when he describes the green movement as "the new secular religion." He says "green imperialists" are a new avatar of the "white man's burden," set to impose their values on the world.

According to Lal, rules restricting trade through the Basel Convention or attempts to ban genetically modified foods are designed to exclude poorer countries from world markets. "I look upon the green agenda as ultimately trying to stop growth in the Third World. And that means condemning three-quarters of the world's population to continuing poverty."

Other voices in the South, however, argue that environmental controls like the CDM are not all bad. Developing countries, say experts, will receive $5 to $17 billion to fund climate-friendly technologies. "The CDM gives us an opportunity to invest in projects that promote sustainable development. If incidentally they also reduce emissions, we shouldn't quarrel with the fact," says Dr R.K.Pachauri, of the Tata Energy Research Institute, New Delhi. "As sovereign nations, we should be confident about choosing projects, like renewable energy projects, that we would in any case want to invest in."

The rules of the CDM have yet to be established, and Pachauri urges developing countries not to squander the chance to influence their formulation.

But Anil Agarwal, director of the Delhi-based Centre for Science and Environment, insists that the CDM has neglected the concerns of poor countries. "The CDM still begs the long-term question about when and how developing countries will take on commitments of their own to reduce emissions."

Not all experts in developing countries, however, shun global environmental controls. Indian economist and newspaper editor Swaminathan S. Aiyar argues that there is "much to be learnt from the rest of the world." Like British colonial rule, which brought with it some desirable elements like democracy, civil and gender rights to India, the new crusaders could impart useful technologies, he writes in the *Times of India*. "Instead of rejecting wholesale what green imperialists say, we need to extract what is of value, and reject the dross."

Though the die are loaded against developing countries during negotiations, part of the fault lies with them, say experts. "The North is aggressive and assertive about what it wants, and comes well-prepared to negotiations," laments Agarwal. "Developing countries are comparatively disorganized and unclear about their objectives."

The environment, Pachauri explains, is a low priority for politicians in developing countries. Lack of cooperation between these nations is another handicap. A dearth of resources and language barriers come in the way. But as developing countries know, they will be hardest hit by global warming. In the end, counsels Agarwal, it is up to them to devise ways to protect our common future.

"Getting into the other's shoes"

Ecuador's former environment minister Yolanda Kakabadse once said that "my heart is in conservation, but my head tells me I must be fair to my country." Today, as head of the World Conservation Union*, she calls for a better understanding between North and South**

INTERVIEW BY SHIRAZ SIDHVA *UNESCO COURIER* JOURNALISt

Do you believe that the current environmental agenda is increasingly driven by rich countries?

A majority of industrialized countries are more active than developing ones in international debates. For one, they have the means: it's so costly to participate that you often find a huge team of highly technical people from developed countries and few representatives from developing ones. Recent debates on climate change and biodiversity have been clear signs that some countries in the South might not have the capacity to enter all the technical debates.

Do you believe that countries like Ecuador are targets of eco-imperialism?

I think the term is overrated. I'm not sure we're sending out the right message with it. There are clearly signs of developed countries trying to impose on others. But on the other hand, developing countries have gained a lot of information about environmental protection over the past decade through these negotiations. There is obviously an interest on everyone's part to see that the natural resources and the environment of the planet are dealt with. And most of these resources are concentrated in the tropical areas, where many developing countries are.

It's not just a North-South issue: sometimes I see signs of "imperialism" within our own countries, when a sector or group tries to impose a management pattern or policy.

As an activist, you campaigned against shrimp farming and had to defend that industry when you were a minister (1998–2000). How did you reconcile that?

Is it fair to tell Ecuador not to harvest shrimps—a major export—because the mangroves are an important eco-system, or should the answer be to explore whether it is possible to harvest shrimps and protect mangroves at the same time? I'm absolutely convinced that you can do both, at least in the case of shrimps. You just move the shrimp pools a little further back from the mangroves and achieve the two goals.

Did you actually manage to do that?

We did in some places, but the majority of the pools had already been built where the mangroves were. It was not just the industry that was to blame: it's only in the past decade that people in the North and South have become aware of the importance of mangrove conservation.

How would you have reacted if other countries had sought to impose a trade ban on certain products from Ecuador?

I would be totally against trade sanctions. We destroyed the mangroves without realizing their value. To come 30 years later and blame us for doing something wrong is not the right attitude. Instead we should come up with a way to stop environmental damage while enhancing the capacity of a country like Ecuador to produce goods for the world market.

Has the relationship between North and South evolved on environmental issues?

There is a better understanding in the North that imposing solutions won't take us anywhere, that teamwork is essential. The South, meanwhile, has realized that it does hold some decision-making power. Developed countries certainly have a better appreciation of poor countries' concerns, partly because more information is available. The building of partnerships between the two sides has also increased.

Do you believe that conflicting interests over the conventions currently being formulated can be resolved?

The big challenge is how to strike a balance between conservation and development—sustainable development. That is what we must fight for. Also, there's no process in the world that doesn't involve conflict. Conflicts aren't necessarily bad, they arise from differences of opinion based on different cultures, expectations, backgrounds, drives and ambitions. We have to invest in generating the capacity worldwide to manage these conflicts and understand each other's points of view better.

What would you advise both side to do?

You have to understand the other culture, and both sides must do this by getting into the other's shoes, which entails looking at values, needs and conditions within any culture. That would improve the relationships of different continents an countries dramatically.

The World Conservation Union (IUCN) is an umbrella organization of nearly a thousand NGOs, based in Gland (Switzerland). http://www.iucn.org

Local difficulties

Greenery is for the poor too, particularly on their own doorstep

W HY should we care about the environment? Ask a European, and he will probably point to global warming. Ask the two little boys playing outside a newsstand in Da Shilan, a shabby neighbourhood in the heart of Beijing, and they will tell you about the city's notoriously foul air: "It's bad—like a virus!"

Given all the media coverage in the rich world, people there might believe that global scares are the chief environmental problems facing humanity today. They would be wrong. Partha Dasgupta, an economics professor at Cambridge University, thinks the current interest in global, future-oriented problems has "drawn attention away from the economic misery and ecological degradation endemic in large parts of the world today. Disaster is not something for which the poorest have to wait; it is a frequent occurrence."

Every year in developing countries, a million people die from urban air pollution and twice that number from exposure to stove smoke inside their homes. Another 3m unfortunates die prematurely every year from water-related diseases. All told, premature deaths and illnesses arising from environmental factors account for about a fifth of all diseases in poor countries, bigger than any other preventable factor, including malnutrition. The problem is so serious that Ian Johnson, the World Bank's vice-president for the environment, tells his colleagues, with a touch of irony, that he is really the bank's vice-president for health: "I say tackling the underlying environmental causes of health problems will do a lot more good than just more hospitals and drugs."

The link between environment and poverty is central to that great race for sustainability. It is a pity, then, that several powerful fallacies keep getting in the way of sensible debate. One popular myth is that trade and economic growth make poor countries' environmental problems worse. Growth, it is said, brings with it urbanisation, higher energy consumption and industrialisation—all factors that contribute to pollution and pose health risks.

In a static world, that would be true, because every new factory causes extra pollution. But in the real world, economic growth unleashes many dynamic forces that, in the longer run, more than offset that extra pollution. As chart 5 " Dangers old

and new" makes clear, traditional environmental risks (such as water-borne diseases) cause far more health problems in poor countries than modern environmental risks (such as industrial pollution).

Rigged rules

However, this is not to say that trade and economic growth will solve all environmental problems. Among the reasons for doubt are the "perverse" conditions under which world trade is carried on, argues Oxfam. The British charity thinks the rules of trade are "unfairly rigged against the poor", and cites in evidence the enormous subsidies lavished by rich countries on industries such as agriculture, as well as trade protection offered to manufacturing industries such as textiles. These measures hurt the environment because they force the world's poorest countries to rely heavily on commodities—a particularly energy-intensive and ungreen sector.

Mr Dasgupta argues that this distortion of trade amounts to a massive subsidy of rich-world consumption paid by the world's poorest people. The most persuasive critique of all goes as follows: "Economic growth is not sufficient for turning environmental degradation around. If economic incentives facing producers and consumers do not change with higher incomes, pollution will continue to grow unabated with the growing scale of economic activity." Those words come not from some anti-globalist green group, but from the World Trade Organisation.

Another common view is that poor countries, being unable to afford greenery, should pollute now and clean up later. Certainly poor countries should not be made to adopt American or European environmental standards. But there is evidence to suggest that poor countries can and should try to tackle some environmental problems now, rather than wait till they have become richer.

This so-called "smart growth" strategy contradicts conventional wisdom. For many years, economists have observed that as agrarian societies industrialised, pollution increased at first, but as the societies grew wealthier it declined again. The trouble

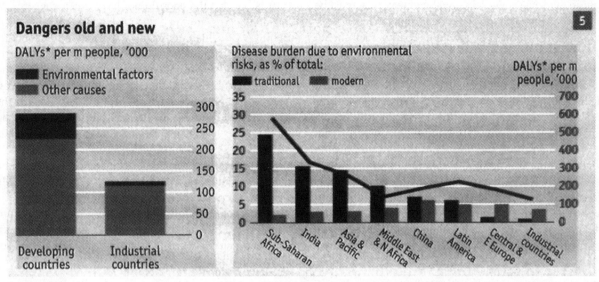

Source: World Bank

*Disability Adjusted Life Years - number of years lived with disability and years lost to premature death.

is that this applies only to some pollutants, such as sulphur dioxide, but not to others, such as carbon dioxide. Even more troublesome, those smooth curves going up, then down, turn out to be misleading. They are what you get when you plot data for poor and rich countries together at a given moment in time, but actual levels of various pollutants in any individual country plotted over time wiggle around a lot more. This suggests that the familiar bell-shaped curve reflects no immutable law, and that intelligent government policies might well help to reduce pollution levels even while countries are still relatively poor.

Developing countries are getting the message. From Mexico to the Philippines, they are now trying to curb the worst of the air and water pollution that typically accompanies industrialisation. China, for example, was persuaded by outside experts that it was losing so much potential economic output through health troubles caused by pollution (according to one World Bank study, somewhere between 3.5% and 7.7% of GDP) that tackling it was cheaper than ignoring it.

One powerful—and until recently ignored—weapon in the fight for a better environment is local people. Old-fashioned paternalists in the capitals of developing countries used to argue that poor villagers could not be relied on to look after natural resources. In fact, much academic research has shown that the poor are more often victims than perpetrators of resource depletion: it tends to be rich locals or outsiders who are responsible for the worst exploitation.

Local people usually have a better knowledge of local ecological conditions than experts in faraway capitals, as well as a direct interest in improving the quality of life in their village. A good example of this comes from the bone-dry state of Rajasthan in India, where local activism and indigenous know-how about rainwater "harvesting" provided the people with reliable water supplies—something the government had failed to do. In Bangladesh, villages with active community groups or concerned mullahs proved greener than less active neighbouring villages.

Community-based forestry initiatives from Bolivia to Nepal have shown that local people can be good custodians of nature. Several hundred million of the world's poorest people live in and around forests. Giving those villagers an incentive to preserve forests by allowing sustainable levels of harvesting, it turns out, is a far better way to save those forests than erecting tall fences around them.

To harness local energies effectively, it is particularly important to give local people secure property rights, argues Mr Dasgupta. In most parts of the developing world, control over resources at the village level is ill-defined. This often means that local elites usurp a disproportionate share of those resources, and that individuals have little incentive to maintain and upgrade forests or agricultural land. Authorities in Thailand tried to remedy this problem by distributing 5.5m land titles over a 20-year period. Agricultural output increased, access to credit improved and the value of the land shot up.

Name and shame

Another powerful tool for improving the local environment is the free flow of information. As local democracy flourishes, ordinary people are pressing for greater environmental disclosure by companies. In some countries, such as Indonesia, governments have adopted a "sunshine" policy that involves naming and shaming companies that do not meet environmental regulations. It seems to achieve results.

Bringing greenery to the grass roots is good, but on its own it will not avert perceived threats to global "public goods" such as the climate or biodiversity. Paul Portney of Resources for the Future explains: "Brazilian villagers may think very carefully and unselfishly about their future descendants, but there's no reason for them to care about and protect species or habitats that no future generation of Brazilians will care about."

That is why rich countries must do more than make pious noises about global threats to the environment. If they believe

165

that scientific evidence suggests a credible threat, they must be willing to pay poor countries to protect such things as their tropical forests. Rather than thinking of this as charity, they should see it as payment for environmental services (say, for carbon storage) or as a form of insurance.

In the case of biodiversity, such payments could even be seen as a trade in luxury goods: rich countries would pay poor countries to look after creatures that only the rich care about. Indeed, private green groups are already buying up biodiversity "hot spots" to protect them. One such initiative, led by Conservation International and the International Union for the Conservation of Nature (IUCN), put the cost of buying and preserving 25 hot spots exceptionally rich in species diversity at less than $30 billion. Sceptics say it will cost more, as hot spots will need buffer zones of "sustainable harvesting" around them. Whatever the right figure, such creative approaches are more likely to achieve results than bullying the poor into conservation.

It is not that the poor do not have green concerns, but that those concerns are very different from those of the rich. In Beijing's Da Shilan, for instance, the air is full of soot from the many tiny coal boilers. Unlike most of the neighbouring districts, which have recently converted from coal to natural gas, this area has been considered too poor to make the transition. Yet ask Liu Shihua, a shopkeeper who has lived in the same spot for over 20 years, and he insists he would readily pay a bit more for the cleaner air that would come from using natural gas. So would his neighbours.

To discover the best reason why poor countries should not ignore pollution, ask those two little boys outside Mr Liu's shop what colour the sky is. "Grey!" says one tyke, as if it were the most obvious thing in the world. "No, stupid, it's blue!" retorts the other. The children deserve blue skies and clean air. And now there is reason to think they will see them in their lifetime.

From *The Economist*, July 6, 2002, pp. 11-13. © 2002 by The Economist, Ltd. Distributed by the New York Times Special Features. Reprinted by permission.

POVERTY

AND ENVIRONMENTAL DEGRADATION:
CHALLENGES WITHIN THE GLOBAL ECONOMY

by Akin L. Mabogunje

THE LINK BETWEEN DEEPENING POVERTY AND ENVIRONMENTAL degradation confronts anyone living in a developing country on a daily basis. Perhaps most striking is the increased visibility and extent of both of these phenomena since the end of the Second World War, despite the organized efforts of the United Nations and related international agencies to promote global development in a series of well-publicized social and economic development efforts. Although some countries have made significant progress in this respect, and some individual groups and social classes have escaped poverty, millions remain mired in desperation.

According to the *World Development Report 2000/2001*, 1.2 billion of the world's 6 billion people live on less than $1 per day, and 2.8 billion people, or almost half of the of the world's population, live on less than $2 per day. In 1998, at least 40 percent of the population in South Asia and more than 46 percent in sub-Saharan Africa was living on less than $1 per day.[1]

However, poverty can no longer be adequately defined in terms of income alone—it must be recognized as a multifaceted phenomenon. In an attempt to represent the complexity of poverty, the United Nations Development Programme (UNDP) distinguishes between income poverty and human poverty.[2] According to UNDP, income poverty occurs when the income level of an individual falls below a nationally defined poverty line. Income-based measures of poverty attempt to express the failure of economic resources to meet basic minimum needs—especially food; they also facilitate comparative assessments of countries' progress in poverty reduction. For example, the World Bank has established an international poverty line at $1 per day per person for the purpose of international comparison.

UNDP defines human poverty as the denial or deprivation of opportunities and choices that would enable an individual "to lead a long, healthy, creative life and to enjoy a decent standard of living, freedom, dignity, self-respect and the respect of others."[3] To measure human poverty, UNDP proposes three indices: The first relates to an individual's vulnerability to death at a relatively early age and is measured by the percentage of the population expected to die before the age of 40; the second relates to an individual's exclusion from the world of reading and communication and is measured by the percentage of adults who are illiterate; the third index relates to the standard of living and is measured by the percentage of people with access to health services and safe water and the percentage of malnourished children less than 5 years old.

The failure of these definitions to relate poverty to the environment reflects a shortcoming in the approach to solving these problems. In an address to the high session of the Economic and Social Council of the United Nations in June 1993, Boutros Boutros-Ghali, then UN secretary-general, demonstrated an effort to change this approach when he emphasized that poverty is only one aspect of the generally dehumanizing phenomenon of deprivation:

> *Deprivation is a multidimensional concept. In the sphere of economics, deprivation manifests itself as poverty; in politics, as marginalization; in social relations, as discrimination; in culture, as rootlessness; in ecology, as vulnerability. The different forms of deprivation reinforce one another Often the same household, the same region, the same country is the victim of all these forms of deprivation. We must attack deprivation in all its forms. None of the other dimensions of deprivation, however, can be tackled unless we address the problem of poverty and unemployment.*[4]

This conceptualization places poverty in a broader web of deprivation. Because the poor are often cut off from the decisionmaking process of their communities, discriminated against by other stations in society, cut off from abiding roots in the community, and relegated to occupy environmentally unsafe areas of societal space, the solutions to their dilemma require a multifaceted approach based on an understanding of broader deprivation.

Globalization, Poverty, and the Environment

Perhaps the single most important development in the world today is what is generally referred to as "globalization." Globalization is partly a result of the tremendous advances in information technology that have, in effect, shrunk the world and linked distant parts of the Earth, creating global relationships. Globalization is also a result of the expanding reach of the capitalist mode of production. Changes in technology and manufacturing organization have fostered the emergence of transnational corporations that have been able to amass wealth within and beyond individual nation-states such that their roles in the economy of their countries and of other countries rival or exceed those of nation-states. Communication technologies have enabled enormous financial resources to be moved from one end of the world to another in a matter of minutes. The instantaneous transfer of vast economic resources has the potential to make or break the economic fortunes of countries and affect the lives and employment opportunities of large numbers of people. Therefore, nation-states are forced to vigorously compete for foreign investment to enhance the rate of growth of their economies. To attract these investments, nation-states must achieve minimum levels of infrastructure development and, more importantly, maintain a certain degree of economic, social, and political stability.

Manufacturing operations have evolved from the classic model (epitomized by the vehicle production operations of Henry Ford in the early twentieth century), in which a huge factory produces all the components as well as the end-product, to an increasingly flexible method of production whereby components are produced in different countries and then assembled at another location close to the market site.

Therefore, although globalization exacerbates poverty in some places and among some groups, it also has a democratizing potential that may be essential to breaking out of poverty. Developing economies cannot get out of poverty without attracting transnational corporations, but at the same time, they cannot attract these corporations unless they have achieved a certain level of development. Because many developing-country economies are in the early stage of the transformation to a free-market, capitalist mode, the conditions necessary to attract international investment are difficult to fulfill, particularly for South Asia and sub-Saharan Africa.

Integrating Developing Economies into the Global Market

In the early stage of capitalism, it is crucial to treat the different factors of production—land, labor, capital, and Entrepreneurship—as commodities, so they can be brought to the free, self-regulating market of supply and demand. To enter the global economy, the poor of the developing world must transform themselves from self-employed peasants to wage-earning labor. However, the price for their labor must be determined by supply and demand. Although labor is treated as a discrete commodity, in reality it is an attribute of human beings. Other commodities can be shoved about, used indiscriminately, and even left unused, but labor cannot be treated in this manner without severe human consequences. For this reason, when the capitalist mode of production emerged in Europe, masses of people were thrown into abject and humiliating poverty. Developing economies are likely to face similar consequences as they are integrated into the global capitalist market.

Commenting on the capitalist system in England in the first half of the nineteenth century, Karl Polanyi noted,

The system, in disposing of a man's labour power, would incidentally dispose of the physical, psychological and moral entity, "man," attached to that tag. Robbed of the protective covering of cultural institutions, human beings would per-

ish from the effects of social exposure. They would die as victims of acute social dislocation through vice, perversion, crime and starvation. Nature would be reduced to its elements, neighbourhoods and landscapes defiled, rivers polluted, military safety jeopardized, the power to produce food and raw materials destroyed.... Undoubtedly, labour, land and money markets are essential to a market economy. But no society could stand the effects of such a system of crude fictions even for the shortest stretch of time, unless its human natural substance as well as its business organization was protected against the ravages of this satanic mill.[5]

In the face of such social consequences, it is no wonder that societies in Western Europe and later in North America, from the middle of the twentieth century until today, were forced to protect individuals from full exposure to the ravages of the free-market economy. The protections came in the form of trade unions and centralized representative governments committed to restraining the socially disruptive potency of capitalism through regulations such as tariff laws, factory laws, social security and pension laws, labor codes, and a host of other social welfare legislation.

Globalization has extended the reach of capitalism and has thus forced countries in the initial phase of capitalism to confront the poverty among their populations. Such populations are caught in producing raw materials for a global market that is becoming increasingly sophisticated and has the option to use industrially produced substitutes, such as the replacement of copper wire with fiber optics. Developing countries with nascent industrial production have to compete with cheaper and better substitutes from the developed countries. Developing countries' capacity to compete in the emerging areas of technological innovation is constrained by weaknesses in their educational system and their institutional capabilities. Thus, prices for their primary production show a decline in real value.

This situation has not been helped by developed countries' tendency to protect the value of their own agricultural products with tariffs, quotas, and export subsidies. This practice cuts into the international trade of developing countries. Although agricultural exports are important for the foreign-exchange earnings of many developing countries, where more than two-thirds of poor people live in rural areas, world trade in agricultural products grew between 1985 and 1994 at only 1.8 percent per annum compared with the expansion of 5.8 percent per annum for manufactured products during the same period.[6]

In addition to globalization, political instability and regional conflicts have been major factors in deepening poverty in many developing countries. Between 1987 and 1997, more than 85 percent of the world's armed conflicts were civil wars fought within the borders of individual countries.[7] Fourteen such conflicts were in African countries, including Sudan, Somalia, Angola, Rwanda, Burundi, Liberia, and Sierra Leone. Asia recorded 14 conflicts in Cambodia, Vietnam, Sri Lanka, and Indonesia, while Europe contended with the breakup of the former Yugoslavia. Although the populations caught in these conflicts come from different socioeconomic classes, the casualties occur predominantly among the poor. A significant number of those whose assets (social and material) and sources of livelihood were destroyed and who have been displaced as a result of armed conflict are often added to the ranks of the poor. Population displacement creates masses of refugees, disrupts markets and other forms of economic and social institutions, and diverts human resources and public expenditure away from productive activities. In 1998, it was estimated that there were 12.4 million international refugees and 18 million internally displaced people, almost half of them in Africa.[8]

Poverty and Environmental Degradation

Despite the constraints globalization places on economic growth and the insecurity that arises from regional armed

conflicts, advances in health sciences—especially in epidemiology—have led to a human population explosion. Between 1960 and 2000, the world's population grew from less than 3 billion to some 6 billion. World population reached 6.1 billion in mid-2000 and is currently growing at an annual rate of 1.2 percent (about 77 million people). The United Nations estimates that by 2050, world population will reach between 7.9 billion and 10.9 billion people.[9] Population in developed countries is expected to change little during the next 50 years and is even expected to decrease in some countries. However, in the developing world, population is expected to increase by 3.3 million between 2000 and 2050.[10]

A remarkable shift of global population from rural to urban areas has occurred in developed and developing regions of the world. In fact, by 2030, the urban population is expected to be twice the size of the rural population globally.[11] The shift in population distribution from rural to urban areas has been accompanied by a shift in the concentration of the poor. Poverty in urban centers has been increasing more rapidly than in rural areas. According to a United Nations estimate, 600 million people in urban areas in developing countries (almost 28 percent of the developing world's urban population) cannot meet their basic needs for shelter, water, and health. In fact, about half the urban population in poor countries is living below official poverty levels.[12] This number is expected to rise phenomenally over the next few decades. Rapid population growth and urbanization, coupled with the need to produce for export, has negatively affected the environment in at least seven ways:

Deforestation. The agricultural practices in developing countries are still relatively primitive, depending largely on bush-fallow cultivation, with the repeated clearing and burning of shrub and forest to make room for food crops. Equally significant is the deforestation arising from the need for firewood. For instance, it is estimated, that firewood and brush provide about 52 percent of the domestic energy supply in sub-Saharan Africa; charcoal, another forest

product, is also major source of domestic energy.[13]

Desertification. Overcultivation and overgrazing on marginal lands are the major causes of desertification. Although desertification results from many factors and occurs in a variety of environments, rangelands are particularly at risk because they are often found in arid and semiarid regions. In the tropical grassland regions that often border deserts, overgrazing is a potent cause of desertification because feeding the livestock population (which is rapidly expanding to meet increased demand) requires frontier expansion. Desertification also arises from the removal of wood for fuel and the salinization of croplands caused by poorly managed irrigation.

Biodiversity loss. The wide range of ecosystems on which the poor eke out a living has been degraded, and the ecosystems' diverse communities of plants and animals have been put at risk in the process. According to the World Resources Institute, most scientists agree that between 5 and 10 percent of closed tropical forest species will become extinct each decade at current rates of forest loss and disturbance. This loss amounts to about 100 species a day.[14] Indeed, about one-third of the forests that existed in 1950 have been cleared, primarily for agriculture, grazing, or firewood collection.[15] The U.S. National Academy of Sciences estimates that more than 50 percent of all the Earth's species live in tropical rain-forests: A typical four-square-mile patch of rainforest contains as many as 1,500 species of flowering plants, 750 tree species, 125 mammal species, 400 bird species, 100 reptile species, 60 amphibian species, and 150 butterfly species.[16]

Erosion. Population pressure has led to a reduction in the fallow period and to overcultivation of cropland, particularly in developing countries, where poverty ensures that bush-fallow farming predominates. As a result of overcultivation and forest clearing, soil erosion has become widespread. For instance, in Ethiopia, annual topsoil losses of up to 296 metric tons per hectare have been recorded on relatively steep slopes. Even in countries with somewhat moderate slopes, erosion can proceed rapidly where such areas are unprotected by vegetation. In West Africa, losses of 30 to 55 metric tons per hectare have been noted on slopes of only 1 to 2 percent.[17] In regions with unstable sedimentary rock formations, such as in southeastern Nigeria, gully erosion devastates a considerable area of land. Wind erosion is also significant in drier, marginal lands close to the deserts.

Urban pollution. Urban pollution represents an increasing feature of cities and metropolitan areas in developing countries. It begins with the difficult shelter conditions of squatter settlements, which consist of makeshift huts on land to which the poor have no ownership rights and which usually lack adequate water supply and sanitation facilities. Air pollution becomes a serious concern in such areas. The dependence of the poor on biomass fuels for cooking and other domestic uses increases the concentration of suspended particulates, which often reach levels that exceed World Health Organization (WHO) standards in areas where the poor are concentrated. The need of the poor for cheap means of transport within urban areas has encouraged the proliferation of highly polluting transportation modes such as single-stroke engine motorcyles. Poorly maintained secondhand vehicles heighten the level of air pollution in most cities of developing countries.[18]

Water pollution. Contaminated drinking water transmits diseases such as diarrhea, typhoid, and cholera. In developing countries, diarrheal diseases are believed to have killed about 3 million children annually in the early 1990s and 1 million adults and children older than 5 years annually in the mid-1980s.[19] The lack of solid waste management in squatter settlements is also visibly disturbing. These areas generally receive minimal garbage collection service or none at all. For example, in 1993, in Dhaka, Bangladesh, 90 percent of the slum areas did not have regular garbage collection services.[20] The problems resulting from such conditions are obvious—odors, disease vectors, pests that are attracted to garbage (including rats, mosquitoes, and flies), and the overflowing drainage channels clogged with garbage. Leachate from decomposing and putrefying garbage con-

taminates water sources. Because the poorest areas of cities are generally those that receive the fewest sanitation services, the uncollected solid wastes usually include a significant proportion of fecal matter.[21]

Water pollution is also a serious problem in areas where farmers have been given fertilizers and pesticides to increase agricultural productivity. In India, pollution caused by the leaching of nitrogen fertilizers has been detected in the ground water in many areas. In parts of India's Haryana State, for example, well water with nitrate concentrations ranging from 114 milligrams per liter (mg/L) to 1,800 mg/L (far above the 45 mg/L national standard) have been reported.[22] Pesticides that governments have supplied to peasant farmers contaminate sources of ground water, endanger local water supplies, and pollute aquatic systems. Indeed, according to a 1990 estimate published by WHO, occupational pesticide poisoning may affect as many as 25 million people, or 3 percent of the agricultural workforce each year in developing countries. In Africa alone, where some 80 percent of the populace is involved in agriculture, as many as 11 million cases of acute pesticide exposures occur each year.[23]

Climate change. The melting snows of Kilimanjaro provide dramatic evidence that climate change is already affecting the Earth.[24] The Third Assessment Report of the Intergovernmental Panel on Climate Change projects that African countries are especially vulnerable to climate change because much of its agriculture is rain-fed, and it experiences a high frequency of droughts and floods. In particular, grain yields are projected to decline. Coastal settlements in such regions as Egypt, southeastern Africa, the Gulf of Guinea, Senegal, and Gambia will be affected by rising sea levels and coastal erosion. All over the world, the range of infectious disease will likely increase, and significant extinctions of plant and animal species are projected. Desertification is expected to worsen, and most importantly, the numbers and impact of extreme droughts and floods are expected to grow.[25] All of these projections are made worse by the limited

ability of the poor to adapt to climate change.[26]

Poverty, Environmental Hazards, and Natural Disasters

The relationship between poverty and the environment is an obvious feature of life in the developing world. The poor degrade the environment in various ways, while the environment—in the form of environmental hazards and natural disasters—takes a particularly devastating toll on the poor. A review of 30 case studies in developing countries (14 from in Africa) found three major spirals of impoverishment and environmental decline driven by two forces external to the case study locales— development/commercialization and natural hazard events—and two internal to the communities studied—population growth and existing poverty.[27] In the first spiral, poor people were displaced from their resources by richer claimants or by competition for existing land or employment. They were driven by development activities, commercialization, and population growth. For these displaced people, division of the remaining resources followed, or else forced migration to other areas that are usually more marginal. In the second, driven by population growth and poverty, meager resources were further divided to meet the needs of generations or the exigencies of poverty. Remaining resources were then degraded by excessive use of divided lands or inappropriate use of environments unable to sustain the requisite resource use. And in the third, driven by poverty and natural hazard events, poor families were unable to maintain protective works against natural hazards of disease, drought, flood, soil erosion, landslides, and pests or to restore resources damaged by these hazards.[28]

Thus, although the poor in urban and rural areas tend to have a negative impact on the environment, they are also the most vulnerable to environmental hazards and natural disasters. Environmental hazards represent ever-present dangers of life-threatening proportions, and natural disasters tend to be episodic and of varying duration. In many developing countries, inadequate attention is given to environmental management in areas occupied by the poor, and therefore, they are exposed to numerous environmental hazards. For example, carbon-monoxide poisoning, one of the most serious environmental hazards, arises from the use of biomass fuels in poorly ventilated dwellings.[29]

The environmental vulnerability of the poor is most pronounced when natural disaster strikes. Earthquakes, droughts, floods, landslides, volcanic eruptions, windstorms, and forest fires devastate the poor and greatly diminish their chances of escaping poverty. Developing countries, especially in densely populated regions, are known to suffer the brunt of natural disasters. According to the World Bank, between 1990 and 1998, 94 percent of the world's 568 major disasters and more than 97 percent of all natural disaster-related deaths occurred in developing countries.[30] In Bangladesh alone, three storms, four floods, one tsunami, and two cyclones killed more than 400,000 people and affected another 42 million. In southern Africa in 1991–92, Malawi, South Africa, Zambia, and Zimbabwe experienced severe droughts. Between 1995 and 1999, in Latin America and the Caribbean, major natural disasters associated with El Niño, Hurricane Mitch, Hurricane Georges, and the Quindio earthquake in Colombia claimed thousands of lives and caused billions of dollars of damage.

Because the poor often live in insubstantial, makeshift, overcrowded conditions in disaster-prone areas, they are the primary victims of natural disasters. Given that labor is the only economic asset for the majority of the poor, any injury, disability, or loss of life resulting from such vulnerability results in a major loss of economic power. For example, Thomas Reardon, professor of Population Resources, Economic Development, and Applied Microeconomics at the University of California at Davis, and Edward J. Taylor, professor of International Agricultural Development and Agribusiness/Marketing at Michigan State University, observed that the 1984 drought in Burkina Faso resulted in a 50-percent decline in the income of the poorest third

of the rural population residing in the difficult agroclimatic zone of the Sahelian Savanna; those living in the Sudan Savanna (an easier agroclimatic zone) suffered only a 7-percent decline in income.[31] Similarly, Rob Vos, Margarita Velasco, and Edgar de Labastida estimated that in Ecuador, El Niño may have increased the incidence of poverty in affected areas by more than 10 percent.[32]

Women and children are especially vulnerable to such hazards. Indeed, female heads of households tend to suffer more during natural disasters than male heads of households, not only because of their much smaller asset base but also because of customary usages and social relations in many of the communities in developing countries. John Hoddinott, a research fellow at the International Food Policy Research Institute, and Bill Kinsey, professor in the Institute of Development Studies at the University of Zimbabwe, noted, for instance, that although the 1994–95 drought in Zimbabwe had no impact on the health of the men, women and young children were adversely affected.[33] Although the effect of the drought on womens' health was temporary and largely in terms of their loss of body mass (which they regained with the good rains in the following year), the consequence for children, especially those under 2 years of age, was more permanent—an average reduction of 1.5 to 2.0 centimeters of linear growth.

Other hazards derive from disease vectors that are fostered by unsanitary surroundings and settlement on polluted or poorly drained sites. For instance, the sandfly that transmits leishmaniases (a group of parasitic diseases) breeds in piles of refuse or in pit latrines, and the mosquito *culex quinquefasciatus*, a vector for bancroftian filariasis (or elephantiasis), breeds in open or cracked septic tanks, flooded pit latrines, and drains. D. Sapir noted that leptospirosis (a bacterial disease) outbreaks have been associated with an increase in the level of water in the poorly drained areas occupied by the poor in Sao Paulo and Rio de Janeiro. The disease passes to human beings through water contaminated with the urine of infected rats or certain domestic animals.[34] Indeed, many of the diseases

passed on by insect vectors that were once predominantly rural are now threatening urban areas. The poor are also exposed to toxic and radioactive waste, when, for instance, their economic desperation forces them to scavenge through landfill sites.

HIV/AIDS is another factor that increases the vulnerability of the poor to environmental threats. The Joint United Nations Programme on HIV/AIDS (UNAIDS) estimates that more than 40 million people worldwide are infected with HIV. About one-third of those infected are between ages 15 and 24. Almost three-quarters of the people currently infected with HIV live in sub-Saharan Africa.[35] Although HIV/AIDS is primarily discussed as a health threat, its destructive effects are also clear in economic, social, and environmental sectors, particularly because the disease makes it difficult for infected people to escape poverty. People with the disease demonstrate decreased productivity, and high mortality rates often result in single-parent families or in orphans, who have little capacity to support themselves. The increased desperation that results from these scenarios makes it unlikely that infected people will benefit from education about improved planning and development that will help improve environmental practices. Thus, HIV/AIDS stresses the environment in much the same way it stresses other sectors.

The Challenge of Poverty Reduction and Environmental Improvement

The spiral of degradation of the environment and vulnerability to environmental hazards and natural disasters characteristic of the poor in the developing world cannot be allowed to continue. Strategic policies and programs are critically needed to address these problems. One factor that underscores the exigent nature of the problem relates to the globalization process. The information technology that fueled the initial poverty-inducing consequences of globalization can also serve as a tool to change the relationship between states and civil society in virtually all countries of the developing world: The instantaneous transmission of live events across the globe can empower societies to demand rights from their governments. Already, the expanded availability of information has spawned the growth of worldwide, voluntary non-governmental organizations (NGOs) that catalyze, mobilize, and give direction to civil society. Availability of information also forces government to be more responsive to the needs and interests of its population. NGOs make it difficult for states to ignore problems of income distribution, environmental conditions, poverty, employment, and housing for the vast majority of their population.

Good governance is central to issues of poverty reduction as well as improvement in environmental quality. Good governance requires three basic conditions: decentralization (the authority structures must be decentralized and devolved); inclusiveness (decisionmaking processes must be participatory and all-inclusive); and accountability (government strategies and activities must be transparent and accountable to the populace).

Decentralization begins with the relation of local governments to the higher levels of government and the lower systems of administrative delegation within the administrative unit, and down to the neighborhood levels, especially those populated largely by the poor. This means a review of the responsibilities delegated to local governments and their implications for the welfare of citizens at the ward and the neighborhood levels. Such delegation must be accompanied by commensurate powers and resources so that local governments can effectively carry out the responsibilities assigned to them. It also entails creating transparent systems, which ensures that local taxation and other statutory transfers of resources are used in a manner to create necessary opportunities for poverty reduction. This involves providing necessary social services (e.g., schools and health services), physical infrastructure (e.g., roads, water, and electricity), environmental improvement (e.g., sanitation, potable water, solid waste disposal, and air pollution reduction) and creating jobs, credit, and market opportunities.

Good governance also emphasizes a participatory process of decisionmaking.

It encourages the involvement of all citizens, especially disadvantaged groups—such as women, the urban poor, and the disabled—in the affairs of the community. An emerging strategy for ensuring the efficacy of this process is to engage all stakeholders, including the poor, in the preparation of the budget of local governments both in urban and rural areas. Participatory budgeting ensures that priorities are determined on the basis of reconciling the divergent interests of all sections of the populace for the common good. It facilitates the collective commitment of all, not only to the long-term, strategic vision of sustainable human development but also to the more immediate goals of poverty reduction. Thus, the community as a whole becomes engaged in the design, implementation, and monitoring of local priorities. Such a participatory process of decisionmaking not only facilitates partnership with private-sector organizations but also results in at least three major beneficial consequences: efficiency, equity, and sustainability in the planning and management of community affairs.

Efficiency in community effort comes first, because participation—especially by the poor—is a first and important factor in their empowerment. When all segments of a population are involved in decisionmaking, they will be more likely to fulfill their civic responsibilities. Where the poor contribute to local development of needed services, infrastructure, and environmental improvements, greater efficiency can be achieved in the overall delivery of such services. This entails not only making their own tax contribution to the local revenue base but also ensuring that they have the resources to do so. Experiments in which banks provide credit to the poor have shown that, rather than subsidized credit, the poor require access to noncollaterized credit, entailing minimal transaction costs to lenders and borrowers, while allowing for charging market-based interest rates. This is especially true when accompanied by fair and predictable legal and regulatory frameworks, which recognize and legitimize the predominantly informal characteristics of their economic activities. The many ways in which the poor make, sell, or trade things hardly register in conven-

tional economic measures, but in aggregate, it constitutes a significant sector of the economy.

The equity effects of involving the poor in decisionmaking are no less substantial. Participation promotes the development of human resources and ensures the provision of facilities for primary education, primary health care, nutrition, family planning, employment and livelihood, shelter, safe drinking water, sanitation, and other basic services. This is crucial not only for improving the labor productivity and income of the poor but also for giving their children opportunities to escape from the trap of poverty. In particular, participation enhances the status of poor women and encourages their upward mobility by giving them a voice. It promotes actions that reduce the vulnerability of the poor to natural and human-made disasters and hazards in the environment.

Perhaps the most important consequence of an all-inclusive and participatory decisionmaking process is its potential to foster sustainable development (i.e., meeting the needs of the present without compromising the ability of future generations to meet their needs).[36] Because of the emphasis on consultation and the reconciliation of divergent interests of various groups, it is possible to reach agreement on issues such as environmental planning and management of cities. Participation promotes strong local democracies and fosters the ethic of civic responsibility among all segments of the community, including the poor. The heightened civic engagement that results encourages the development of a consensual vision for the community's future. Realizing such a vision engenders concern with the quality of the environment and the well-being of present and future generations.

Accountability is the third major feature of good governance. A fundamental tenet of good governance is that local authorities must be accountable to their citizens. Access of all citizens to all relevant information is critical for citizens to understand what is happening in the local government and who is benefiting from the decisions and actions of the government. Therefore, there is a need for some institutional framework to ensure that local governments are directly accountable to their electorates. These frameworks can be fostered through regularly organized and open consultations of all citizens and public feedback mechanisms such as an ombudsman, hotlines, complaint offices and procedures, local government report cards, and procedures for public petitioning and/or public interest litigation.

In any country, strategies for alleviating and reducing poverty have greater chances of succeeding under conditions of national stability and sustained economic growth. Consequently, poverty-reduction measures must relate to how well the macroeconomic and sectoral policies of national governments are oriented to promoting increased productivity and output in the economy. Such policies must not only stimulate economic growth but must also create more employment opportunities. Moreover, in order to ensure the enhanced participation of the poor in economic growth, special attention must be given to reforming laws and regulations that impede access to land, credit, and public infrastructure and services. Improving the access of the poor to land through secure tenure is one way of enhancing their sense of stewardship of the portion of the Earth's surface they occupy.

Conclusion

Any discussion of poverty and the environment must be posited within a political economy framework that focuses on the globalization process and its emphasis on the social relations of production and reproduction. Globalization compels different countries to approach problems of their environment and their poor as resources to be developed. Thus, the 1992 Earth Summit in Rio, in pursuing its Agenda 21 for sustainable development, identified the role local authorities could play in support of Agenda.[37] This was a recognition that local authorities, as the level of governance closest to the people, are able to play the pivotal role of educating, mobilizing, and responding to the public to promote sustainable development.[38]

Despite this recognition, the record shows that not many local governments in developing countries are aware of their role in meeting the expectations of Agenda 21. A few, especially in Latin America and Asia, have developed their own Agenda 21. In the last few years, the International Council for Local Environment Initiatives (ICLEI) has been furthering these initiatives in various countries. Particularly with respect to developing countries, ICLEI noted that lack of resources and technical capacity within most local authorities has been a major constraint in carrying out this agenda. In many countries, this problem is further compounded by local governments, which are often restricted by central governments in raising finances and in other activities.

Clearly, the issue of local governance has to be given greater emphasis in any program to reduce poverty and enhance the quality of the environment. Equally important in this regard is the need to strengthen the role of networks of NGOs committed to mobilizing, educating, and defending the interests of the poor. Several such networks are gaining global visibility. In 1997, one network representing the alliance of grassroots organizations, researchers, and international organizations formed Women in Informal Employment: Globalizing and Organizing, whose mission is to promote better statistics, research, and policy in support of poor women.[39] Another network, known as HomeNet, was created in the mid-1990s by unions, grassroots organizations, and NGOs working with home-based workers and street vendors in developing and developed countries. HomeNet is concerned with the adverse impact of globalization on the livelihoods of poor women in the informal economy.

In the decades ahead, much work remains to be done. Despite the increasing proactive interventionism from different tiers of governments and the network of NGOs, it is important to recognize our limited understanding of the complex and dynamic interaction between poverty and the environment. Because the collective impact of the poor on the environment is nonlinear, complex, and often does not show up immediately, it is necessary to begin to reconceptualize research procedures and methodologies to account for the role of various social actors.

It is important to develop effective systems and mechanisms that can better provide needed scientific and technical knowledge to support local efforts to address the needs of both the environment and the poor. The emerging paradigm of sustainability science encourages processes of coproduction of knowledge in which scholars and stakeholders, including the poor, interact to define important questions, relevant evidence, and convincing arguments that are scientifically sound and rooted in social understanding.

Sustainability science emphasizes that when the stakeholders are involved in the production of such requisite knowledge, they will become active agents of sustainable and equitable development.[40] Only in this way is it possible to deal effectively with the twin problems of poverty and environmental degradation in developing countries and to begin the arduous task of attempting to eliminate poverty in the context of sustainable development.

NOTES

1. World Bank. *World Development Report 2000/2001: Attacking Poverty* (New York: Oxford University Press, 2000), 3.
2. *UNDP* (United Nations Development Programme) *Human Development Report 1997* (New York: Oxford University Press, 1997), 3.
3. Ibid., page 15.
4. United Nations. 1993.
5. K. Polanyi, *The Great Transformation: The Political and Economic Origins of Our Time* (New York: Holt, Rinehart and Winston, 1944), 75.
6. World Bank. note I above, 180.
7. D. A. Pottebaum, *Economic and Social Implications of War and Conflict* (Ithaca, N.Y.: Cornell University. Agricultural Economics Department, 1999).
8. World Bank, note 1 above. page 127.
9. United Nations Population Division, *World Population Prospects: The 2000 Revision*, draft, 28 February 2001.
10. Ibid.
11. United Nations Centre for Human Settlements (UNCHS), *An Urbanizing World: Global Report on Human Settlements 1996* (Oxford: Oxford University Press, 1996).
12. A. Marshall, ed., *The State of the World Population 1996: Changing Places: Population, Development and the Urban*

Future (United Nations Population Fund. 1996). Available at http://www.un-fpa.org/swp/1996/SWP96MN.HTM.

13. World Resources Institute (WRI), *World Resources 1994–95: A Guide to the Global Environment* (New York: Oxford University Press, 1994), 10.

14. WRI, available at http://www.wri.org/biodivtropical/html.

15. P. H. Raven and T. Williams, eds., *Human Nature and Society: The Quest for a Sustainable World*, Proceedings of the 1997 Forum on Biodiversity, Board on Biology, National Research Council (National Academy Press. 2000).

16. Rainforest Web.Org. Available at http://www.rainforestweb.org_information/biodiversity/?state=more.

17. WRI, *World Resources 1987: An Assessment of the Resource Base that Supports the Global Economy* (New York: Basic nooks Inc., 1987), 3.

18. United Nations Environment Programme (UNEP) and World Health Organization (WHO), "Air Pollution in the World's Megacities," *Environment*, March 1994, 4–13, 25–37; and UNEP and WHO, "Monitoring the Global Environment: An Assessment of Urban Air Quality." *Environment*, October 1998, 6–13, 26–37.

19. UNCHS, *An Urbanizing World: Global Report on Human Settlements 1996* (Oxford: Oxford University Press, 1996), 171.

20. Economic and Social Commission for Asia and the Pacific (ESCAP), *State of Urbanization in Asia and the Pacific*, ST/ESCAP/1300 (New York: United Nations, 1992).

21. UNCHS, note 19 above, page 270.

22. India, *Status of Ground Water Pollution in India, Central Ground Water Board, Ministry of Water Resources* (Lucknow, Uttar Pradesh: Ground Water Pollution Directorate, 1991), 32.

23. J. Jeyaratnam, "Acute Pesticide Poisoning: A Major Global Health Problem," *World Health Statistics Quarterly* 43 (1990): 139–43.

24. J. McCarthy and M. McKenna, "How Earth's Ice Is Changing," *Environment*, December 2000, 8–18.

25. Intergovernmental Panel on Climate Change, *Climate Change 2000: Impacts, Adaptation, and Vulnerability* (New York: Cambridge University Press, 2001), 14.

26. R. W. Kates, "Cautionary Tales: Adaptation and the Global Poor." *Climatic Change 2000* 45, no. 1 (2000): 5–17.

27. R. W. Kates and V. Haarmann, "Where the Poor Live: Are the Assumptions Correct?" *Environment*, May 1992, 4–11, 25–28.

28. Ibid.

29. G. McGranahan and J. Songsore, "Wealth, Health, and the Urban Household: Weighing Environmental Burdens in Accra, Jakarta, and São Paulo," *Environment*, July/August 1994, 4–11, 40–45.

30. World Bank, note 1 above, page 182.

31. T. Reardon and J. E. Taylor, "Agroclimatic Shock, Income Inequality and Poverty: Evidence from Burkina Faso," *World Development* 24, no. 5 (1996): 901–14.

32. R. Vos, M. Velasco, and E. de Labastida, *Economic and Social Effects of El Niño in Ecuador, 1997–1998* (Washington, D.C.: Inter-America Development Bank, 1999).

33. J. Hoddinott and B. Kinsey, *Child Growth in the Time of Drought* (Washington, D.C.: International Food Policy Research Institute, 1998).

34. D. Sapir, *Infectious Disease Epidemics and Urbanization: A Critical Review of the Issues*, Paper prepared for the WHO Commission on Health and Environment (Geneva: WHO, Division of Environmental Health, 1990).

35. Joint United Nations Progmmme on HIV/AIDS and WHO, "AIDS Epidemic Update—December 2001." December 2001. Available at http://www.unaids.org/epidemic_update/report_dec01/index*full.

36. World Commission on Environment and Development, *Our Common Future* (Oxford: Oxford University Press, 1987).

37. For more on Agenda 21, see P. Haas, M. A. Levy, and E. A. Parson. "Appraising the Earth Summit: How Should We Judge UNCED's Success?" *Environment* October 1992, 6–11, 26–33.

38. N. A. Robinson, ed., *Agenda 21: Earth's Action Plan* (New York: Oceania, 1996), 63.

39. World Bank, note 1 above, page 187.

40. R. W. Kates et al., "Sustainability Science," *Science*, 27 April 2001, 641–42. Also available at http://www.sustainabilityscience.org.

Akin L. Mabogunje is a geographer whose research and policy efforts center on urban and regional development. He is chairman of the Development Policy Centre, Ibadan, Nigeria, and coconvener of the international initiative on science and technology for sustainability. A foreign associate of the U.S. National Academy of Sciences, he was recently honored with Nigeria's highest distinction: Commander of the Order of the Niger. He is also a contributing editor of *Environment*.

This article is taken from a paper presented at the 2001 Open Meeting of the Human Dimensions of Global Environmental Change Research Community in Rio de Janeiro, Brazil, on 6–8 October 2001 under the auspices of the Brazilian Academy of Sciences.

Endangered Humans

How global land conservation efforts are creating a growing class of invisible refugees

By Charles C. Geisler

Shortly after William the Conqueror won the Battle of Hastings in 1066, he evicted almost 2,000 local Saxons and established a nearly 100,000-acre hunting preserve. Some 800 years later in North America, the U.S. government granted protected status to Yellowstone, Yosemite, and Glacier national parks. Again, as in England, the native peoples were evicted, forbidden to hunt or gather on their ancestral lands, or simply eliminated.

With ever more compelling reasons to protect nature—invaluable ecosystems, biodiversity, and genetic libraries—these stories are repeating themselves today on a global scale. Some 70 percent of the planet's protected areas are inhabited by human beings, and these local residents are widely viewed as a menace to environmental conservation. Thus, a new breed of refugee is in the making.

The recent worldwide growth of parks and protected areas is impressive. According to the Switzerland-based World Conservation Union, nearly 29,000 protected areas now shield some 2.1 billion acres of land from a series of residential and economic uses. These territories compose 6.4 percent of the earth's land, or about half of the world's croplands, and are roughly the size of the continental United States plus half of Alaska. Most of this protection is recent. From less than 1,000 protected areas in 1950, the count grew to 3,500 in 1985 before ballooning to 29,000 today. The most ardent conservationists seek to multiply today's base several times. If such global "greenlining" continues without concern for the rights of resident populations, its gains could take an enormous human toll.

Africa offers a telling example of greenlining with manifest social costs. In 1985, Africa had 443 publicly protected areas encompassing 217 million acres of land. Facing international pressure, virtually all African countries have since increased their protected land base. Today, over 1,000 protected areas account for nearly 380 million acres of African land, with 7 countries claiming protected status for more than 10 percent of their land base. In 14 African countries, more land is greenlined than cultivated, and the poorer countries in Africa today have on average more land set aside for conservation than the continent's more affluent nations.

How many people have these conservation efforts displaced? A precise count of conservation refugees in Africa and elsewhere remains elusive, in part due to diverse definitions of a "protected area," enforcement problems, and recidivism among refugees. In Africa, well-known cases of mass eviction have occurred in Uganda, Botswana, Cameroon, Madagascar, South Africa, Togo, Zimbabwe, Rwanda, and Democratic Republic of the Congo, affecting nearly half a million people. For example, Tanzania's Masailand is now dotted with national parks that have displaced more than 60,000 farmers and pastoralists from their ancestral lands. Indirect measurements—such as multiplying the area under protection by a low range of possible human densities—yield estimates of 900,000 to 14.4 million people. If accurate, these upper bounds mean that conservation refugees in Africa could roughly equal the global refugee population of 14.5 million people currently calculated by the U.N. High Commission for Refugees.

Global conservation is surely a worthy cause, but its sometimes insensitive implementation raises serious environmental justice questions. Indeed, the very notion of "environmental justice" typically applies to infringement on environmental rights by ecologically blind development. But when conservation erases the rights of resident peoples, the expansion of protected areas resembles urban renewal or megaproject development—both notorious for displacing human communities in the name of some perplexing public interest. Whose public interest does greenlining serve? Seldom the interests of conservation refugees, who remain invisible in conservation planning debates.

This invisibility has multiple causes. The first involves the nature of refugee reality: Authorities usually deny refugee problems until they take on crisis proportions, and the official definitions of a refugee exclude many forms of displacement. Policymakers and environmentalists periodically disregard the social and cultural impact of protected areas, perceiving conservation as the opposite of "development" and portraying local residents as intruders. A final reason is class bias. Environmental refugees in

Africa and elsewhere tend to be poor and powerless. It is the wealthy inhabitants of the planet who benefit most from greenlining—enjoying exotic vacation destinations, new targets for their tax-deductible largess, windfall gains in value for their high-end properties in or near protected zones, and what Harvard entomologist Edward O. Wilson calls "biophilia," or a deeply felt loyalty to the earth's biota. Local inhabitants are rarely so fortunate. Many live on marginal lands in marginal places with marginal rights to remain in their communities. Their contributions to the ecosystem are taken for granted and appropriated with little compensation. Simply put, conservation refugees are invisible because visibility raises the price of conservation.

Green consciousness, vital as it is to human survival, must broaden its vision of global welfare as it broadens its reach. Nearly half of the planet's most species-rich areas contain human populations suffering severe economic disadvantages. The tropics, where biodiversity flourishes most, are home to nearly 60 percent of the world's most destitute people. While global conservation does not cause poverty, neither should it exacerbate poverty. The poor should not be asked to disproportionately subsidize the expansion of conservation. They, too, must have voice and choice. If conservationists are to retain the mantle of justice, they must find alternatives to involuntary and uncompensated human displacement.

Charles C. Geisler is professor of rural sociology at Cornell University.

From *Foreign Policy*, May/June 2002, pp. 80-81. © 2002 by Charles C. Geisler.

A Dirty Dilemma: The Hazardous Waste Trade

Zada Lipman

Since the 1980s, exporters of hazardous waste have targeted developing countries. Some of this waste is destined for dumping or disposal, while other waste is directed to resource recovery, recycling, or reuse. To protect developing countries from the dangers associated with hazardous waste, the international community adopted the Basel Convention on the Transboundary Movements of Hazardous Wastes and their Disposal, which first regulated and then banned the trade of hazardous waste. Although lauded as a landmark for global democracy and environmental justice, the ban has created a dilemma for developing countries with large recycling industries that rely on hazardous waste imports for their continued operation.

Environmental problems arising from the disposal of hazardous waste in developing countries did not gain international attention until the late 1980s, when several incidents of dumping were reported in African nations. One of the most serious cases occurred in 1987. Several thousand tons of highly toxic and radioactive waste, labeled "substances relating to the building trade," were exported from Italy to Koko, Nigeria, and stored in drums in a backyard. Many of these drums were damaged and leaking; workers packing the drums into containers for retransport to Italy suffered severe chemical burns and partial paralysis, and land within a 500-meter radius of the dump site was declared unsafe. The Italian government eventually accepted the return of the waste, and the Nigerian government has since imposed the death penalty on the waste importers. In 1988, Guinea-Bissau was offered a US$600 million contract—four times its gross national product—to dispose of 15 million tons of toxic waste over five years. The contract was never concluded because of public concern within Guinea-Bissau, but many similar arrangements were reported in the 1980s in countries such as Namibia, Guinea, Sierra Leone, and Haiti. In some cases, dumping took place with the consent of the government in question, while in other cases it was part of an illegal operation. Since then, numerous incidents of dumping in developing countries have been reported throughout the world.

Logic of the Market

Although precise estimates of the worldwide generation of hazardous waste are difficult to obtain, the United Na-tions Environment Programme (UNEP) estimated in 1992 that approximately 400 million metric tons of hazardous waste were generated annually, with 80 percent of this waste coming from countries in the Organisation for Economic Cooperation and Development (OECD). This figure is likely to be significantly higher today.

The disposal of hazardous waste has become a major issue for countries that are large waste-generators. Before the dangers associated with disposal were understood, most of this waste was deposited in landfills, causing serious problems for surrounding areas. A well-documented example is the "Valley of the Drums" in Kentucky, a seven-acre site with 17,000 drums of hazardous waste that has contaminated nearby soil and water. As a result of incidents like this, most developed countries introduced stringent environmental and safety measures for the disposal of hazardous waste. This trend led to increasingly limited and costly disposal options in developed countries.

Developing countries became targets for waste generators—mostly developed countries—since they provided disposal options for a mere fraction of the equivalent cost in the state of origin. According to a study by Katharina Kummer in *International Management of Hazardous Wastes*, disposal costs for hazardous waste in developing countries in 1988 ranged from US$2.50 to US$50 per ton, compared with costs of US$100 to US$2,000 per ton in OECD countries. The cost of incineration was even higher, at US$10,000 for one ton of hazardous waste in the United Kingdom. The lower disposal costs in developing countries generally stem from low or nonexistent environmental standards, less stringent laws, and an absence of public opposition due to a lack of information concerning the dangers involved. Given these considerations, the economic logic for exporting hazardous waste to developing countries is indisputable.

The Basel Convention

International concerns about the export of hazardous waste to developing countries led to the negotiation of the 1989 Basel Convention, which became binding in 1992. As of August 2001, 148 countries had ratified the Convention. Unfortunately, the United States, which generates approximately 60 percet of the world's hazardous waste, has yet to ratify.

The Basel Convention itself does not ban the transboundary movements of hazardous waste, except to Antarctica. Rather, it seeks to control and limit the movement of waste based on a process of prior informed consent. Hazardous waste exports cannot proceed unless the pertinent authorities in the recipient and transit countries are notified in advance and provide written consent. Any movement of hazardous waste without a movement document or prior notification is illegal under the Basel Convention. This Convention requirement applies to both hazardous waste exported for final disposal and waste exported for recycling. The Convention also requires parties to prohibit the import of hazardous wastes when it is likely that the waste will not be managed in an environmentally sound manner.

The Basel Convention initially focused on protecting developing countries from hazardous waste dumping by developed countries. But by 1992, at the First Meeting of the Conference of the Parties (COP-1), concerns had already shifted to hazardous waste traded for recycling or recovery. Jim Puckett of the Basel Action Network estimates that from 1980 to 1988, only 36 percent of hazardous waste exports to developing countries were destined for recycling. In 1992 these exports had risen to 88 percent, and in 2001 they are likely to constitute over 95 percent of all waste exports.

The lack of a distinction between "waste" and "products" in the Convention and its vague criteria for "hazardous" allowed the continued export of hazardous waste to developing countries for recycling on the basis that the toxic substances exported were commodities rather than wastes. Some exports of waste are for "sham" recycling, but even when recycling takes place, this waste presents environmental and health risks to developing countries that lack the technology to handle the waste safely. For instance, Greenpeace and the Basel Action Network have not encountered a single hazardous waste recycling facility in a non-OECD country that does not cause serious pollution. In addition, Greenpeace estimates that more than 2.5 million tons of hazardous waste were exported to developing countries between 1989 and 1994. Dissatisfaction with the Basel Convention resulted in over 100 developing countries unilaterally imposing regional or national waste-import bans.

The control regime imposed by the Basel Convention was a compromise between developing countries, which favored a total ban on the transboundary movements of hazardous waste, and developed countries, which wanted a more flexible control regime. In 1994, developing countries finally gained sufficient international support to achieve a total ban on hazardous waste exports, when the parties to the Basel Convention decided to ban all exports of hazardous wastes from OECD to non-OECD (largely developing) countries. To ensure that the ban was legally binding, it was adopted as an amendment to the Convention in 1995. At the same time, a newly added paragraph recognized that exports of hazardous waste to developing countries are highly unlikely to be managed in an environmentally sound manner consistent with the Convention. Clearly, protection of developing countries was the rationale for this amendment. However, the ban has not yet received the requisite number of ratifications to come into force. As of August 2001, there were 26 ratifications—out of the minimum of 62—and only half of these were from developing countries, which had been the chief initiators and presumed benefactors of the ban. However, the European Union has ratified the amendment and introduced regulations to ban exports to developing countries, a decision that binds its 15 member states.

Destiny's Landfill?

Developing countries with large recycling and reclamation industries are concerned that they will be deprived of resources if the ban comes into force. This concern is exacerbated by the uncertainty as to which types of waste are subject to the ban. A major weakness of the Basel Convention is its failure to provide clear definitions of hazardous waste. Waste characterized as hazardous and subject to the ban, as well as waste not covered by the Convention (including a large percentage of internationally traded metals and secondary raw materials), were classified into two lists and adopted as Annexes to the Convention in 1998, which made them legally binding. Waste is primarily classified according to its degree of hazard, with its value as a secondary raw material also taken into account. Wastes that have yet to be classified include items of economic importance such as zinc, lead and copper compounds, lithium, and spent catalysts.

The ban will likely have a considerable impact on recycling industries in developing countries. In particular, a future decision to classify lead-acid batteries as hazardous will affect countries such as India and the Philippines, which rely on imported lead-acid batteries for a significant proportion of their lead requirements. The demand for lead in developing countries in Southeast Asia is also increasing due to rising demand for batteries in motor vehicles, telecommunications, and computer equipment. As Jonathan Kreuger points out in his book, *International Trade and the Basel Convention*, if the ban proceeds, lead ingots will have to be bought to supplement the output of the domestic recycling industry which will itself become reliant on domestic supplies or on imports from other non-OECD countries. A 1999 United Nations Conference on Trade and Development (UNCTAD) study into lead-acid batteries found that a Philippine secondary lead smelter that provides 80 percent of the country's refined lead output may need to close if feedstock requirements were to become unavailable.

Of course, any financial benefits that the recycling of hazardous waste may provide should be offset against the costs to human health and to the environment. Most developing countries lack the capacity to handle hazardous waste safely. The UNCTAD study identified thousands of small battery-reconditioning shops in major cities throughout the Philippines that were located in busy streets, often adjacent to fast-food vendors. Workers did not wear protective clothing and often dismantled batteries with their bare hands. Reconditioners routinely dumped diluted sulfuric acid down street drains or behind their premises. Lead plates were then sold to licensed smelters, and the residues from the smelters, which can have a lead content of over 90 percent, were dumped in a river, in the countryside, or behind the smelters. Greenpeace research in the Philippines into imports of lead-acid batteries has revealed that even legitimate hazardous-waste recycling operations promoted by the Philippine government are in many cases creating residual waste more toxic than the original product.

Mercury waste is also subject to the ban. The danger associated with the reprocessing of mercury wastes is well illustrated by a notorious incident in South Africa. Thor Chemicals, a British company, established one of the world's largest mercury-reprocessing plants in South Africa and began importing mercury wastes in 1986. Over the next eight years, the plant imported thousands of tons of waste from the United States and Europe. In 1988, mercury contamination 1,000 times higher than the World Health Organization's standards was discovered in a river about 50 kilometers from the Thor plant. Subsequent samples of soil at the reprocessing site revealed high levels of mercury contamination. Mercury, which can be absorbed into the body through food, air, or skin contact, is linked to many neurological problems causing symptoms such as trembling, loss of muscle control, headaches, mental confusion, nausea, and hair loss. Long-term exposure can lead to a coma and eventually to death. Workers at the Thor plant claim that they were not warned about the dangers of working with mercury. They continued to work at the plant because unemployment was high and Thor "paid the highest wages." This comment encapsulates the painful dilemma for workers in developing countries—poverty or pollution.

By 1992, two Thor workers had died of mercury poisoning, while many others were permanently disabled. Tests conducted on workers in 1992 revealed that almost 30 percent were at risk of severe mercury poisoning. The families of the deceased workers sued Thor in a British court and were awarded almost US$2 million. In comparison, compensation for injured workers has been paltry. In 1997, 20 former Thor workers were paid approximately US$1.3 million in an out-of-court settlement by Thor's head office in Britain. A second class-action lawsuit in 1999 involving 20 workers resulted in an award of approximately US$400,000 split amongst the plaintiffs—an amount barely sufficient to cover all the medical expenses. In the interim, the plant site remains an ecological time bomb. Although the plant was shut down in 1994, 1,000 tons of stockpiled waste remains on the site in leaking barrels. A South African Commission of Inquiry has recommended the incineration of the stockpiled wastes at standards far below those in developed countries. Environmental groups have opposed the recommendations of the Commission and have called on companies that originally exported the waste to South Africa to reclaim it.

Out of the Waste Land

A ban on hazardous-waste exports to developing countries is the simplest control measure to implement, as well as being morally justifiable. Countries that benefit from industrialization should also bear the full weight of its burdens. This principle is exemplified by the "polluter pays" principle, by which countries that are the primary generators of hazardous waste have the responsibility to deal with the waste at its source rather than exporting it to developing countries. As the case studies illustrate, most developing countries do not have efficient hazardous-waste disposal facilities. They also lack the requisite skill to evaluate the risks posed by hazardous waste or, where unsafe disposals have taken place, to institute monitoring systems and remedial strategies to mitigate widespread contamination and loss of workers' lives and health. Greenpeace believes that the ban has succeeded in preventing a global environmental disaster and claims that hazardous waste exports to developing countries have diminished dramatically since the ban was adopted, notwithstanding the fact that it has not yet entered into force.

However, many problems with the ban have yet to be resolved. For example, not all developing countries are in favor of a total ban. Attempts by developed countries to impose a ban on countries that oppose it not only suggest paternalism, but also infringe on the sovereign right of these countries to consent to the import of hazardous waste. Cash-starved economies and corruption at the government level also leave open the possibility of illegal imports of hazardous wastes. Regardless of the legality of such actions, developing countries may continue to import hazardous waste as long as it remains profitable. A ban on imports of hazardous waste may put thousands of workers in developing countries out of work; it also denies developing countries the opportunity to import cheap materials. There is an undeniable tension between the justice of banning hazardous waste exports that cause harm to people in a developing country and the realization that doing so may endanger the very livelihood of these people.

Although it is impossible to ignore the ethical problems that arise in imposing a ban on exports of hazardous waste to developing countries, the alternative is an escalating waste colonization process that would further degrade the health of impoverished people and stressed environments. This process is well-illustrated by developments in the global shipbreaking industry, which is not regulated by the Basel Convention and its ban amendment. This industry has relocated from the developed world to countries such as India, Bangladesh, and China. Thousands of workers toil under the most arduous conditions without any safety precautions. Apart from the dangers posed by exposure to toxic chemicals, it is estimated that one in four of these workers will contract cancer from the asbestos on board these waste vessels.

A Way Forward

International investment is necessary to assist developing countries in establishing their own industries and to relieve them from their dependency on hazardous waste imports. As developing countries continue to industrialize, they will require additional funding and technology to build environmentally safe recycling and disposal plants for any waste that they generate. Developed countries have a responsibility under the Basel Convention to provide training for developing countries in managing hazardous waste and to facilitate technology transfers. The underlying rationale is that if developing countries are trained to adopt clean production technology from the outset, they can avoid the mistakes made by developed countries.

Technology transfers, however, may not be a panacea for problems associated with the handling of hazardous waste. Greenpeace has observed that the challenges faced by developing nations are not just a matter of know how and technology; the successful export of the developed world's environmental knowledge would, instead, also require the export of an entire social structure. This claim may be an overstatement. It is more likely that as developing countries improve economically, social and environmental reforms will follow. The requirement that exporters adhere to global environmental standards for admission to international markets may provide the necessary incentive for developing countries to upgrade their environmental performance. Yet to impose these requirements without providing sufficient funding and technology will result in greater poverty and hardship for these countries.

At the same time, it is essential for all responsible developed countries to ratify the ban amendment to ensure its lasting success. It is also in the long-term interest of developing countries to ratify the ban. The developed world must provide more assistance in order to persuade developing countries to give up a hazardous livelihood chosen only out of the fear of having no livelihood at all.

ZADA LIPMAN is Associate Director of the Centre for Environmental Law, Macquarie University, Australia and a Barrister of the Supreme Court of New South Wales.

DEATH STALKS A CONTINENT

In the dry timber of African societies, AIDS was a spark. The conflagration it set off continues to kill millions. Here's why

By Johanna McGeary

IMAGINE YOUR LIFE THIS WAY. You get up in the morning and breakfast with your three kids. One is already doomed to die in infancy. Your husband works 200 miles away, comes home twice a year and sleeps around in between. You risk your life in every act of sexual intercourse. You go to work past a house where a teenager lives alone tending young siblings without any source of income. At another house, the wife was branded a whore when she asked her husband to use a condom, beaten silly and thrown into the streets. Over there lies a man desperately sick without access to a doctor or clinic or medicine or food or blankets or even a kind word. At work you eat with colleagues, and every third one is already fatally ill. You whisper about a friend who admitted she had the plague and whose neighbors stoned her to death. Your leisure is occupied by the funerals you attend every Saturday. You go to bed fearing adults your age will not live into their 40s. You and your neighbors and your political and popular leaders act as if nothing is happening.

Across the southern quadrant of Africa, this nightmare is real. The word not spoken is AIDS, and here at ground zero of humanity's deadliest cataclysm, the ultimate tragedy is that so many people don't know—or don't want to know—what is happening.

As the HIV virus sweeps mercilessly through these lands—the fiercest trial Africa has yet endured—a few try to address the terrible depredation. The rest of society looks away. Flesh

and muscle melt from the bones of the sick in packed hospital wards and lonely bush kraals. Corpses stack up in morgues until those on top crush the identity from the faces underneath. Raw earth mounds scar the landscape, grave after grave without name or number. Bereft children grieve for parents lost in their prime, for siblings scattered to the winds.

The victims don't cry out. Doctors and obituaries do not give the killer its name. Families recoil in shame. Leaders shirk responsibility. The stubborn silence heralds victory for the disease: denial cannot keep the virus at bay.

The developed world is largely silent too. AIDS in Africa has never commanded the full-bore response the West has brought to other, sometimes lesser, travails. We pay sporadic attention, turning on the spotlight when an international conference occurs, then turning it off. Good-hearted donors donate; governments acknowledge that more needs to be done. But think how different the effort would be if what is happening here were happening in the West.

By now you've seen pictures of the sick, the dead, the orphans. You've heard appalling numbers: the number of new infections, the number of the dead, the number who are sick without care, the number walking around already fated to die.

But to comprehend the full horror AIDS has visited on Africa, listen to the woman we have dubbed Laetitia Hambahlane in

Durban or the boy Tsepho Phale in Francistown or the woman who calls herself Thandiwe in Bulawayo or Louis Chikoka, a long-distance trucker. You begin to understand how AIDS has struck Africa—with a biblical virulence that will claim tens of millions of lives—when you hear about shame and stigma and ignorance and poverty and sexual violence and migrant labor and promiscuity and political paralysis and the terrible silence that surrounds all this dying. It is a measure of the silence that some asked us not to print their real names to protect their privacy.

HALF A MILLION AFRICAN CHILDREN WERE INFECTED WITH HIV LAST YEAR

Theirs is a story about what happens when a disease leaps the confines of medicine to invade the body politic, infecting not just individuals but an entire society. As AIDS migrated to man in Africa, it mutated into a complex plague with confounding social, economic and political mechanics that locked together to accelerate the virus' progress. The region's social dynamics colluded to spread the disease and help block effective intervention.

We have come to three countries abutting one another at the bottom of Africa—Botswana, South Africa, Zimbabwe—the heart of the heart of the epidemic. For nearly a decade, these nations suffered a hidden invasion of infection that concealed the dimension of the coming calamity. Now the omnipresent dying reveals the shocking scale of the devastation.

AIDS in Africa bears little resemblance to the American epidemic, limited to specific high-risk groups and brought under control through intensive education, vigorous political action and expensive drug therapy. Here the disease has bred a Darwinian perversion. Society's fittest, not its frailest, are the ones who die—adults spirited away, leaving the old and the children behind. You cannot define risk groups: everyone who is sexually active is at risk. Babies too, unwittingly infected by mothers. Barely a single family remains untouched. Most do not know how or when they caught the virus, many never know they have it, many who do know don't tell anyone as they lie dying. Africa can provide no treatment for those with AIDS.

They will all die, of tuberculosis, pneumonia, meningitis, diarrhea, whatever overcomes their ruined immune systems first. And the statistics, grim as they are, may be too low. There is no broad-scale AIDS testing: infection rates are calculated mainly from the presence of HIV in pregnant women. Death certificates in these countries do not record AIDS as the cause. "Whatever stats we have are not reliable," warns Mary Crewe of the University of Pretoria's Center for the Study of AIDS. "Everybody's guessing."

THE TB PATIENT

CASE NO. 309 IN THE TUGELA FERRY HOME-CARE PROGRAM shivers violently on the wooden planks someone has knocked into a bed, a frayed blanket pulled right up to his nose. He has the flushed skin, overbright eyes and careful breathing of the tubercular. He is alone, and it is chilly within the crumbling mud

walls of his hut at Msinga Top, a windswept outcrop high above the Tugela River in South Africa's KwaZulu-Natal province. The spectacular view of hills and veld would gladden a well man, but the 22-year-old we will call Fundisi Khumalo, though he does not know it, has AIDS, and his eyes seem to focus inward on his simple fear.

Before he can speak, his throat clutches in gasping spasms. Sharp pains rack his chest; his breath comes in shallow gasps. The vomiting is better today. But constipation has doubled up his knees, and he is too weak to go outside to relieve himself. He can't remember when he last ate. He can't remember how long he's been sick—"a long time, maybe since six months ago." Khumalo knows he has TB, and he believes it is just TB. "I am only thinking of that," he answers when we ask why he is so ill.

But the fear never leaves his eyes. He worked in a hair salon in Johannesburg, lived in a men's hostel in one of the cheap townships, had "a few" girlfriends. He knew other young men in the hostel who were on-and-off sick. When they fell too ill to work anymore, like him, they straggled home to rural villages like Msinga Top. But where Khumalo would not go is the hospital. "Why?" he says. "You are sick there, you die there."

"He's right, you know," says Dr. Tony Moll, who has driven us up the dirt track from the 350-bed hospital he heads in Tugela Ferry. "We have no medicines for AIDS. So many hospitals tell them, 'You've got AIDS. We can't help you. Go home and die.'" No one wants to be tested either, he adds, unless treatment is available. "If the choice is to know and get nothing," he says, "they don't want to know."

Here and in scattered homesteads all over rural Africa, the dying people say the sickness afflicting their families and neighbors is just the familiar consequence of their eternal poverty. Or it is the work of witchcraft. You have done something bad and have been bewitched. Your neighbor's jealousy has invaded you. You have not appeased the spirits of your ancestors, and they have cursed you. Some in South Africa believe the disease was introduced by the white population as a way to control black Africans after the end of apartheid.

Ignorance about AIDS remains profound. But because of the funerals, southern Africans can't help seeing that something more systematic and sinister lurks out there. Every Saturday and often Sundays too, neighbors trudge to the cemeteries for costly burial rites for the young and the middle-aged who are suddenly dying so much faster than the old. Families say it was pneumonia, TB, malaria that killed their son, their wife, their baby. "But you starting to hear the truth," says Durban home-care volunteer Busi Magwazi. "In the church, in the graveyard, they saying, 'Yes, she died of AIDS.' Oh, people talking about it even if the families don't admit it." Ignorance is the crucial reason the epidemic has run out of control. Surveys say many Africans here are becoming aware there is a sexually transmitted disease called AIDS that is incurable. But they don't think the risk applies to them. And their vague knowledge does not translate into changes in their sexual behavior. It's easy to see why so many don't yet sense the danger when few talk openly about the disease. And Africans are beset by so plentiful a roster of perils—famine, war, the violence of desperation or ethnic hatred, the regular illnesses of poverty, the dangers inside mines or on the roads—that the delayed risk of AIDS ranks low.

A CONTINENT IN PERIL

17 million Africans have died since the AIDS epidemic began in the late 1970s, more than 3.7 million of them children. An additional 12 million children have been orphaned by AIDS. An estimated 8.8% of adults in Africa are infected with HIV/AIDS, and in the following seven countries, at least 1 adult in 5 is living with HIV

1. Botswana

Though it has the highest per capita GDP, it also has the highest estimated adult infection rate—**36%**. 24,000 die each year. 66,000 children have lost their mother or both parents to the disease.

2. Swaziland

More than **25%** of adults have HIV/AIDS in this small country. 12,000 children have been orphaned, and 7,100 adults and children die each year.

3. Zimbabwe

One-quarter of the adult population is infected here. 160,000 adults and children died in 1999, and 900,000 children have been orphaned. Because of AIDS, life expectancy is 43.

4. Lesotho

24% of the adults are infected with HIV/AIDS. 35,000 children have been orphaned, and 16,000 adults and children die each year.

5. Zambia

20% of the adult population is infected, 1 in 4 adults in the cities. 650,000 children have been orphaned, and 99,000 Zambians died in 1999.

6. South Africa

This country has the largest number of people living with HIV/AIDS, about **20%** of its adult population, up from 13% in 1997. 420,000 children have been orphaned, and 250,000 people die each year from the disease.

7. Namibia

19.5% of the adult population is living with HIV. 57% of the infected are women. 67,000 children are AIDS orphans, and 18,000 adults and children die each year.

Source: UNAIDS

THE OUTCAST

TO ACKNOWLEDGE AIDS IN YOURSELF IS TO BE BRANDED AS monstrous. Laetitia Hambahlane (not her real name) is 51 and sick with AIDS. So is her brother. She admits it; he doesn't. In her mother's broken-down house in the mean streets of Umlazi township, though, Laetitia's mother hovers over her son, nursing him, protecting him, resolutely denying he has anything but TB, though his sister claims the sure symptoms of AIDS mark him. Laetitia is the outcast, first from her family, then from her society.

For years Laetitia worked as a domestic servant in Durban and dutifully sent all her wages home to her mother. She fell in love a number of times and bore four children. "I loved that last man," she recalls. "After he left, I had no one, no sex." That was 1992, but Laetitia already had HIV.

She fell sick in 1996, and her employers sent her to a private doctor who couldn't diagnose an illness. He tested her blood and found she was HIV positive. "I wish I'd died right then," she says, as tears spill down her sunken cheeks. "I asked the doctor, 'Have you got medicine?' He said no. I said, 'Can't you keep me alive?' " The doctor could do nothing and sent her away. "I couldn't face the word," she says. "I couldn't sleep at night. I sat on my bed, thinking, praying. I did not see anyone day or night. I ask God, Why?"

Laetitia's employers fired her without asking her exact diagnosis. For weeks she could not muster the courage to tell anyone. Then she told her children, and they were ashamed and frightened. Then, harder still, she told her mother. Her mother raged about the loss of money if Laetitia could not work again. She was so angry she ordered Laetitia out of the house. When her daughter wouldn't leave, the mother threatened to sell the house to get rid of her daughter. Then she walled off her daughter's room with plywood partitions, leaving the daughter a pariah, alone in a cramped, dark space without windows and only a flimsy door opening into the alley. Laetitia must earn the pennies to feed herself and her children by peddling beer, cigarettes and candy from a shopping cart in her room, when people are brave enough to stop by her door. "Sometimes they buy, sometimes not," she says. "That is how I'm surviving."

Her mother will not talk to her. "If you are not even accepted by your own family," says Magwazi, the volunteer home-care giver from Durban's Sinoziso project who visits Laetitia, "then others will not accept you." When Laetitia ventures outdoors, neighbors snub her, tough boys snatch her purse, children taunt her. Her own kids are tired of the sickness and don't like to help her anymore. "When I can't get up, they don't bring me food," she laments. One day local youths barged into her room, cursed her as a witch and a whore and beat her. When she told the police, the youths returned, threatening to burn down the house.

But it is her mother's rejection that wounds Laetitia most. "She is hiding it about my brother," she cries. "Why will she do nothing for me?" Her hands pick restlessly at the quilt covering her paper-thin frame. "I know my mother will not bury me properly. I know she will not take care of my kids when I am gone."

Jabulani Syabusi would use his real name, but he needs to protect his brother. He teaches school in a red, dusty district of KwaZulu-Natal. People here know the disease is all around them, but no one speaks of it. He eyes the scattered huts that make up his little settlement on an arid bluff. "We can count 20 who died just here as far as we can see. I personally don't remember any family that told it was AIDS," he says. "They hide it if they do know."

Syabusi's own family is no different. His younger brother is also a teacher who has just come home from Durban too sick to work anymore. He says he has tuberculosis, but after six months the tablets he is taking have done nothing to cure him. Syabusi's wife Nomsange, a nurse, is concerned that her 36-year-old brother-in-law may have something worse. Syabusi finally asked the doctor tending his brother what is wrong. The doctor said the information is confidential and will not tell him. Neither will his brother. "My brother is not brave enough to tell me," says Syabusi, as he stares sadly toward the house next door, where his only sibling lies ill. "And I am not brave enough to ask him."

Kennedy Fugewane, a cheerful, elderly volunteer counselor, sits in an empty U.S.-funded clinic that offers fast, pinprick blood tests in Francistown, Botswana, pondering how to break through the silence. This city suffers one of the world's highest infection rates, but people deny the disease because HIV is linked with sex. "We don't reveal anything," he says. "But people are so stigmatized even if they walk in the door." Africans feel they must keep private anything to do with sex. "If a man comes here, people will say he is running around," says Fugewane, though he acknowledges that men never do come. "If a woman comes, people will say she is loose. If anyone says they got HIV, they will be despised."

Pretoria University's Mary Crewe says, "It is presumed if you get AIDS, you have done something wrong." HIV labels you as living an immoral life. Embarrassment about sexuality looms more important than future health risks. "We have no language to talk candidly about sex," she says, "so we have no civil language to talk about AIDS." Volunteers like Fugewane try to reach out with flyers, workshops, youth meetings and free condoms, but they are frustrated by a culture that values its dignity over saving lives. "People here don't have the courage to come forward and say, 'Let me know my HIV status,'" he sighs, much less the courage to do something about it. "Maybe one day…"

Doctors bow to social pressure and legal strictures not to record AIDS on death certificates. "I write TB or meningitis or diarrhea but never AIDS," says South Africa's Dr. Moll. "It's a public document, and families would hate it if anyone knew." Several years ago, doctors were barred even from recording compromised immunity or HIV status on a medical file; now they can record the results of blood tests for AIDS on patient charts to protect other health workers. Doctors like Moll have long agitated to apply the same openness to death certificates.

THE TRUCK DRIVER

HERE, MEN HAVE TO MIGRATE TO WORK, INSIDE THEIR COUNTries or across borders. All that mobility sows HIV far and wide, as Louis Chikoka is the first to recognize. He regularly drives the highway that is Botswana's economic lifeline and its curse. The road runs for 350 miles through desolate bush that is the Texas-size country's sole strip of habitable land, home to a large majority of its 1.5 million people. It once brought prospectors to Botswana's rich diamond reefs. Now it's the link for transcontinental truckers like Chikoka who haul goods from South Africa to markets in the continent's center. And now the road brings AIDS.

Chikoka brakes his dusty, diesel-belching Kabwe Transport 18-wheeler to a stop at the dark roadside rest on the edge of Francistown, where the international trade routes converge and at least 43% of adults are HIV-positive. He is a cheerful man even after 12 hard hours behind the wheel freighting rice from Durban. He's been on the road for two weeks and will reach his destination in Congo next Thursday. At 39, he is married, the father of three and a long-haul trucker for 12 years. He's used to it.

Lighting up a cigarette, the jaunty driver is unusually loquacious about sex as he eyes the dim figures circling the rest stop. Chikoka has parked here for a quickie. See that one over there, he points with his cigarette. "Those local ones we call bitches. They always waiting here for short service." Short service? "It's according to how long it takes you to ejaculate," he explains. "We go to the 'bush bedroom' over there [waving at a clump of trees 100 yds. away] or sometimes in the truck. Short service, that costs you 20 rands [$2.84]. They know we drivers always got money."

Chikoka nods his head toward another woman sitting beside a stack of cardboard cartons. "We like better to go to them," he says. They are the "businesswomen," smugglers with gray-market cases of fruit and toilet paper and toys that they need to transport somewhere up the road. "They come to us, and we negotiate privately about carrying their goods." It's a no-cash deal, he says. "They pay their bodies to us." Chikoka shrugs at a suggestion that the practice may be unhealthy. "I been away two weeks, madam. I'm human. I'm a man. I have to have sex."

What he likes best is dry sex. In parts of sub-Saharan Africa, to please men, women sit in basins of bleach or saltwater or stuff astringent herbs, tobacco or fertilizer inside their vagina. The tissue of the lining swells up and natural lubricants dry out. The resulting dry sex is painful and dangerous for women. The drying agents suppress natural bacteria, and friction easily lacerates the tender walls of the vagina. Dry sex increases the risk of HIV infection for women, already two times as likely as men to contract the virus from a single encounter. The women, adds

Chikoka, can charge more for dry sex, 50 or 60 rands ($6.46 to $7.75), enough to pay a child's school fees or to eat for a week.

UNVANQUISHED

A Fighter in a Land of Orphans

Silence and the ignorance it promotes have fed the AIDS epidemic in Africa perhaps more than any other factors. In Malawi, where until the end of dictator Hastings Banda's rule in 1994 women were barred from wearing short skirts and men could be jailed for having long hair, public discussion of AIDS was forbidden. According to the government, AIDS didn't exist inside Malawi. Catherine Phiri, 38, knew otherwise. She tested positive in 1990, after her husband had died of the disease. Forced to quit her job as a nurse when colleagues began to gossip, she sought refuge with relatives in the capital, Lilongwe. But they shunned her and eventually forced her to move, this time to Salima on beautiful Lake Malawi. "Even here people gossiped," says Phiri, who's brave, open face is fringed by a head of closely cropped graying hair.

Determined to educate her countrymen, Phiri set up a group that offers counseling, helps place orphans and takes blood that can then be tested in the local hospital. "The community began to see the problem, but it was very difficult to communicate to the government. They didn't want to know.

They do now. According to a lawmaker, AIDS has killed dozens of members of Parliament in the past decade. And Malawi's government has begun to move. President Bakili Muluzi incorporates AIDS education into every public rally. In 1999 he launched a five-year plan to fight the disease, and last July he ordered a crackdown on prostitution (though the government is now thinking of legalizing it). At the least, his awareness campaign appears to be working: 90% of Malawians know about the dangers of AIDS. But that knowledge comes too late for the estimated 8% of HIV-positive citizens—800,000 people in 1999—or the 276,000 children under 15 orphaned by the disease.

Last October, Phiri picked up an award for her efforts from the U.N. But, she says, "I still have people who look at me like trash..." Her voice trails off. "Sometimes when I go to sleep I fear for the future of my children. But I will not run away now. Talking about it: that's what's brave."

—By Simon Robinson/Salima

Chikoka knows his predilection for commercial sex spreads AIDS; he knows his promiscuity could carry the disease home to his wife; he knows people die if they get it. "Yes, HIV is terrible, madam," he says as he crooks a finger toward the businesswoman whose favors he will enjoy that night. "But, madam, sex is natural. Sex is not like beer or smoking. You can stop them.

But unless you castrate the men, you can't stop sex—and then we all die anyway."

Millions of men share Chikoka's sexually active lifestyle, fostered by the region's dependence on migrant labor. Men desperate to earn a few dollars leave their women at hardscrabble rural homesteads to go where the work is: the mines, the cities, the road. They're housed together in isolated males-only hostels but have easy access to prostitutes or a "town wife" with whom they soon pick up a second family and an ordinary STD and HIV. Then they go home to wives and girlfriends a few times a year, carrying the virus they do not know they have. The pattern is so dominant that rates of infection in many rural areas across the southern cone match urban numbers.

IN SOME AFRICAN COUNTRIES, THE INFECTION RATE OF TEEN GIRLS IS FOUR TIMES THAT OF BOYS

If HIV zeros in disproportionately on poor migrants, it does not skip over the educated or the well paid. Soldiers, doctors, policemen, teachers, district administrators are also routinely separated from families by a civil-service system that sends them alone to remote rural posts, where they have money and women have no men. A regular paycheck procures more access to extramarital sex. Result: the vital professions are being devastated.

Schoolmaster Syabusi is afraid there will soon be no more teachers in his rural zone. He has just come home from a memorial for six colleagues who died over the past few months, though no one spoke the word AIDS at the service. "The rate here—they're so many," he says, shaking his head. "They keep on passing it at school." Teachers in southern Africa have one of the highest group infection rates, but they hide their status until the telltale symptoms find them out.

Before then, the men—teachers are mostly men here—can take their pick of sexual partners. Plenty of women in bush villages need extra cash, often to pay school fees, and female students know they can profit from a teacher's favor. So the schoolmasters buy a bit of sex with lonely wives and trade a bit of sex with willing pupils for A's. Some students consider it an honor to sleep with the teacher, a badge of superiority. The girls brag about it to their peers, preening in their ability to snag an older man. "The teachers are the worst," says Jabulani Siwela, an AIDS worker in Zimbabwe who saw frequent teacher-student sex in his Bulawayo high school. They see a girl they like; they ask her to stay after class; they have a nice time. "It's dead easy," he says. "These are men who know better, but they still do it all the time."

THE PROSTITUTE

THE WORKINGWOMAN WE MEET DIRECTS OUR CAR TO A reedy field fringing the gritty eastern townships of Bulawayo, Zimbabwe. She doesn't want neighbors to see her being interviewed. She is afraid her family will find out she is a prostitute, so we will call her Thandiwe. She looked quite prim and proper in her green calf-length dress as she waited for johns outside 109 Tongogaro Street in the center of downtown. So, for that matter, do the dozens of other women cruising the city's dim street corners: not a mini or bustier or bared navel in sight. Zimbabwe is in many ways a prim and proper society that frowns on commercial sex work and the public display of too much skin.

FINANCIAL AID

A Lending Tree

Getting ahead in Africa is tough. Banks lend money only to the middle class and the wealthy. Poor Africans—meaning most Africans—stay poor. It's even harder if you're sick. Without savings to fall back on, many HIV-positive parents pull their kids out of school. They can't afford the fees and end up selling their few possessions to feed the family. When they die, their kids are left with nothing.

Though not directly targeted at people with AIDS, microcredit schemes go some way toward fixing that problem. The schemes work like minibanks, lending small amounts—often as little as $100—to traders or farmers. Because they lack the infrastructure of banks and don't charge fees, most charge an interest rate of as much as 1% a week and repayment rates of over 99%—much better than that for banks in Africa, or in most places.

Many microcredit schemes encourage clients to set aside some of the extra income generated by the loan as savings. This can be used for medical bills or to pay school fees if the parents get sick. "Without the loans I would have had to look for another way to make money," says Florence Muriungi, 40, who sings in a Kampala jazz band and whose husband died of AIDS four years ago. Muriungi, who cares for eight children—five of her own and three her sister left when she too died of AIDS—uses the money to pay school fees in advance and fix her band's equipment. Her singing generates enough money for her to repay the loans and save a bit.

Seventeen of the 21 women at a weekly meeting of regular borrowers in Uganda care for AIDS orphans. Five are AIDS widows. "I used to buy just one or two bunches of bananas to sell. Now I buy 40, 50, 60," says Elizabeth Baluka, 47, the group's secretary. "Every week I put aside a little bit of money to help my children slowly by slowly."

—By Simon Robinson/Kampala

That doesn't stop Thandiwe from earning a better living turning tricks than she ever could doing honest work. Desperate for a job, she slipped illegally into South Africa in 1992. She cleaned floors in a Johannesburg restaurant, where she met a cook from back home who was also illegal. They had two daughters, and they got married; he was gunned down one night at work.

She brought his body home for burial and was sent to her in-laws to be "cleansed." This common practice gives a dead husband's brother the right, even the duty, to sleep with the widow. Thandiwe tested negative for HIV in 1998, but if she were positive, the ritual cleansing would have served only to pass on the disease. Then her in-laws wanted to keep her two daughters because their own children had died, and marry her off to an old uncle who lived far out in the bush. She fled.

Alone, Thandiwe grew desperate. "I couldn't let my babies starve." One day she met a friend from school. "She told me she was a sex worker. She said, 'Why you suffer? Let's go to a place where we can get quick bucks.'" Thandiwe hangs her head. "I went. I was afraid. But now I go every night."

She goes to Tongogaro Street, where the rich clients are, tucking a few condoms in her handbag every evening as the sun sets and returning home strictly by 10 so that she won't have to service a taxi-van driver to get a ride back. Thandiwe tells her family she works an evening shift, just not at what. "I get 200 zim [$5] for sex," she says, more for special services. She uses two condoms per client, sometimes three. "If they say no, I say no." But then sometimes resentful johns hit her. It's pay-and-go until she has pocketed 1,000 or 1,500 Zimbabwe dollars and can go home—with more cash than her impoverished neighbors ever see in their roughneck shantytown, flush enough to buy a TV and fleece jammies for her girls and meat for their supper.

"I am ashamed," she murmurs. She has stopped going to church. "Every day I ask myself, 'When will I stop this business?' The answer is, 'If I could get a job'…" Her voice trails off hopelessly. "At the present moment, I have no option, no other option." As trucker Chikoka bluntly puts it, "They give sex to eat. They got no man; they got no work; but they got kids, and they got to eat." Two of Thandiwe's friends in the sex trade are dying of AIDS, but what can she do? "I just hope I won't get it."

In fact, casual sex of every kind is commonplace here. Prostitutes are just the ones who admit they do it for cash. Everywhere there's premarital sex, sex as recreation. Obligatory sex and its abusive counterpart, coercive sex. Transactional sex: sex as a gift, sugar-daddy sex. Extramarital sex, second families, multiple partners. The nature of AIDS is to feast on promiscuity.

79% OF THOSE WHO DIED OF AIDS LAST YEAR WERE AFRICAN

Rare is the man who even knows his HIV status: males widely refuse testing even when they fall ill. And many men who suspect they are HIV positive embrace a flawed logic: if I'm already infected, I can sleep around because I can't get it again. But women are the ones who progress to full-blown AIDS first and die fastest, and the underlying cause is not just sex but power. Wives and girlfriends and even prostitutes in this part of the world can't easily say no to sex on a man's terms. It matters little what comes into play, whether it is culture or tradition or the pathology of violence or issues of male identity or the subservient status of women.

Beneath a translucent scalp, the plates of Gertrude Dhlamini's cranium etch a geography of pain. Her illness is obvious in the thin, stretched skin under which veins throb with the shingles that have blinded her left eye and scarred that side of her face. At 39, she looks 70. The agonizing thrush, a kind of fungus, that paralyzed her throat has ebbed enough to enable her to swallow a spoon or two of warm gruel, but most of the nourishment flows away in constant diarrhea. She struggles to keep her hand from scratching restlessly at the scaly rash flushing her other cheek. She is not ashamed to proclaim her illness to the world. "It must be told," she says.

Gertrude is thrice rejected. At 19 she bore a son to a boyfriend who soon left her, taking away the child. A second boyfriend got her pregnant in 1994 but disappeared in anger when their daughter was born sickly with HIV. A doctor told Gertrude it was her fault, so she blamed herself that little Noluthando was never well in the two years she survived. Gertrude never told the doctor the baby's father had slept with other women. "I was afraid to," she says, "though I sincerely believe he gave the sickness to me." Now, she says, "I have rent him from my heart. And I will never have another man in my life."

Gertrude begged her relatives to take her in, but when she revealed the name of her illness, they berated her. They made her the household drudge, telling her never to touch their food or their cooking pots. They gave her a bowl and a spoon strictly for her own use. After a few months, they threw her out.

Gertrude sits upright on a donated bed in a cardboard shack in a rough Durban township that is now the compass of her world. Perhaps 10 ft. square, the little windowless room contains a bed, one sheet and blanket, a change of clothes and a tiny cooking ring, but she has no money for paraffin to heat the food that a home-care worker brings. She must fetch water and use a toilet down the hill. "Everything I have," she says, "is a gift." Now the school that owns the land under her hut wants to turn it into a playground and she worries about where she will go. Gertrude rubs and rubs at her raw cheek. "I pray and pray to God," she says, "not to take my soul while I am alone in this room."

Women like Gertrude were brought up to be subservient to men. Especially in matters of sex, the man is always in charge. Women feel powerless to change sexual behavior. Even when a woman wants to protect herself, she usually can't: it is not uncommon for men to beat partners who refuse intercourse or request a condom. "Real men" don't use them, so women who want their partners to must fight deeply ingrained taboos. Talk to him about donning a rubber sheath and be prepared for accusations, abuse or abandonment.

A nurse in Durban, coming home from an AIDS training class, suggested that her mate should put on a condom, as a kind of homework exercise. He grabbed a pot and banged loudly on it with a knife, calling all the neighbors into his house. He pointed the knife at his wife and demanded: "Where was she between 4 p.m. and now? Why is she suddenly suggesting this? What has changed after 20 years that she wants a condom?"

Schoolteacher Syabusi is an educated man, fully cognizant of the AIDS threat. Yet even he bristles when asked if he uses a condom. "Humph," he says with a fine snort. "That question is nonnegotiable." So despite extensive distribution of free condoms, they often go unused. Astonishing myths have sprung up. If you don one, your erection can't grow. Free condoms must be too cheap to be safe: they have been stored too long, kept too hot, kept too cold. Condoms fill up with germs, so they spread AIDS. Condoms from overseas bring the disease with them. Foreign governments that donate condoms put holes in them so that Africans will die. Education programs find it hard to compete with the power of the grapevine.

THE CHILD IN NO. 17

IN CRIB NO. 17 OF THE SPARTAN BUT CROWDED CHILDREN'S ward at the Church of Scotland Hospital in KwaZulu-Natal, a tiny, staring child lies dying. She is three and has hardly known a day of good health. Now her skin wrinkles around her body like an oversize suit, and her twig-size bones can barely hold her vertical as nurses search for a vein to take blood. In the frail arms hooked up to transfusion tubes, her veins have collapsed. The nurses palpate a threadlike vessel on the child's forehead. She mews like a wounded animal as one tightens a rubber band around her head to raise the vein. Tears pour unnoticed from her mother's eyes as she watches the needle tap-tap at her daughter's temple. Each time the whimpering child lifts a wan hand to brush away the pain, her mother gently lowers it. Drop by drop, the nurses manage to collect 1 cc of blood in five minutes.

The child in crib No. 17 has had TB, oral thrush, chronic diarrhea, malnutrition, severe vomiting. The vial of blood reveals her real ailment, AIDS, but the disease is not listed on her chart, and her mother says she has no idea why her child is so ill. She breast-fed her for two years, but once the little girl was weaned, she could not keep solid food down. For a long time, her mother thought something was wrong with the food. Now the child is afflicted with so many symptoms that her mother had to bring her to the hospital, from which sick babies rarely return.

VIRGINITY TESTING IS BACK The practice of virginity testing used to be part of traditional Zulu rites. It is regaining popularity among anxious mothers who believe that if their daughters remain virgins, they won't get AIDS.

She hopes, she prays her child will get better, and like all the mothers who stay with their children at the hospital, she tends her lovingly, constantly changing filthy diapers, smoothing sheets, pressing a little nourishment between listless lips, trying to tease a smile from the vacant, staring face. Her husband works in Johannesburg, where he lives in a men's squatter camp. He comes home twice a year. She is 25. She has heard of

AIDS but does not know it is transmitted by sex, does not know if she or her husband has it. She is afraid this child will die soon, and she is afraid to have more babies. But she is afraid too to raise the subject with her husband. "He would not agree to that," she says shyly. "He would never agree to have no more babies."

Dr. Annick DeBaets, 32, is a volunteer from Belgium. In the two years she has spent here in Tugela Ferry, she has learned all about how hard it is to break the cycle of HIV transmission from mother to infant. The door to this 48-cot ward is literally a revolving one: sick babies come in, receive doses of rudimentary antibiotics, vitamins, food; go home for a week or a month; then come back as ill as ever. Most, she says, die in the first or second year. If she could just follow up with really intensive care, believes Dr. DeBaets, many of the wizened infants crowding three to a crib could live longer, healthier lives. "But it's very discouraging. We simply don't have the time, money or facilities for anything but minimal care."

Much has been written about what South African Judge Edwin Cameron, himself HIV positive, calls his country's "grievous ineptitude" in the face of the burgeoning epidemic. Nowhere has that been more evident than in the government's failure to provide drugs that could prevent pregnant women from passing HIV to their babies. The government has said it can't afford the 300-rand-per-dose, 28-dose regimen of AZT that neighboring nations like Botswana dole out, using funds and drugs from foreign donors. The late South African presidential spokesman Parks Mankahlana even suggested publicly that it was not cost effective to save these children when their mothers were already doomed to die: "We don't want a generation of orphans."

Yet these children—70,000 are born HIV positive in South Africa alone every year—could be protected from the disease for about $4 each with another simple, cheap drug called nevirapine. Until last month, the South African government steadfastly refused to license or finance the use of nevirapine despite the manufacturer's promise to donate the drug for five years, claiming that its "toxic" side effects are not yet known. This spring, however, the drug will finally be distributed to leading public hospitals in the country, though only on a limited basis at first.

The mother at crib No. 17 is not concerned with potential side effects. She sits on the floor cradling her daughter, crooning over and over, "Get well, my child, get well." The baby stares back without blinking. "It's sad, so sad, so sad," the mother says. The child died three days later.

The children who are left when parents die only add another complex dimension to Africa's epidemic. At 17, Tsepho Phale has been head of an indigent household of three young boys in the dusty township of Monarch, outside Francistown, for two years. He never met his father, his mother died of AIDS, and the grieving children possess only a raw concrete shell of a house. The doorways have no doors; the window frames no glass. There is not a stick of furniture. The boys sleep on piled-up blankets, their few clothes dangling from nails. In the room that passes for a kitchen, two paraffin burners sit on the dirt floor alongside the month's food: four cabbages, a bag of oranges and one of potatoes, three sacks of flour, some yeast, two jars of oil

and two cartons of milk. Next to a dirty stack of plastic pans lies the mealy meal and rice that will provide their main sustenance for the month. A couple of bars of soap and two rolls of toilet paper also have to last the month. Tsepho has just brought these rations home from the social-service center where the "orphan grants" are doled out.

Tsepho has been robbed of a childhood that was grim even before his mother fell sick. She supported the family by "buying and selling things," he says, but she never earned more than a pittance. When his middle brother was knocked down by a car and left physically and mentally disabled, Tsepho's mother used the insurance money to build this house, so she would have one thing of value to leave her children. As the walls went up, she fell sick. Tsepho had to nurse her, bathe her, attend to her bodily functions, try to feed her. Her one fear as she lay dying was that her rural relatives would try to steal the house. She wrote a letter bequeathing it to her sons and bade Tsepho hide it.

As her body lay on the concrete floor awaiting burial, the relatives argued openly about how they would divide up the profits when they sold her dwelling. Tsepho gave the district commissioner's office the letter, preventing his mother's family from grabbing the house. Fine, said his relations; if you think you're a man, you look after your brothers. They have contributed nothing to the boys' welfare since. "It's as if we don't exist anymore either," says Tsepho. Now he struggles to keep house for the others, doing the cooking, cleaning, laundry and shopping.

The boys look at the future with despair. "It is very bleak," says Tsepho, kicking aimlessly at a bare wall. He had to quit school, has no job, will probably never get one. "I've given up my dreams. I have no hope."

Orphans have traditionally been cared for the African way: relatives absorb the children of the dead into their extended families. Some still try, but communities like Tsepho's are becoming saturated with orphans, and families can't afford to take on another kid, leaving thousands alone.

Now many must fend for themselves, struggling to survive. The trauma of losing parents is compounded by the burden of becoming a breadwinner. Most orphans sink into penury, drop out of school, suffer malnutrition, ostracism, psychic distress. Their makeshift households scramble to live on pitiful handouts—from overstretched relatives, a kind neighbor, a state grant—or they beg and steal in the streets. The orphans' present desperation forecloses a brighter future. "They hardly ever succeed in having a life," says Siphelile Kaseke, 22, a counselor at an AIDS orphans' camp near Bulawayo. Without education, girls fall into prostitution, and older boys migrate illegally to South Africa, leaving the younger ones to go on the streets.

1 IN 4 SOUTH AFRICAN WOMEN AGES 20 TO 29 IS INFECTED WITH HIV

EVERY DAY SPENT IN THIS PART OF AFRICA IS ACUTELY DEPRESSING: there is so little countervailing hope to all the stories of the dead and the doomed. "More than anywhere else in the world, AIDS in Africa was met with apathy," says Suzanne LeClerc-

Madlala, a lecturer at the University of Natal. The consequences of the silence march on: infection soars, stigma hardens, denial hastens death, and the chasm between knowledge and behavior widens. The present disaster could be dwarfed by the woes that loom if Africa's epidemic rages on. The human losses could wreck the region's frail economies, break down civil societies and incite political instability.

In the face of that, every day good people are doing good things. Like Dr. Moll, who uses his after-job time and his own fund raising to run an extensive volunteer home-care program in KwaZulu-Natal. And Busi Magwazi, who, along with dozens of others, tends the sick for nothing in the Durban-based Sinoziso project. And Patricia Bakwinya, who started her Shining Stars orphan-care program in Francistown with her own zeal and no money, to help youngsters like Tsepho Phale. And countless individuals who give their time and devotion to ease southern Africa's plight.

But these efforts can help only thousands; they cannot turn the tide. The region is caught in a double bind. Without treatment, those with HIV will sicken and die; without prevention, the spread of infection cannot be checked. Southern Africa has no other means available to break the vicious cycle, except to change everyone's sexual behavior—and that isn't happening.

The essential missing ingredient is leadership. Neither the countries of the region nor those of the wealthy world have been able or willing to provide it.

South Africa, comparatively well off, comparatively well educated, has blundered tragically for years. AIDS invaded just when apartheid ended, and a government absorbed in massive transition relegated the disease to a back page. An attempt at a national education campaign wasted millions on a farcical musical. The premature release of a local wonder drug ended in scandal when the drug turned out to be made of industrial solvent. Those fiascoes left the government skittish about embracing expensive programs, inspiring a 1998 decision not to provide AZT to HIV-positive pregnant women. Zimbabwe too suffers savagely from feckless leadership. Even in Botswana, where the will to act is gathering strength, the resources to follow through have to come from foreign hands.

AIDS' grip here is so pervasive and so complex that all societies—theirs and ours—must rally round to break it. These countries are too poor to doctor themselves. The drugs that could begin to break the cycle will not be available here until global pharmaceutical companies find ways to provide them inexpensively. The health-care systems required to prescribe and monitor complicated triple-cocktail regimens won't exist unless rich countries help foot the bill. If there is ever to be a vaccine, the West will have to finance its discovery and provide it to the poor. The cure for this epidemic is not national but international.

The deep silence that makes African leaders and societies want to deny the problem, the corruption and incompetence that render them helpless is something the West cannot fix. But the fact that they are poor is not. The wealthy world must help with its zeal and its cash if southern Africa is ever to be freed of the AIDS plague.

A UGANDAN TALE

Not Afraid to Speak Out

Major Rubaramira Ruranga knows something about fighting. During Idi Amin's reign of terror in Uganda in the 1970s, Ruranga worked as a spy for rebels fighting the dictator. After Amin's ouster, the military man studied political intelligence in Cuba before returning to find a new dictator at the helm and a blood war raging. Hoping for change, Ruranga supplied his old rebel friends with more secrets, this time from within the President's office. When he was discovered, he fled to the bush to "fight the struggle with guns."

The turmoil in Uganda was fueling the spread of another enemy—AIDS. Like many rebel soldiers, Ruranga was on the move constantly to avoid detection. "You never see your wife, and so you get to a new place and meet someone else," he says. "I had sex without protection with a few women." Doctors found he was HIV positive in 1989. "They told me I would die in two to three years, so I started preparing for when I was away. I told my kids, my wife. Worked on finishing the house for them. I gave up hope." But as he learned about AIDS, his attitude changed. After talking to American and European AIDS activists—some had lived with the disease for 15 years or more—"I realized I was not going to die in a few years. I was reborn, determined to live."

He began fighting again. After announcing his HIV status at a rally on World AIDS Day in 1993—an extraordinarily brave act in Africa, where few activists, let alone army officers, ever admit to having HIV—he set up a network for those living with HIV/AIDS in Uganda, "so that people had somewhere to go to talk to friends." And while Uganda has done more to slow the spread of AIDS than any other country—in some places the rate of infection has dropped by half—"we can always do better," says Ruranga. "Why are we able to buy guns and bullets to kill people and we are not able to buy drugs to save people?" The fight continues.

—By Simon Robinson/Kampala

An Epidemic of Neglect

Neglected Diseases and

the Health Burden in Poor Countries

By Rachel Cohen

CARDIOVASCULAR DISEASE AND CANCER, baldness and impotence—in recent decades, innovative drug treatments have helped mitigate or cure everything from life-threatening disease to the unpleasant marks of aging. But this "health revolution," which has resulted in considerable gains in life expectancy and health improvements in some parts of the world, has left most of the world's population behind.

Drug research and development (R&D) for diseases that disproportionately affect poor people in developing countries diseases is at a virtual standstill. According to a study by Patrice Trouiller and colleagues, soon to be published in the Lancet, of the 1,393 new drugs approved between 1975 and 1999, only 16 (or just over 1 percent) were specifically developed for tropical diseases and tuberculosis, diseases that account for 11.4 percent of the global disease burden.

Doctors Without Borders/Médecins Sans Frontières (MSF), the international humanitarian organization that was awarded the 1999 Nobel Peace Prize, has called attention to the problem of "neglected diseases." These are seriously disabling or life-threatening diseases—such as malaria, tuberculosis (TB), Human African trypanosomiasis (sleeping sickness), American trypanosomiasis (Chagas' disease), and visceral leishmaniasis (kala azar)—which mainly affect people in developing countries, for which treatment options are inadequate or do not exist, and for which R&D is insufficient or non-existent.

Those stricken with neglected diseases do not constitute a valuable enough market to stimulate adequate R&D for new medicines by the multinational pharmaceutical industry.

With a global system that relies on private companies to bring drugs to market, this situation has left much of the world without prospect for treatment and cure of the diseases that take a dreadful toll on their populations.

However, there are now, for the first time, glimmers of hope on the horizon. With public health campaigners drawing some international attention to the problem of neglected diseases, a number of diverse initiatives have been launched to spur R&D into these illnesses. But these new efforts are unproven and poorly funded, and some are pursuing controversial partnerships with industry. Their ultimate, or even partial, success is far from assured.

THE BURDEN OF NEGLECTED DISEASES

According to the World Health Organization, 14 million people die each year from communicable diseases such as malaria, TB, sleeping sickness and kala azar. An estimated 97 percent of deaths from communicable diseases occur in developing countries, with the poorest people in those nations disproportionately affected. Infectious and parasitic diseases account for 25 percent of the disease burden in low-and middle-income countries, compared to only 3 percent in high-income countries. According to the World Bank, eliminating communicable diseases would dramatically close the mortality gap between the richest 20 percent of the world population and the poorest 20 percent.

In many instances, the ravages of communicable diseases are worsening. Old diseases are reappearing and drug resistance is spreading rapidly, rendering useless medicines that were once effective. Some of the medicines used today were invented decades ago and are highly toxic. For some diseases, there is no treatment to offer: no effective medicine has been found and nobody is looking for new possibilities.

Sleeping sickness is a primary example. A parasitic disease that affects 36 countries in sub-Saharan Africa, an estimated 300,000 to 500,000 people are currently infected with this disease, and the figure appears to be on the rise. Left untreated, sleeping sickness is 100 percent fatal. It kills more than 65,000 people each year.

In many places, melarsoprol has been the only treatment available for second stage sleeping sickness. Developed in 1949, melarsoprol is an arsenic derivative so toxic that it kills one in 20 patients. Production of eflornithine, a safer and far more effective medicine, was discontinued in 1995 because it was not considered profitable by its manufacturer (now Aventis).

Production of eflornithine resumed in 2001, after Bristol-Myers Squibb (BMS) discovered that a topical ointment containing eflornithine as the active ingredient was useful in removing unwanted facial hair in women. BMS received approval to market the ointment for cosmetic use in 2000.

Following a major behind-the-scenes advocacy effort by MSF and other health advocates, Aventis agreed to resume production of eflornithine, donating enough of the drug to meet global sleeping sickness need for the next five years. Aventis also committed to working with WHO to find a long-term solution for the manufacture of the drug, and to invest in R&D, for example, into an oral formulation of eflornithine that will be easier to administer in field conditions. Eflornithine has yet to reach all the hospitals and clinics in Africa that desperately need it in order to begin treating people, however.

A more well-known neglected disease in industrialized countries, because it affects travelers to, as well as residents in, tropical countries, is malaria. Every year, an estimated 300 million to 500 million cases of malaria occur in more than 90 countries worldwide. Ninety percent of cases occur in Africa. One child dies every 30 seconds from malaria. Chloroquine, a medicine developed in 1934, continues to be the only treatment available for many people with malaria, despite growing resistance to it. Unfortunately, high resistance in many parts of the world has rendered it virtually useless—in some regions, the chloroquine resistance rate is up to 100 percent, and the parasite has now also developed resistance to some newer drugs, including sulfadoxinepyrimethamine (SP) and mefloquine.

Unlike some other neglected diseases, effective treatments do exist for malaria, most notably combination therapies containing artemisinin, which is derived from a Chinese plant. But governments are hesitant to change national malaria protocols to include these new medicines because they are much more costly than traditional antimalarials like chloroquine and SP. In many places,

doctors have no choice but to prescribe chloroquine or SP, often knowing these drugs will not work. Meanwhile, there is not enough in the research pipeline, particularly for drug-resistant malaria.

Leishmaniasis is another severe neglected disease. Around 2 million people become ill with leishmaniasis every year, and an estimated 12 million people are currently affected by leishmaniasis in its different forms. The most severe form, visceral leishmaniasis or kala azar, is a fatal disease that affects 500,000 and kills more than 65,000 annually.

Sodium stibogluconate (SSG), also known by its brand name Pentostam and marketed by GlaxoSmithKline, is the most common medicine available for kala azar. Developed in the 1930s, it is expensive, difficult to administer and requires hospitalization for one month. Many people affected live in conflict zones where it is hard for medical teams to reach patients, and the long course of treatment can be interrupted if the hospitals are caught in fighting. For them, a month's hospital stay is an impossibility. Resistance to SSG is on the rise and the few other drugs that can be used are highly toxic and very expensive.

Clinical tests in India are proving encouraging for miltefosine and paromomycin, but these results have yet to be finalized for registration.

MARKET FAILURES

Over the last few decades, major progress in molecular biology and biotechnology has enabled the development of increasingly sophisticated medicines to cure a wide variety of diseases.

Meanwhile, global expenditures on health R&D have increased dramatically. However, R&D funds are focused on diseases of the rich. Only 10 percent of global health research is devoted to conditions that account for 90 percent of the global disease burden—an imbalance that has been referred to as the 10/90 disequilibrium.

"Heavy reliance on an increasingly consolidated and highly competitive multinational drug industry to generate new medicines has left the development of life-saving drugs subject to the forces of the market economy," says Els Torreele, co-chair of the Drugs for Neglected Diseases (DND) Working Group, an independent, international, interdisciplinary team formed by MSF.

"Pharmaceutical companies simply have no financial incentive to invest in the development of new treatments for diseases that affect people who can't afford the medicines." she says.

People in developing countries, who make up about 80 percent of the world population, only represent about 20 percent of worldwide medicine sales. All of sub-Saharan Africa totals less than 2 percent of the global pharmaceutical market.

Industry leaders acknowledge their narrow focus. "The United States has become the must-win market for every pharmaceutical company," Fred Hassan, CEO of Pharmacia Corporation, told the World Conference on Clinical Pharmacology and Therapeutics in July 2000. "In addition, there are just six or seven other critical markets, including Japan and key countries in Europe.

"This does not mean ignoring other markets," Hassan said. "But it does mean strategically concentrating resources and top management attention on success in the key market. Again, this is very different from our industry's approach in the past, which focused on therapeutic areas across geographical regions."

Following this model, the pharmaceutical industry has prospered, with companies often having sales of hundreds of millions or even billions of dollars a year on a single drug. Pharmaceutical companies top the U.S. industry performance list for return on investment, with a 39 percent return for shareholders, according to Fortune magazine.

This model has not worked to develop new drugs for diseases that affect the world's poor, however. A Harvard University survey of 11 top-grossing pharmaceutical companies found that none had brought a drug to market in the last five years for three of the most neglected diseases—sleeping sickness, kala azar and Chagas' disease—and that there is hardly anything currently in the research pipeline for these diseases. Overall R&D budgets for the responding companies ranged from $500 million to more than $1 billion per year.

A recent survey on new medicines in development conducted by the U.S. drug industry lobby group PhRMA reports that of the 137 medicines for infectious diseases in the pipeline during 2000, only one mentioned sleeping sickness as an indication, and only one mentioned malaria. There were no new medicines in the pipeline for tuberculosis or leishmaniasis. PhRMA's 2000 "New Medicines in Development" list shows eight drugs in development for impotence and erectile dysfunction, seven for obesity and four for sleep disorders.

FROM MARKETS TO PUBLIC HEALTH

With purchasing power defining the industry research agenda and priorities, there is little prospect for the pharmaceutical industry addressing neglected diseases.

But public health activists say doing nothing is unacceptable. Governments, which should have a broader outlook than corporations, they say, have also failed.

"Having left drug development to the most profitable industry in the world, governments have shirked their responsibility to meet the health needs that the market does not care about," says Dr. Bernard Pecoul, director of MSF's Campaign for Access to Essential Medicines.

"Patents are granted with the aim of stimulating innovation," says Pecoul, "but a market monopoly is irrelevant when the market is non-existent. No amount of market tinkering will stimulate drug companies to invest

in drug development for patients with kala azar and sleeping sickness in Sudan, Congo or India, who are too poor to ever represent a 'viable market.'"

In this vacuum, advocates for neglected diseases are searching for solutions.

Some recent initiatives have sought to find novel approaches for stimulating research into neglected diseases. Public-private partnerships, such as the Global Alliance for TB Drug Development (GATB) and the Medicines for Malaria Venture (MMV), are emerging, seeking to induce public and private sector research.

The single biggest change in funding for neglected diseases over the past few years has come not from private industry or the public sector, but from the increased commitment of foundations. The Bill and Melinda Gates Foundation has become a major force in neglected disease drug development. The Gates Foundation has given $25 million (over five years) to both GATB and MMV.

It is too early to say how successful these efforts will be. Dr. James Orbinksi, co-chair of the DND Working Group, worries that many of the public-private partnership initiatives still largely depend on the fact that there are "overlap" markets in wealthy countries—as for malaria and TB—that can be leveraged.

"Ensuring the development of life-saving medicine," says Orbinski, "should not be based on philanthropic donations or a coincidence of government and industry interests. These strategies, while extremely important, still rely partially on market forces—tourists get malaria, and TB affects the western world. This does not work for the most neglected diseases like sleeping sickness and kala azar. The only way to solve this chronic health crisis is to change the way we think about medicines for neglected diseases. They must be considered public goods, not simply consumer products."

(The Gates Foundation, to its credit, has in at least one instance broken away from the overlap market model, and is providing funding for kala azar vaccine research.)

The DND Working Group is looking to a new model for stimulating R&D for the most neglected diseases. With MSF's preliminary support, the DND Working Group is carrying out a feasibility study for a not-for-profit Drugs for Neglected Diseases Initiative (DNDI), which Orbinski describes as a "partnership for public response." The DNDI would aim to develop new, effective, field-relevant medicines for the most neglected diseases, and would be firmly committed to ensuring that these drugs are affordable for those who need them.

"Ensuring the development and availability of treatments for these diseases will require a paradigm shift from market-driven to needs-driven drug development, with public responsibility and leadership in both developing and developed countries," says Dr. Yves Champey, the chair of the Drugs for Neglected Diseases Initiative Task Force, which is heading up the feasibility study.

According to the DND Working Group, the DNDI, should it be created, will differ from market-driven drug development in two critical respects: it will be guided by a clear focus on the most neglected diseases and it will be driven by need rather than return on investment.

In addition to identifying and developing new chemical entities, the DNDI will place special attention on developing and registering novel combinations of existing drugs, and reformulating existing products to improve use. These relatively simple and low-cost efforts, it is hoped, may yield major advances to address neglected diseases.

A complementary focus of DNDI will be to invest in sustainable capacity and leadership in developing countries at all stages of R&D. Public leadership from developed countries will also be vital to the DNDI.

"National self-interest and profit prospects can no longer be the single motor of medical research and devel-

opment," says Dr. Pecoul. "Without serious political commitment and a clear realization that there is a public duty to commit to health issues at a global level, advances in science and medicine will contribute nothing to alleviating the suffering and meeting the critical medical needs of the millions who die of neglected diseases in the developing world."

Rachel Cohen is advocacy liaison for Médecins Sans Frontières' Access to Essential Medicines Campaign.

From *Multinational Monitor*, June 2002, pp. 9-12. © 2002 by Multinational Monitor. Reprinted by permission.

UNIT 6

Women and Development

Unit Selections

Key Points to Consider

- What factors can lead to enhancing the status of women?

- What prevents wider educational opportunities for women in the developing world?

- What factors have contributed to a larger role for women in the Middle East?

- Why might a larger role for women help to resolve conflicts?

- What effect have foreign aid programs had on gender?

 **Links: www.dushkin.com/online/**
These sites are annotated in the World Wide Web pages.

African Women Global Network
http://www.osu.edu/org/awognet/main.html

WIDNET: Women in Development NETwork
http://www.focusintl.com/widnet.htm

Women's Issues—3rd World
http://www.women3rdworld.about.com/newsissues/women3rdworld/mbody.htm

WomenWatch/Regional and Country Information
http://www.un.org/womenwatch/

There is widespread recognition of the crucial role that women play in the development process. Women are critical to the success of family planning programs, bear much of the responsibility for food production, account for an increasing share of wage labor in developing countries, and are acutely aware of the consequences of environmental degradation. Despite their important contributions, however, women lag behind men in access to health care, nutrition, and education while continuing to face formidable social, economic, and political barriers.

Women's lives in the developing world are invariably difficult. Often female children are valued less than male offspring, resulting in higher female infant and child mortality rates. In extreme cases, this undervaluing leads to female infanticide. Those females who do survive face lives characterized by poor nutrition and health, multiple pregnancies, hard physical labor, discrimination, and perhaps violence.

Clearly, women are central to any successful population policy. Evidence shows that educated women have fewer and healthier children. This connection between education and population indicates that greater emphasis should be placed on educating women. In reality, female school enrollments are lower than those of males for reasons having to do with state priorities, family resources that are insufficient to educate both boys and girls, female socialization, and cultural factors. Although education is probably the largest single contributor to enhancing the status of women and thereby promoting development, access to education is still limited for many women. Sixty percent of the children worldwide who are not enrolled in school are girls. Higher status for women also has benefits in terms of improved health, better wages, and greater influence in decision making.

Women make up a significant portion of the agricultural workforce. They are heavily involved in food production from planting to cultivation, harvesting, and marketing. Despite their agricultural contribution, women frequently do not have adequate access to advances in agricultural technology or the benefits of extension and training programs. They are also discriminated against in land ownership. As a result, important opportunities to improve food production are lost when women are not given access to technology, training, credit, and land ownership commensurate with their agricultural role.

The industrialization that has accompanied the globalization of production has meant more employment opportunities for women, but often these are low-tech, low-wage jobs. The lower labor costs in the developing world that attract manufacturing facilities are a mixed blessing for women. Increasingly, women are recruited to fill these production jobs because wage differentials allow employers to pay women less. On the other hand, expanding opportunities for women in these positions contributes to family income. The informal sector, where jobs are smaller-scale, more traditional, and labor-intensive, has also attracted more women. These jobs are often their only source of employment due to family responsibilities or discrimination. Clearly, women also play a critical role in economic expansion in developing countries. Nevertheless, women are often the first to feel the effects of an economic slowdown.

The consequences of the structural adjustment programs that many developing countries have had to adopt have also fallen disproportionately on women. As employment opportunities have declined because of austerity measures, women have lost jobs in the formal sector and faced increased competition from males in the informal sector. Cuts in spending on health care and education also affect women, who already receive fewer of these benefits. Currency devaluations further erode the purchasing power of women.

Because of the gender division of labor, women are often more aware of the consequences of environmental degradation. Depletion of resources such as forests, soil, and water are much more likely to be felt by women who are responsible for collecting firewood and water and who raise most of the crops. As a result, women are an essential component of successful environmental protection policies but are often overlooked in planning environmental projects.

Enhancing the status of women has been the primary focus of recent international conferences. The 1994 International Conference on Population and Development focused attention on women's health and reproductive rights and the crucial role that these issues play in controlling population growth. The 1995 Fourth World Conference on Women held in Beijing, China, proclaimed women's rights to be synonymous with human rights. These developments represent a turning point in women's struggle for equal rights. International conferences not only focus attention on gender issues, but also provide additional opportunities for developing leadership and encouraging grassroots efforts to realize the goal of enhancing the status of women. Greater political involvement may also increase women's participation in efforts to resolve conflict. Some argue that women have certain qualities that are more likely to facilitate compromise and the peaceful resolution of conflicts.

There are some indications that women have made progress in certain regions of the developing world. In the Middle East, several factors have contributed to the erosion of some of the restrictions on women. The countries of the Southern African Development Community have made substantial but uneven progress in meeting the targets for women's political participation set at the 1995 Beijing Conference on Women. In India, a constitutional amendment has set aside one-third of village council and chief positions for women, and a percentage of those positions must go to lower-caste women. In Latin America, women have made significant gains in attaining equal pay, access to reproductive health, and protection from violence, but there are differences between urban and rural women as well as differences in women's status among countries in the region.

Empowering Women

Lori S. Ashford

Thanks to the growing activism and influence of women's rights advocates around the world, the situation of women has moved to the forefront of both national and international population policy debates. Since the Cairo and Beijing conferences, there has been greater discussion of gender issues, and of the differences in men's and women's socially defined roles. Governments and donor agencies increasingly acknowledge that women's inferior status hinders development and support policies and programs to reduce gender inequalities. Those concerned about the negative effects of population growth also see a connection between enhancing a woman's status within the family and society and increasing her control over childbearing.

> ## No government could deny that women deserve higher status and better opportunities.

The agreements adopted at the population and development conference (ICPD) in Cairo and the women's conference in Beijing called for equal participation and partnerships between men and women in nearly all areas of public and private life. As innovative—even revolutionary—as these notions seem, they met with little dissent during the conferences themselves. In spite of the diversity of cultures represented, no government could deny that women deserve higher status and better opportunities. Beyond the conference halls, however, putting these ideas into practice requires overcoming deeply rooted cultural values and ways of life.

Understanding Gender

Gender refers to the different roles that men and women play in society, and the relative power they wield. Gender roles vary from one country to another, but almost everywhere, women face disadvantages relative to men in the social, economic, and political spheres of life. Where men are viewed as the principal decisionmakers, women often hold a subordinate position in negotiations about limiting family size, contraceptive use, managing family resources, protecting family health, or seeking jobs.

Inequalities between men and women are closely linked to women's health–making the issue of gender pertinent to discussions on how to improve reproductive health. Gender differences affect women's health and well-being throughout the life cycle:[1]

- Before or at birth, parents who prefer boys may put girls at risk of sex-selective abortions (where technology is available to identify the sex) or infanticide.
- Where food is scarce, girls often eat last, and usually less than boys.
- Girls may be less likely than boys to receive health care when they are ill.
- In some countries, mainly in Africa, girls are subjected to female genital cutting.
- Adolescent girls may be pressured into having sex at an early age—within an arranged marriage, by adolescent boys proving their virility, or by older men looking for partners not infected with STIs.

Table 1

Fertility and Access to Reproductive Health Care Among the Poorest and Richest Women, Selected Countries, 1990s

Country	Births per woman (TFR)*		Prenatal care (Percent of pregnant women)		Births attended by skilled staff (Percent of deliveries)	
	Poorest fifth	Richest fifth	Poorest fifth	Richest fifth	Poorest fifth	Richest fifth
Bolivia	7.4	2.1	39	95	20	98
Cameroon	6.2	4.8	53	99	32	95
Guatemala	8.0	2.4	35	90	9	92
India	4.1	2.1	25	89	12	79
Indonesia	3.3	2.0	74	99	21	89
Morocco	6.7	2.3	8	74	5	78
Vietnam	3.1	1.6	50	92	49	99

*TFR (total fertility rate) is the average total number of births per woman given prevailing birth rates.

Note: Women were ranked according to their household assets.

Source: A. Tinker, K. Finn, and J. Epp, *Improving Women's Health: Issues and Interventions* (World Bank, June 2000): 10.

- Married women may be pressured by husbands or families to have more children than they prefer, and women may be unable to seek or use contraception.

- Married and unmarried women may be unable to deny sexual advances or persuade partners to use a condom, thereby exposing themselves to the risk of STIs.

- In all societies, women are more likely than men to experience domestic violence. Women may sustain injuries from physical abuse by male partners or family members, and the fear of abuse can make women less willing to resist the demands of their husbands or families.

A mixture of cultural and social factors explain women's lack of power in protecting themselves or their daughters from these health threats. These factors include women's limited exposure to information and new ideas, ignorance of good health practices, limited physical mobility, and lack of control over money and resources.[2] In some South Asian and Middle Eastern countries, for example, women's use of health care services is inhibited by cultural restrictions on women traveling alone or being treated by male health care providers.

The Poverty Connection

Gender disadvantages intertwine with poverty. Poverty is strongly linked to poor health, and women represent a disproportionate share of the poor. In all regions of the world, including wealthier regions, reproductive health is worse among the poor (see Table 1). Women in the poorest households have the highest fertility, poorest nutrition, and most limited access to skilled pregnancy and delivery care, which contribute to higher maternal and infant death rates. Women's disadvantaged position also perpetuates poor health, an inadequate diet, early entry into motherhood, frequent pregnancies, and a continued cycle of poverty.[3]

Women's low socioeconomic status also makes them more vulnerable to physical and sexual abuse. Unequal power in sexual relationships exposes women to coerced sex (see Box 1), unwanted pregnancies, and sexually transmitted infections. Impoverishment can also lead some women into commercial sex.[4] Thus, women's access to and control over resources can give women greater control of their sexuality, which is fundamental to controlling their fertility and improving their health.

Reducing Gender Inequalities

Recent reports from the World Bank show that reducing gender inequalities can bring about greater economic prosperity and help reduce poverty. One study found that a 1 percent increase in women's secondary schooling results in a 0.3 percent increase in economic growth.[5] In addition, the strong links between women's status, health, and fertility rates make gender equality a critical strategy for policies to improve health and stabilize population growth.

Women's control over their sexuality is central to population and health concerns.

The empowerment of women is seen as a key avenue for reducing the differences between the sexes that exist in nearly all societies. Empowerment refers to "the process by which the powerless gain greater control over the circumstances of their lives."[6] It means not only greater control over resources but also greater self-confidence and the ability to make decisions on an equal basis with men. Empowerment of women also requires that men are aware of gender inequalities and are willing to question traditional definitions of masculinity.

In many societies, these concepts may be threatening to men, who are accustomed to having authority in the household, the community, the economy, and national politics. The concepts may also be frightening to women, who may fear the implications of these changes for their personal lives.[7] For these reasons, and because concepts of women's rights, empowerment, and gender equality are still relatively new in many places, progress in advancing women's rights has been modest.

Box 1

Violence Against Women

Violence against women is the issue that perhaps best illustrates the connection between women's rights and women's health, and the tragic consequences of women's inferior position. Once thought to be only a private matter, violence against women has gained visibility as a serious public policy and public health concern.

Violence against women (also referred to as gender-based violence) occurs in nearly all societies, within the home or in the wider community, and it is largely unpoliced. It may include female infanticide, incest, child prostitution, rape, wife-beating, sexual harassment, wartime violence, or harmful traditional practices such as forced early marriage, female genital cutting, and widow or bride burning. A recent study published by the Center for Health and Gender Equity and Johns Hopkins University estimated that one in three women worldwide suffers from some form of gender-based violence.

Domestic violence is the most common form of gender-based violence, and it is most often perpetrated by a boyfriend or husband against a woman (see figure). Psychological abuse almost always accompanies physical abuse, and the majority of women who are abused by their partners are abused many times. Many women tolerate the abuse because they fear retaliation by their spouse or extended family, or both, if they protest. Women's vulnerability to violence is reinforced by their economic dependence on men, widespread cultural acceptance of domestic violence, and a lack of laws and enforcement mechanisms to combat it.

Although women's control over their sexuality is central to population and health concerns, the extent to which sexual activity

is forced or coerced has only recently been addressed. Most coerced sex takes place between people who know each other—spouses, family members, or acquaintances. One-quarter to one-half of domestic violence cases involve forced sex. Coercion also takes place against children and adolescents in more developed and less developed countries. Statistics on rape suggest that between one-third and two-thirds of rape victims around the world are younger than 16.

What is the connection between gender-based violence and reproductive health? Violence against women is rooted in unequal power between men and women. It affects women's physical, mental, economic, and social well-being. It can lead to a range of health problems. Since girls are more often subjected to coerced sex than boys, they are at risk of becoming infected with sexually transmitted infections (STIs) at a younger age than are boys. Some STIs can lead to pelvic inflammatory disease, infertility, and AIDS. Forced and unprotected sex also leads to unintended pregnancies, abortions, and unwanted children. The experience of abuse puts women at greater risk of mental health problems, including depression, suicide, and alcohol and drug abuse. Ultimately, these outcomes have negative consequences for the whole society, not just the women who are victims of such violence.

After a series of international conferences and conventions in the 1990s called for eliminating all forms of violence against women, many countries strengthened laws and enforcement mechanisms related to domestic violence. Much of the pressure to change laws and community standards has come from nongovernmental organizations, and particularly women's groups.

These groups are at the forefront of efforts to combat violence against women through grassroots activism, lobbying, and working with women survivors of violence. Ending the violence requires community-level action and, ultimately, changes in the values that lead to the subjugation of women.

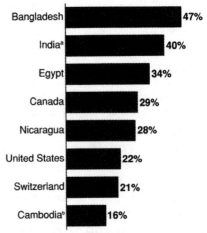

Women Reporting Physical Assault by Male Partner, Selected Studies, 1990s

Bangladesh	47%
India[a]	40%
Egypt	34%
Canada	29%
Nicaragua	28%
United States	22%
Switzerland	21%
Cambodia[b]	16%

[a] Six states only.
[b] Phnom Penh and six provinces.

Source: L. Heise, M. Ellsberg, and M. Gottemoeller, *Population Reports* Series L, no. 11 (1999): table 1.

References

Lori Heise, Mary Ellsberg, and Megan Gottemoeller, "Ending Violence Against Women," *Population Reports* Series L, no. 11 (Baltimore: Johns Hopkins University, 1999); UNFPA, *The State of World Population 2000: Lives Together, Worlds Apart* (New York: UNFPA, 2000); and Yvette Collymore, ed., *Conveying Concerns: Women Report on Gender-Based Violence* (Washington, DC: Population Reference Bureau, 2000).

Women's Education

The World Bank calls women's education the "single most influential investment that can be made in the developing world." Many governments now support women's education not only to foster economic growth, but also to promote smaller families, increase modern contraceptive use, and improve child health. Educating women is an important end in itself, and it is also a long-term strategy for advancing women's reproductive health. The Cairo conference called for universal access to and completion of primary education, and for reducing the gender gap—

differences in boys' and girls' enrollment—in secondary education.

Worldwide, more men than women are literate (80 percent, compared with 64 percent). While nearly all boys and girls in more developed countries are enrolled in both primary and secondary school, women in less developed countries complete fewer years of education than men, on average, and are more likely to be illiterate.

According to the UN, primary and secondary school enrollments increased for both girls and boys during the 1990s in almost all world regions. The gender gap in school enrollments closed somewhat in recent years but remains pronounced at the secondary school level (see Table 2). Girls are more likely than boys to discontinue their schooling for a number of reasons: household duties; early marriage and childbearing; parents' perceptions that education is more beneficial for sons; worries about girls' safety as they travel to schools away from their villages; and limited job opportunities for women in sectors that require higher education. In some settings, gender bias among teachers and sexual harassment may lead to higher dropout rates among young women.[8]

Table 2

Secondary School Enrollment by Sex in World Regions, 1980 and 1990s

| | Percent enrolled* | | | |
| | 1980 | | 1990s | |
Region	Boys	Girls	Boys	Girls
More developed countries	88	89	99	102
Less developed countries	43	30	57	48
Northern Africa	47	29	63	57
Sub-Saharan Africa	19	10	29	23
Western Asia	49	31	63	48
South-Central Asia	38	20	55	37
Southeast Asia	40	35	53	49
East Asia	59	45	77	70
Central America	46	42	56	57
Caribbean	–	–	49	55
South America	38	42	–	–

–Not available.

* The percent enrolled is the ratio of the total number enrolled in secondary school (regardless of age) to the number of secondary-school-age children, or the gross enrollment ratio. Data from the 1990s are the latest available, generally between 1990 and 1996.

Source: A. Boyd, C. Haub, and D. Cornelius, *The World's Youth 2000* (Population Reference Bureau, 2000), based on national figures from UNESCO, *Statistical Yearbook 1999* (1999).

During the 1990s, the countries seeing the greatest gains in closing the gender gap were regions that had the lowest enrollments in the past: Northern Africa, sub-Saharan Africa, Southern Asia and Western Asia. Nevertheless, in the regions where almost a third of the world's women live (Southern Asia and sub-Saharan Africa), girls are much less likely than boys to attend secondary school. The populations of these two regions are among the world's fastest growing, which suggests that the number of illiterate women in these regions will continue to grow.[9]

Research over the last 20 years has shown that women with more education make a later transition to adulthood and have smaller, healthier families. Women with more education usually have their first sexual experience later, marry later, want smaller families, and are more likely to use contraception and other health care than their less educated peers (see Figure 1). In many less developed countries, women with no schooling

have about twice as many children as do women with 10 or more years of school.[10] Expanding educational opportunities for women has been embraced as a means to lower national fertility rates and to slow population growth.

Figure 1

Women's Education and Childbearing, Selected Countries, 1995–1999

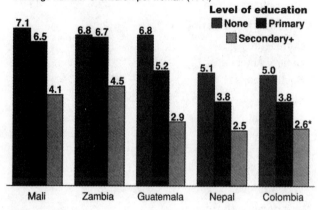

Average number of children per woman (TFR)

*Secondary level only.
Note: The TFR is the average total number of children born per woman given prevailing birth rates.

Source: Demographic and Health Surveys, Final Country Reports. Available online at: www.measuredhs.com.

Women's Work

Employment is another way that women can elevate their status: It enables them to earn income and have more control over resources. It can also increase women's involvement in the public sphere and help enhance decisionmaking skills. Women's participation in the labor force increased during the 1990s, but they are still at the lower end of the labor market in pay and authority.

Women make up an increasing share of the labor force in almost all regions of the world (see Table 3). Several factors explain this: Women's improved ability to limit and space pregnancies has enabled them to spend less time on child care and more on work outside the home; attitudes toward the employment of women have become more accepting; and new policies on family and child care ensure more flexibility and therefore favor working women. In addition, economic growth, and particularly the expansion of service industries (like finance, communications, and tourism), which tend to employ large numbers of women, has increased women's labor force opportunities. And finally, programs that have made credit available for small enterprises have benefited women.[10]

Table 3

Womens Share of the Labor Force, 1980 and 1997

Region	Women as percent of labor force	
	1980	1997
Africa		
Northern Africa	20	26
Sub-Saharan Africa	42	43
Latin America & Caribbean		
Caribbean	38	43
Central America	27	33
South America	27	38
Asia		
East Asia	40	43
Southeast Asia	41	43
Southern Asia	31	33
Central Asia	47	46
Western Asia	23	27
More developed regions		
Eastern Europe	45	45
Western Europe	36	42
Other more developed regions	39	44

Source: United Nations, *The World's Women 2000: Trends and Statistics* (2000): 110.

Figure 2

Women's Wages as a Percentage of Men's Wages, in Manufacturing, Selected Countries, 1992–1997

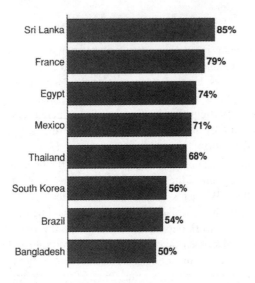

Source: United Nations, *The World's Women 2000: Trends and Statistics* (2000): chart 5.23.

These trends are positive, but equality in the work force is still a long way from reality. Women typically occupy lower-paid and lower-status jobs than men; women's unemployment rates are higher than men's; and far more women than men work in the "informal sector," occupations like street vending and market work, where wages are very low. Even when women work in the same sector as men, wages are typically lower (see Figure 2).

In addition, more women are remaining in the work force during their reproductive years, leading to a "dual burden": working outside the home while at the same time doing a larger share of work in the home than men—such as childrearing, cooking, and cleaning. The few studies that are available on how women's time is used (in more developed countries) show that women spend 50 percent to 70 percent as much time as men on paid work, but almost twice as much or more time as men on unpaid work.[11] A multicountry study on women's lives and family planning in less developed regions found that many women reported additional stress because of these dual responsibilities, and that many women in these situations would prefer not to work outside the home.[12]

The Cairo population and development conference and the Beijing women's conference called on governments to reduce disparities between men and women in the work force and to provide additional support to working women, such as maternity leave, child-care assistance, and other flexible arrangements. One area that has received much visibility and high-level political support is that of providing microcredit—small low-interest loans—that allow women to start their own small businesses. Experience has shown that women are good "credit risks" and repayment rates are high. These programs tend to be small, however, and they do not address the underlying social and cultural reasons for women's economic disadvantages.

Changing Family Dynamics

Women's strong attachment to family and responsibility for the household define much of their adult lives. Women usually marry at a younger age than do men, and the age gap between spouses tends to be wider in low-income countries. The age differences between spouses help perpetuate women's weaker authority. In societies where childbearing starts soon after marriage, women's opportunities to pursue careers or additional schooling, or to develop contacts beyond the family, are limited.

In many societies, laws pertaining to the family put women at a disadvantage and reinforce their dependence on men: Women may be unable to inherit land or other property, divorce their husbands, or get custody of their children if they can divorce. The gender gap in legal rights is gradually narrowing in some countries. Egypt revised its laws in 2000 to allow women similar divorce rights as men, for example, and Morocco revised its personal status laws in the 1990s to enhance women's rights in marriage—including polygamous marriages—and divorce.[13]

Women's Political Leadership

Women's right to equal participation in political life is guaranteed by a number of international conventions, most notably the

Convention on the Elimination of All Forms of Discrimination Against Women (CEDAW), which was adopted in 1979 but has not been ratified by all governments. In practice, in all countries, women are underrepresented at every level of government, especially in high-level executive positions and legislative bodies.

Figure 3
Legislative Seats Held by Women, Selected Countries, 1999

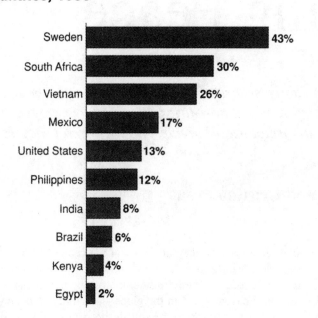

Source: United Nations, *The World's Women 2000: Trends and Statistics* (2000): table 6.A.

Worldwide, women held only 11 percent of seats in parliaments and congresses in 1999, up from 9 percent in 1987.[14] In the United States, sub-Saharan Africa, Latin America and the Caribbean, and East and Southeast Asia, these percentages ranged between 10 percent and 13 percent in 1999. In Northern Africa and South and West Asia, women held just 3 percent to 5 percent of seats in legislative bodies (see Figure 3). There are some exceptions: Women in Vietnam, South Africa, Mozambique, Cuba, and Argentina hold 25 percent to 30 percent of parliamentary seats.

In the 1990s, a number of countries, including India, Uganda, Brazil, and the Philippines, formally set aside a percentage of seats on national and local bodies to be held by women. In South Africa, the proportion of women in parliament

rose from 1 percent to 30 percent following the establishment of a new constitution.

One drawback of mandated quotas is that women may be placed on election ballots and voted into office before they have developed the political and technical skills necessary to govern effectively. Until these reforms become more widespread and meaningful, women's lack of political power will limit their influence on laws and policies that affect their rights, as well as on priority-setting in health care and other public services.

Notes

1. Program for Appropriate Technology (PATH), *Reproductive Health Outlook*. Website accessed online at: www.rho.org, on Jan. 23, 2001.
2. AbouZahr, "Background Paper on Reproductive Health": 16
3. Ann Tinker, K. Finn, and Joanne E. Epp, *Improving Women's Health: Issues and Interventions* (Washington, DC: The World Bank, June 2000): 10.
4. Ibid.: 11.
5. David Dollar and Roberta Gotti, "Gender Inequality, Income and Growth: Are Good Times Good for Women?" *Policy Research Reports on Gender and Development*, Working Paper Series No. 1 (Washington, DC: World Bank, 1999).
6. Gita Sen and Srilatha Batliwala, "Empowering Women for Reproductive Rights," in *Women's Empowerment and Demographic Processes*, eds., Harriet B. Presser and Gita Sen (Oxford, England: Oxford University Press, 2000): 18.
7. Ibid.
8. Caroline Bledsoe, John Casterline, J. Johnson-Kuhn, and John Haaga, *Critical Perspectives on Schooling and Fertility in the Developing World* (Washington, DC: National Academy Press, 1999): 81–104.
9. United Nations, *The World's Women 2000: Trends and Statistics* (New York: United Nations, 2000): 85–86.
10. United Nations, *Linkages Between Population and Education* (New York: United Nations, 1997): 12.
11. United Nations, *The World's Women 2000: Trends and Statistics*: 110.
12. Ibid.: 126.
13. Barbara Barnett and Jane Stein, *Women's Voices, Women's Lives: The Impact of Family Planning* (Research Triangle Park, NC: Family Health International, 1998).
14. UNFPA, *State of World Population 2000*: 55; and Radouane Belouali and Najib Guédira, *Reproductive Health in Policy and Practice: Morocco* (Washington, DC: Population Reference Bureau, 1998): 8–10.

Women Waging Peace

You can't end wars simply by declaring peace. "Inclusive security" rests on the principle that fundamental social changes are necessary to prevent renewed hostilities. Women have proven time and again their unique ability to bridge seemingly insurmountable divides. So why aren't they at the negotiating table?

By Swanee Hunt and Cristina Posa

Allowing men who plan wars to plan peace is a bad habit. But international negotiators and policymakers can break that habit by including peace promoters, not just warriors, at the negotiating table. More often than not, those peace promoters are women. Certainly, some extraordinary men have changed the course of history with their peacemaking; likewise, a few belligerent women have made it to the top of the political ladder or, at the grass-roots level, have taken the roles of suicide bombers or soldiers. Exceptions aside, however, women are often the most powerful voices for moderation in times of conflict. While most men come to the negotiating table directly from the war room and battlefield, women usually arrive straight out of civil activism and—take a deep breath—family care.

Yet, traditional thinking about war and peace either ignores women or regards them as victims. This oversight costs the world dearly. The wars of the last decade have gripped the public conscience largely because civilians were not merely caught in the crossfire; they were targeted, deliberately and brutally, by military strategists. Just as warfare has become "inclusive"—with civilian deaths more common than soldiers'—so too must our approach toward ending conflict. Today, the goal is not simply the absence of war, but the creation of sustainable peace by fostering fundamental societal changes. In this respect, the United States and other countries could take a lesson from Canada, whose innovative "human security" initiative—by making human beings and their communities, rather than states, its point of reference—focuses on safety and protection, particularly of the most vulnerable segments of a population.

The concept of "inclusive security," a diverse, citizen-driven approach to global stability, emphasizes women's agency, not their vulnerability. Rather than motivated by gender fairness, this concept is driven by efficiency: Women are crucial to inclu-

sive security since they are often at the center of nongovernmental organizations (NGOs), popular protests, electoral referendums, and other citizen-empowering movements whose influence has grown with the global spread of democracy. An inclusive security approach expands the array of tools available to police, military, and diplomatic structures by adding collaboration with local efforts to achieve peace. Every effort to bridge divides, even if unsuccessful, has value, both in lessons learned and links to be built on later. Local actors with crucial experience resolving conflicts, organizing political movements, managing relief efforts, or working with military forces bring that experience into ongoing peace processes.

International organizations are slowly recognizing the indispensable role that women play in preventing war and sustaining peace. On October 31, 2000, the United Nations Security Council issued Resolution 1325 urging the secretary-general to expand the role of women in U.N. field-based operations, especially among military observers, civilian police, human rights workers, and humanitarian personnel. The Organization for Security and Co-operation in Europe (OSCE) is working to move women off the gender sidelines and into the everyday activities of the organization—particularly in the Office for Democratic Institutions and Human Rights, which has been useful in monitoring elections and human rights throughout Europe and the former Soviet Union. Last November, the European Parliament passed a hard-hitting resolution calling on European Union members (and the European Commission and Council) to promote the equal participation of women in diplomatic conflict resolution; to ensure that women fill at least 40 percent of all reconciliation, peacekeeping, peace-enforcement, peace-building, and conflict-prevention posts; and to support the creation and strengthening of NGOs (including women's organizations) that focus on conflict prevention, peace building, and post-conflict reconstruction.

Ironically, women's status as second-class citizens is a source of empowerment, since it has made women adept at finding innovative ways to cope with problems.

But such strides by international organizations have done little to correct the deplorable extent to which local women have been relegated to the margins of police, military, and diplomatic efforts. Consider that Bosnian women were not invited to participate in the Dayton talks, which ended the war in Bosnia, even though during the conflict 40 women's associations remained organized and active across ethnic lines. Not surprisingly, this exclusion has subsequently characterized—and undermined—the implementation of the Dayton accord. During a 1997 trip to Bosnia, U.S. President Bill Clinton, Secretary of State Madeleine Albright, and National Security Advisor Samuel Berger had a miserable meeting with intransigent politicians elected under the ethnic-based requirements of Dayton. During the same period, First Lady Hillary Rodham Clinton engaged a dozen women from across the country who shared story after story of their courageous and remarkably effective work to restore their communities. At the end of the day, a grim Berger faced the press, offering no encouraging word from the meetings with the political dinosaurs. The first lady's meeting with the energetic women activists was never mentioned.

We can ignore women's work as peacemakers, or we can harness its full force across a wide range of activities relevant to the security sphere: bridging the divide between groups in conflict, influencing local security forces, collaborating with international organizations, and seeking political office.

BRIDGING THE DIVIDE

The idea of women as peacemakers is not political correctness run amok. Social science research supports the stereotype of women as generally more collaborative than men and thus more inclined toward consensus and compromise. Ironically, women's status as second-class citizens is a source of empowerment, since it has made women adept at finding innovative ways to cope with problems. Because women are not ensconced within the mainstream, those in power consider them less threatening, allowing women to work unimpeded and "below the radar screen." Since they usually have not been behind a rifle, women, in contrast to men, have less psychological distance to reach across a conflict line. (They are also more accepted on the "other side," because it is assumed that they did not do any of the actual killing.) Women often choose an identity, notably that of mothers, that cuts across international borders and ethnic enclaves. Given their roles as family nurturers,

women have a huge investment in the stability of their communities. And since women know their communities, they can predict the acceptance of peace initiatives, as well as broker agreements in their own neighborhoods.

As U.N. Secretary-General Kofi Annan remarked in October 2000 to the Security Council, "For generations, women have served as peace educators, both in their families and in their societies. They have proved instrumental in building bridges rather than walls." Women have been able to bridge the divide even in situations where leaders have deemed conflict resolution futile in the face of so-called intractable ethnic hatreds. Striking examples of women making the impossible possible come from Sudan, a country splintered by decades of civil war. In the south, women working together in the New Sudan Council of Churches conducted their own version of shuttle diplomacy—perhaps without the panache of jetting between capitals—and organized the Wunlit tribal summit in February 1999 to bring an end to bloody hostilities between the Dinka and Nuer peoples. As a result, the Wunlit Covenant guaranteed peace between the Dinka and the Nuer, who agreed to share rights to water, fishing, and grazing land, which had been key points of disagreement. The covenant also returned prisoners and guaranteed freedom of movement for members of both tribes.

On another continent, women have bridged the seemingly insurmountable differences between India and Pakistan by organizing huge rallies to unite citizens from both countries. Since 1994, the Pakistan-India People's Forum for Peace and Democracy has worked to overcome the hysterics of the nationalist media and jingoistic governing elites by holding annual conventions where Indians and Pakistanis can affirm their shared histories, forge networks, and act together on specific initiatives. In 1995, for instance, activists joined forces on behalf of fishers and their children who were languishing in each side's jails because they had strayed across maritime boundaries. As a result, the adversarial governments released the prisoners and their boats.

In addition to laying the foundation for broader accords by tackling the smaller, everyday problems that keep people apart, women have also taken the initiative in drafting principles for comprehensive settlements. The platform of Jerusalem Link, a federation of Palestinian and Israeli women's groups, served as a blueprint for negotiations over the final status of Jerusalem during the Oslo process. Former President Clinton, the week of the failed Camp David talks in July 2000, remarked simply, "If we'd had women at Camp David, we'd have an agreement."

Sometimes conflict resolution requires unshackling the media. Journalists can nourish a fair and tolerant vision of society or feed the public poisonous, one-sided, and untruthful accounts of the "news" that stimulate violent conflict. Supreme Allied Commander of Europe Wesley Clark understood as much when he ordered NATO to bomb transmitters in Kosovo to prevent the Milosevic media machine from spewing ever more inflammatory rhetoric. One of the founders of the independent Kosovo radio station RTV-21 realized that there were "many instances of male colleagues reporting with anger, which served to raise the tensions rather than lower them." As a result, RTV-

21 now runs workshops in radio, print, and TV journalism to cultivate a core of female journalists with a noninflammatory style. The OSCE and the BBC, which train promising local journalists in Kosovo and Bosnia, would do well to seek out women, who generally bring with them a reputation for moderation in unstable situations.

Nelson Mandela suggested at last summer's Arusha peace talks that if Burundian men began fighting again, their women should withhold "conjugal rights" (like cooking, he added).

INFLUENCING SECURITY FORCES

The influence of women on warriors dates back to the ancient Greek play *Lysistrata*. Borrowing from that play's story, former South African President Nelson Mandela suggested at last summer's Arusha peace talks on the conflict in Burundi that if Burundian men began fighting again, their women should withhold "conjugal rights" (like cooking, he added).

Women can also act as a valuable interface between their countries' security forces (police and military) and the public, especially in cases when rapid response is necessary to head off violence. Women in Northern Ireland, for example, have helped calm the often deadly "marching season" by facilitating mediations between Protestant unionists and Catholic nationalists. The women bring together key members of each community, many of whom are released prisoners, as mediators to calm tensions. This circle of mediators works with local police throughout the marching season, meeting quietly and maintaining contacts on a 24-hour basis. This intervention provides a powerful extension of the limited tools of the local police and security forces.

Likewise, an early goal of the Sudanese Women's Voice for Peace was to meet and talk with the military leaders of the various rebel armies. These contacts secured women's access to areas controlled by the revolutionary movements, a critical variable in the success or failure of humanitarian efforts in war zones. Women have also worked with the military to search for missing people, a common element in the cycle of violence. In Colombia, for example, women were so persistent in their demands for information regarding 150 people abducted from a church in 1999 that the army eventually gave them space on a military base for an information and strategy center. The military worked alongside the women and their families trying to track down the missing people. In short, through moral suasion, local women often have influence where outsiders, such as international human rights agencies, do not.

That influence may have allowed a female investigative reporter like Maria Cristina Caballero to go where a man could not go, venturing on horseback alone, eight hours into the jungle to tape a four-hour interview with the head of the paramilitary forces in Colombia. She also interviewed another guerilla leader and published an award-winning comparison of the transcripts, showing where the two mortal enemies shared the same vision. "This [was] bigger than a story," she later said, "this [was] hope for peace." Risking their lives to move back and forth across the divide, women like Caballero perform work that is just as important for regional stabilization as the grandest Plan Colombia.

INTERNATIONAL COLLABORATION

Given the nature of "inclusive" war, security forces are increasingly called upon to ensure the safe passage of humanitarian relief across conflict zones. Women serve as indispensable contacts between civilians, warring parties, and relief organizations. Without women's knowledge of the local scene, the mandate of the military to support NGOs would often be severely hindered, if not impossible.

In rebel-controlled areas of Sudan, women have worked closely with humanitarian organizations to prevent food from being diverted from those who need it most. According to Catherine Loria Duku Jeremano of Oxfam: "The normal pattern was to hand out relief to the men, who were then expected to take it home to be distributed to their family. However, many of the men did what they pleased with the food they received: either selling it directly, often in exchange for alcohol, or giving food to the wives they favored." Sudanese women worked closely with tribal chiefs and relief organizations to establish a system allowing women to pick up the food for their families, despite contrary cultural norms.

In Pristina, Kosovo, Vjosa Dobruna, a pediatric neurologist and human rights leader, is now the joint administrator for civil society for the U.N. Interim Administration Mission in Kosovo (UNMIK). In September 2000, at the request of NATO, she organized a multiethnic strategic planning session to integrate women throughout UNMIK. Before that gathering, women who had played very significant roles in their communities felt shunned by the international organizations that descended on Kosovo following the bombing campaign. Vjosa's conference pulled them back into the mainstream, bringing international players into the conference to hear from local women what stabilizing measures they were planning, rather than the other way around. There, as in Bosnia, the OSCE has created a quota system for elected office, mandating that women comprise one third of each party's candidate list; leaders like Vjosa helped turn that policy into reality.

In addition to helping aid organizations find better ways to distribute relief or helping the U.N. and OSCE implement their ambitious mandates, women also work closely with them to locate and exchange prisoners of war. As the peace processes in Northern Ireland, Bosnia, and the Middle East illustrate, a deadlock on the exchange and release of prisoners can be a major obstacle to achieving a final settlement. Women activists in Armenia and Azerbaijan have worked closely with the International Helsinki Citizens Assembly and the OSCE for the release

The Black and the Green

Grass-roots women's organizations in Israel come in two colors: black and green. The Women in Black, founded in 1988, and the Women in Green, founded in 1993, could not be further apart on the political spectrum, but both claim the mantle of "womanhood" and "motherhood" in the ongoing struggle to end the Israeli-Palestinian conflict.

One month after the Palestinian intifada broke out in December 1988, a small group of women decided to meet every Friday afternoon at a busy Jerusalem intersection wearing all black and holding hand-shaped signs that read: "Stop the Occupation." The weekly gatherings continued and soon spread across Israel to Europe, the United States, and then to Asia.

While the movement was originally dedicated to achieving peace in the Middle East, other groups soon protested against repression in the Balkans and India. For these activists, their status as women lends them a special authority when it comes to demanding peace. In the words of the Asian Women's Human Rights Council: "We are the Women in Black... women, unmasking the many horrific faces of more public 'legitimate' forms of violence—state repression, communalism, ethnic cleansing, nationalism, and wars...."

Today, the Women in Black in Israel continue their nonviolent opposition to the occupation in cooperation with the umbrella group Coalition of Women for a Just Peace. They have been demonstrating against the closures of various Palestinian cities, arguing that the blockades prevent pregnant women from accessing healthcare services and keep students from attending school. The group also calls for the full participation of women in peace negotiations.

While the Women in Black stood in silent protest worldwide, a group of "grandmothers, mothers, wives, and daughters; housewives and professionals; secular and religious" formed the far-right Women in Green in 1993 out of "a shared love, devotion and concern for Israel." Known for the signature green hats they wear at rallies, the Women in Green emerged as a protest to the Oslo accords on the grounds that Israel made too many concessions to Yasir Arafat's Palestinian Liberation Organization. The group opposes returning the Golan Heights to Syria, sharing sovereignty over Jerusalem with the Palestinians, and insists that "Israel remain a Jewish state."

The Women in Green boast some 15,000 members in Israel, and while they have not garnered the global support of the Women in Black, 15,000 Americans have joined their cause. An ardent supporter of Israeli Prime Minister Ariel Sharon, the group seeks to educate the Israeli electorate through weekly street theater and public demonstrations, as well as articles, posters, and newspaper advertisements.

White the groups' messages and methods diverge, their existence and influence demonstrate that women can mobilize support for political change—no matter what color they wear.

—FP

---

of hostages in the disputed region of Nagorno-Karabakh, where tens of thousands of people have been killed. In fact, these women's knowledge of the local players and the situation on the ground would make them indispensable in peace negotiations to end this 13-year-old conflict.

REACHING FOR POLITICAL OFFICE

In 1977, women organizers in Northern Ireland won the Nobel Peace Prize for their nonsectarian public demonstrations. Two decades later, Northern Irish women are showing how diligently women must still work not only to ensure a place at the negotiating table but also to sustain peace by reaching critical mass in political office. In 1996, peace activists Monica McWilliams (now a member of the Northern Ireland Assembly) and May Blood (now a member of the House of Lords) were told that only leaders of the top 10 political parties—all men—would be included in the peace talks. With only six weeks to organize, McWilliams and Blood gathered 10,000 signatures to create a new political party (the Northern Ireland Women's Coalition, or NIWC) and got themselves on the ballot. They were voted into the top 10 and earned a place at the table.

> The grass-roots, get-out-the-vote work of Vox Femina convinced hesitant Yugoslav women to vote for change; those votes contributed to the margin that ousted President Slobodan Milosevic.

The NIWC's efforts paid off. The women drafted key clauses of the Good Friday Agreement regarding the importance of mixed housing, the particular difficulties of young people, and the need for resources to address these problems. The NIWC also lobbied for the early release and reintegration of political prisoners in order to combat social exclusion and pushed for a comprehensive review of the police service so that all members of society would accept it. Clearly, the women's prior work with individuals and families affected by "the Troubles" enabled them to formulate such salient contributions to the agreement. In the subsequent public referendum on the Good Friday Agreement, Mo Mowlam, then British secretary of state for Northern Ireland, attributed the overwhelming success of the YES Campaign to the NIWC's persistent canvassing and lobbying.

Women in the former Yugoslavia are also stepping forward to wrest the reins of political control from extremists (including women, such as ultranationalist Bosnian Serb President Biljana Plavsic) who destroyed their country. Last December, Zorica Trifunovic, founding member of the local Women in Black (an antiwar group formed in Belgrade in October 1991), led a meeting that united 90 women leaders of pro-democracy political campaigns across the former Yugoslavia. According to polling by the National Democratic Institute, the grass-roots, get-out-the-vote work of groups such as Vox Femina (a local NGO that participated in the December meeting) convinced hesitant women to vote for change; those votes contributed to the margin that ousted President Slobodan Milosevic.

International security forces and diplomats will find no better allies than these mobilized mothers, who are tackling the toughest, most hardened hostilities.

Argentina provides another example of women making the transition from protesters to politicians: Several leaders of the Madres de la Plaza de Mayo movement, formed in the 1970s to protest the "disappearances" of their children at the hands of the military regime, have now been elected to political office. And in Russia, the Committee of Soldiers' Mothers—a protest group founded in 1989 demanding their sons' rights amidst cruel conditions in the Russian military—has grown into a powerful organization with 300 chapters and official political status. In January, U.S. Ambassador to Moscow Jim Collins described the committee as a significant factor in countering the most aggressive voices promoting military force in Chechnya. Similar mothers' groups have sprung up across the former Soviet Union and beyond—including the Mothers of Tiananmen Square. International security forces and diplomats will find no better allies than these mobilized mothers, who are tackling the toughest, most hardened hostilities.

YOU'VE COME A LONG WAY, MAYBE

Common sense dictates that women should be central to peacemaking, where they can bring their experience in conflict resolution to bear. Yet, despite all of the instances where women have been able to play a role in peace negotiations, women remain relegated to the sidelines. Part of the problem is structural: Even though more and more women are legislators and soldiers, underrepresentation persists in the highest levels of political and military hierarchies. The presidents, prime ministers, party leaders, cabinet secretaries, and generals who typically negotiate peace settlements are overwhelmingly men. There is also a psychological barrier that precludes women from sitting in on negotiations: Waging war is still thought of as a "man's job," and as such, the task of stopping war often is delegated to men

(although if we could begin to think about the process not in terms of stopping war but promoting peace, women would emerge as the more logical choice). But the key reason behind women's marginalization may be that everyone recognizes just how good women are at forging peace. A U.N. official once stated that, in Africa, women are often excluded from negotiating teams because the war leaders "are afraid the women will compromise" and give away too much.

Some encouraging signs of change, however, are emerging. Rwandan President Paul Kagame, dismayed at his difficulty in attracting international aid to his genocide-ravaged country, recently distinguished Rwanda from the prevailing image of brutality in central Africa by appointing three women to his negotiating team for the conflict in the Democratic Republic of the Congo. In an unusually healthy tit for tat, the Ugandans responded by immediately appointing a woman to their team.

Will those women make a difference? Negotiators sometimes worry that having women participate in the discussion may change the tone of the meeting. They're right: A British participant in the Northern Ireland peace talks insightfully noted that when the parties became bogged down by abstract issues and past offenses, "the women would come and talk about their loved ones, their bereavement, their children and their hopes for the future." These deeply personal comments, rather than being a diversion, helped keep the talks focused. The women's experiences reminded the parties that security for all citizens was what really mattered.

The role of women as peacemakers can be expanded in many ways. Mediators can and should insist on gender balance among negotiators to ensure a peace plan that is workable at the community level. Cultural barriers can be overcome if high-level visitors require that a critical mass (usually one third) of the local interlocutors be women (and not simply present as wives). When drafting principles for negotiation, diplomats should determine whether women's groups have already agreed upon key conflict-bridging principles, and whether their approach can serve as a basis for general negotiations.

Moreover, to foster a larger pool of potential peacemakers, embassies in conflict areas should broaden their regular contact with local women leaders and sponsor women in training programs, both at home and abroad. Governments can also do their part by providing information technology and training to women activists through private and public partnerships. Internet communication allows women peace builders to network among themselves, as well as exchange tactics and strategies with their global counterparts.

"Women understood the cost of the war and were genuinely interested in peace," recalls retired Admiral Jonathan Howe, reflecting on his experience leading the U.N. mission in Somalia in the early 1990s. "They'd had it with their warrior husbands. They were a force willing to say enough is enough. The men were sitting around talking and chewing qat, while the women were working away. They were such a positive force.... You have to look at all elements in society and be ready to tap into those that will be constructive."

Want to Know More?

The Internet is invaluable in enabling the inclusive security approach advocated in this article. The Web offers not only a wealth of information but, just as important, relatively cheap and easy access for citizens worldwide. Most of the women's peace-building activities and strategies explored in this article can be found on the Web site of **Women Waging Peace**—a collaborative venture of Harvard University's John F. Kennedy School of Government and the nonprofit organization Hunt Alternatives, which recognize the essential role and contribution of women in preventing violent conflict, stopping war, reconstructing ravaged societies, and sustaining peace in fragile areas around the world. On the site, women active in conflict areas can communicate with each other without fear of retribution via a secure server. The women submit narratives detailing their strategies, which can then be read on the public Web site. The site also features a video archive of interviews with each of these women. You need a password to view these interviews, so contact Women Waging Peace online or call (617) 868-3910.

The Organization for Security and Co-operation in Europe (OSCE) is an outstanding resource for qualitative and quantitative studies of women's involvement in conflict prevention. Start with the final report of the *OSCE Supplementary Implementation Meeting: Gender Issues* (Vienna: UNIFEM, 1999), posted on the group's Web site. **The United Nations Development Fund for Women** (UNIFEM) also publishes reports on its colorful and easy-to-navigate site. The fund's informative book, *Women at the Peace Table: Making a Difference* (New York: UNIFEM, 2000), available online, features interviews with some of today's most prominent women peacemakers, including Hanan Ashrawi and Mo Mowlam.

For a look at how globalization is changing women's roles in governments, companies, and militaries, read Cynthia Enloe's *Bananas, Beaches and Bases: Making Feminist Sense of International Politics* (Berkeley: University of California Press, 2001). In *Maneuvers: The International Politics of Militarizing Women's Lives* (Berkeley: University of California Press, 2000), Enloe examines the military's effects on women, whether they are soldiers or soldiers' spouses. For a more general discussion of where feminism fits into academia and policymaking, see **"Searching for the Princess? Feminist Perspectives in International Relations"** (*The Harvard International Review*, Fall 1999) by J. Ann Tickner, associate professor of international relations at the University of Southern California.

The Fall 1997 issue of FOREIGN POLICY magazine features two articles that highlight how women worldwide are simultaneously gaining political clout but also bearing the brunt of poverty: **"Women in Power: From Tokenism to Critical Mass"** by Jane S. Jaquette and **"Women in Poverty: A New Global Underclass"** by Mayra Buvinic.

• For links to relevant Web sites, as well as a comprehensive index of related FOREIGN POLICY articles, access **www.foreign policy.com**.

Lasting peace must be homegrown. Inclusive security helps police forces, military leaders, and diplomats do their jobs more effectively by creating coalitions with the people most invested in stability and most adept at building peace. Women working on the ground are eager to join forces. Just let them in.

Swanee Hunt is director of the Women in Public Policy Program at Harvard University's John F. Kennedy School of Government. As the United States' ambassador to Austria (1993–97), she founded the "Vital Voices: Women in Democracy" initiative. Cristina Posa, a former judicial clerk at the United Nations International Criminal Tribunal for the former Yugoslavia, is an attorney at Cleary, Gottlieb, Steen & Hamilton in New York.

Women & Development Aid

Ritu R. Sharma

Key Points

- Economic studies and program evaluations show that considering gender roles and targeting programs to women and girls dramatically enhances economic growth and project effectiveness.
- Women-headed households represent the majority of the poor worldwide. U.S. development programs, which aim to reduce poverty, should logically center on women.
- Despite economic evidence, evaluation results, and directives from Congress, U.S. development assistance programs have largely ignored gender integration.

Over the past 30 years, study after study by academics, development practitioners, and international agencies has demonstrated the seemingly self-evident fact that women are equal to men, and sometimes surpass men, in contributing to social and economic development.

Researchers have also documented the significant economic dividends of investing in women and girls. Studies conducted by the World Bank, United Nations, and various academics have shown that discrimination against women and girls in education, health care, financial services, and human rights dampens overall economic output, productivity,

and growth rates. One World Bank report found that gender inequality in education and employment suppresses Africa's annual per capita growth by 0.8%.

Beyond direct economic impacts, women's increased access to education, health care, and human rights brings a "virtuous" cycle of enhanced child health, improved food production, lower population growth rates, higher incomes, and, of course, better quality of life for women themselves.

In addition to undermining women's potential, discrimination and low status have relegated many women and their children to the ranks of the poor. Women-headed households make up a majority of the poorest of the poor both in developed and developing countries. More than 900 million women live on less than one dollar a day, and the number of rural women living in absolute poverty has risen by 50% over the past 20 years, as opposed to 30% for men.

Advocates, academics, and development practitioners have been working hard for more than thirty years to integrate gender roles—that is, the different roles males and females play in a society—into American aid policy and programming. Yet, despite the evidence that women are active in national development and that investing in women and girls yields a multitude of benefits, U.S. international assistance programs and policy have not caught up with the facts.

In 1970, the women-in-development movement was crystallized by Ester Boserup's groundbreaking book: *Women's Roles in Economic Development*. In her book, she debunked the myth that women are not economic actors, brought to light the extent to which the economies of poor countries are propelled by women, and asserted that programs that considered women's roles would lead to greater contributions to development.

In 1973, Congress passed an amendment that, for the first time, explicitly addressed women's roles in the development process. The Percy Amendment (after its sponsor, Senator Charles Percy) is still in effect. It requires U.S. bilateral assistance programs to enhance the integration of women into the national economies of developing countries, and it instructs the State Department to consider progress on women's issues when making decisions about funding international organizations (e.g., United Nations, World Bank). In 1974, the United States Agency for International Development (USAID) established the Office of Women in Development to assist USAID missions and regional bureaus in integrating women into their various projects in the field.

In 1993, the Government Accounting Office evaluated USAID'S progress in meeting the requirements of the Percy Amendment. The report found that USAID "has only recently begun to consider the role of women in its third-world development strategies, despite the fact that 20 years have passed since Congress directed that AID assistance programs focus on integrating women."

By 1995, First Lady Hillary Rodham Clinton's leadership as head of the U.S. delegation to the UN Conference on Women in Beijing created a flurry of activity within USAID. One outcome was the creation of the Gender Plan of Action (GPA) in 1996, a three-step plan for the total integration of gender dynamics into all USAID activities. This plan was significant in its willingness to use mechanisms that really matter—bids and contracting systems, performance evaluations and promotions, and USAID's annual "Results Review and Resource Request" process—to ensure real change on programming with a gender perspective.

Four years later, the Advisory Committee on Voluntary Foreign Assistance (ACVFA), an independent adviser to the USAID administrator, commissioned an in-depth analysis—including over 500 interviews—of the Gender Plan of Action. The summary report states: "Over 90% of those interviewed in USAID and the PVO/NGO community said that the GPA has not had any measurable impact on Agency operations." This was not due to faults in the plan; it was because the plan was never promoted or implemented by the agency's leadership.

Ritu Sharma   is cofounder and executive director of Women's EDGE.

From *Foreign Policy In Focus Brief*, September 17, 2001, p. 1. Excerpted from Ritu Sharma's "Women and Development Aid" *Foreign Policy in Focus*, Vol. 6, No. 33, September 2001. © 2001 by Interhemispheric Resource Center. Reprinted by permission.

Gender Equality:
A Prerequisite for Sustainable Development

ABSTRACT: This article uses the ten year anniversary of the United Nations Conference on Environment and Development, marked by the World Summit on Sustainable Development in Johannesburg in August 2002, to explore ways in which women have, or have not, become more involved in environmental decision making and whether prerequisites for this involvement (for example equitable education, health care and economic status) are being met. It opens by explaining why women have been identified as a coherent group which should be the focus of attention, by focusing on their relatively poor (compared to men) economic and social status and their disproportionately high involvement in tasks which are unpaid or otherwise undervalued. Women's involvement in Agenda 21 and the resulting document which specifically refers to the need to improve women's health, education, economic position and involvement in decision making is reviewed before considering how far progress has been made in these areas. The article concludes that, while some improvements appear to have been made, these are by no means uniform across space and time and that persistent inequalities continue to exist, and indeed have emerged in some of the new concerns identified by the UN.

SUSAN BUCKINGHAM-HATFIELD

THE EARTH SUMMIT (formerly known as the United Nations Conference on Environment and Development, henceforth referred to as UNCED), held in Rio de Janeiro in 1992, was the first United Nations conference, not specifically concerned with women[1], to agree that women needed to be involved in decision making at all levels in order for environmental sustainability to be achieved. The two main reasons for this inclusion were that, first of all, women are more likely to be affected by negative environmental impacts than men, because of their relatively poor social and economic position, which, in turn, is closely entwined with their domestic role. Second, and not unrelated (as those factors determine the kinds of activity that women engage in, and how they are valued), women are more likely to have a better knowledge of, and

perhaps affinity with, the environment, as they are more likely to be the main meal providers, primary carers and, particularly in the Global South, the dominant food producers (Buckingham-Hatfield, 2000). Mary Mellor (1992) refers to this as "embodiedment", where women are more in touch with the environment through the daily exigencies of feeding and clothing the family, cleaning the home and maintaining family health. Those not so troubled with these roles, she argues, become more abstracted from their environmental connections.

Of all the agreements reached at Rio, Agenda 21 most prominently addressed what needed to be done to achieve equality. Chapter 24 specifically addressed what needed to be achieved to enable women to play a full part in environmental decision making and to eradicate ine-

Objectives
From Chapter 24, Agenda 21
Global action for women towards sustainable and equitable action

The following objectives are proposed for national governments:

1. to implement the Nairobi Forward-Looking Strategies for the Advancement of Women, particularly with regard to women's participation in national eco-system management and control of environmental degradation;
2. to increase the proportion of women decision makers, planners, technical advisors, managers and extension workers in environment and development fields;
3. to consider developing and issuing by the year 2000 a strategy of changes necessary to eliminate constitutional, legal, administrative, cultural, behavioural, social and economic obstacles to women's participation in sustainable development and in public life;
4. to establish by the year 1995 mechanisms at the national, regional and international levels to assess the implementation and impact of development and environment policies and programmes on women and to ensure their contribution and benefits;
5. to assess, review, revise and implement, where appropriate, curricular and other education material, with a view to promoting the dissemination to both men and women of gender relevant knowledge and valuation of women's roles through formal and non-formal education, as well as through training institutions, in collaboration with non-governmental organisations;
6. to formulate and implement clear government policies and national guidelines, strategies and plans for the achievement of equality in all aspects of society, including the promotion of women's literacy, education, nutrition and health and their participation in key decision-making positions and in management of the environment, particularly as it pertains to their access to resources, by facilitating better access to all forms of credit, particularly in the informal sector, taking measures towards ensuring women's access to property rights as well as agricultural inputs and implements;
7. to implement, as a matter of urgency, in accordance with country-specific conditions, measures to ensure that women and men have the same right to decide freely and responsibly the number and spacing of their children and have access to information, education and means, as appropriate, to enable them to exercise this right in keeping with their freedom, dignity and personally held values;
8. to consider adopting, strengthening and enforcing, legislation prohibiting violence against women and to take all necessary administrative, social and educational measures to eliminate violence against women in all its forms.

Source: UN Earth Summit, 1992

qualities, particularly with regard to legal and economic status, education, health and family planning. As the extract from Chapter 24 illustrates, some of the objectives contained deadlines by which action was to have been taken. Of course, although Agenda 21 and a number of other agreements came into force subsequent to UNCED[2], it is very difficult to both enforce and monitor these agreements.

This article considers what the achievements have been, with particular reference to women and the environment, and focuses on the prerequisites required to meaningfully involve women in environmental decision making, such as health care, education and economic status. It considers how much progress (if any) has been made in policy developments, but also in actual improvements in women's status and involvement. It will briefly summarise the purpose of the World Summit on Sustain-

able Development (WSSD) scheduled for August 2002, which is both a review of progress of the UNCED in 1992 and a plan for accelerating the move towards greater environmental sustainability. It will consider how UNCED specifically addressed inequalities of environmental impacts, with particular reference to women. This is important, as UNCED was, in many respects, a landmark international conference in this regard. This article also considers how far, in fact, countries (and localities) have achieved greater gender equity in decision making, by using the proxy indicator of women's political representation, and whether socio-economic indicators suggest that progress has been made in terms of reducing gender inequalities. Finally, it looks at some of the issues that have risen in prominence since 1992, and are being raised as challenges by the UN in 2002, to see what women's experience is/is likely to be in the face of the worsening AIDS/

Violence against women

UNICEF data is used by Jan Jindy Pettman (1996) to demonstrate that violence against women by male partners is the most common crime in the world. Rape is also a common military strategy: in the Yugoslav civil war prior to 1995, estimates from 20,000 to 35,000 have been recorded (Pettman, 1996). Indeed, it was not until the United Nations Conference of Human Rights in Vienna in 1993 adopted the Declaration on the Elimination of Violence Against Women that rape was even considered a war crime (Bunch *et al.*, 2001). It was the Global Campaign for Women's Rights which ensured that this was prominent on the Vienna Conference's agenda (the initial call for the conference did not recognise gender-specific rights), Bunch (2001, p. 218) argued that "women's rights are human rights".

The Grameen Bank

The Grameen Bank was formed to lend money at commercial bank rates to the poorest of households who would otherwise have to rely on the usurious rates of loan sharks. It began as an experimental project in rural Chittagong, Bangladesh, initiated by a professor from Chittagong University. It now receives its funding from the Bangladesh Central Bank, local commercial banks and international donors. Ninety four per cent of the Grameen Bank's borrowers are women, and the Bank has a repayment rate of 98%. Loans are made for individual projects but are issued only to groups of four to five women who take joint responsibility for the loan repayment, which encourages group support and unity. These groups manage the loan and get involved in the community economy through regional federations. The initial loan of US $75 to US $100 is made to the two neediest members of the group who must repay the loan before any other members of the group receive a loan. The loan may be made for any purpose that is approved by the group and the centre chief (each centre is responsible for a number of groups). The loans are repaid in 50 equal installments, and when this is repaid, the borrower may also apply for a larger loan. Members are required to save money which is then converted into Grameen Bank shares. Before receiving a loan, group members are expected to undertake training which helps them manage the loan and their new business. This micro-credit scheme has found women to be the most reliable and conscientious borrowers and, in addition, it has been found to be effective in reducing poverty, increasing children's nutritional intake and in (voluntarily) increasing the use of contraceptives.

Buckingham-Hatfield (2000, pp. 111–2) summarising Hashemi and Morshed (1997).

HIV pandemic, greater globalisation and the accelerating penetration of society by information and communications technologies (ICTs). But first, it may be necessary to argue why women should be the focus of programmes and initiatives to reduce poverty and move towards environmental sustainability.

Why women?

Taken as a whole group, women are the largest disadvantaged group throughout the world. While there are significant differences between women, which should not be minimised (for example, between the Global North and the Global South, between white and black, between mothers and those without children and between the rich and poor), there are marked disadvantages which characterise most women's lives, and have implications for their experience of the environment. For example, world-wide, women are much more likely to suffer violence at the hands of men. Box 2 illustrates the extent of violence worldwide which suggests that, while there has been some progress on Objective 8 of Global action for women towards sustainable and equitable action (see above), there is still a long way to go. In war time, women are increasingly likely to be injured or killed: 80% of war casualties are now civilian, the majority of whom are likely to be women and children, just as they are more likely to be refugees. De Groot (1999) has estimated this as high as 90%. Black (1998) has estimated that refugee women and girls have to travel, on average, one and a half hours per household, per day on top of the journey they would normally expect to make to collect water and firewood.

Economic equalities are as important as legal rights. Arguably, it is difficult, if not impossible, to operationalise legal and civil rights if a person does not have the economic resources to pay for shelter, food, clothing, health care and education. Again, there are world-wide similarities which have not significantly changed in the past decade: hour for hour women earn approximately three-quarters of a man's wage. This is equally true in Belgium (where the female:male hourly earnings ratio in manufacturing jobs is 74.5) and New Zealand (74.5) as in Costa Rica (74.0 for monthly earnings) or Kenya (73.0 also for monthly earnings) (ILO in Anker, 1997).

Minu Hemmati (2001) estimates that 70% of all those who live in absolute poverty (usually considered to be less than US$1 a day) are women. The United Nations recognises that the eradication of poverty must be "one of the fundamental goals of the international community... and it is essential for sustainable development". It goes on to argue that "the empowerment of women is a critical factor for the eradication of poverty" (UN cited in Osborn and Bigg, 1998). This could, however, be seen to be a circular argument, as the empowerment of women is often dependent on their financial independence. Many initiatives demonstrate that where women are able to become economically self-sufficient (for example through micro-credit schemes), they gain the confidence (individually, and also of their neighbours) to take an active part in community affairs (Wickramasinghe, 1997). As the panel below illustrates, the Grameen Bank is a good example of how a micro-credit scheme made available specifically to women has widespread personal, community and environmental benefits. It is also noteworthy that greater financial independence is likely to reduce family size (Wickramasinghe, 1997), while it has been noted that payment direct to women heads of household ensures that the household budget is more likely to be spent on food, clothing and health care, than on cigarettes, alcohol and watches if directed to the male head of household (Chant, 1997).

Although absolute poverty is overwhelmingly concentrated in the Global South, it is worth noting that there are households in the Global North which are sufficiently poor to lack food security and permanent shelter, and that these are more likely to be headed by women. In 1994, Kempson *et al.* reported that half of all mothers living on or just above the income support level in the UK regularly went without food to feed and clothe their children. These concerns, and others, have been the particular focus of a series of UN Conventions on Women, the fourth of which was held in Beijing in 1995. Here, an express link was made between women's legal and political rights, poverty alleviation and environmental sustainability. This, together with UNCED in 1992 marks a linking which now characterises the majority of UN conferences (see also, for example, conferences on Population, Cairo 1994, Habitat, Istanbul 1996, and Social Development Copenhagen 1995). The next section explores how, arguably, the most successful achievement at Rio—Agenda 21—addressed environmental sustainability not only from an "outcomes" perspective, but also by considering the processes signatories would have to engage in to achieve these outcomes. In particular, this involves previously excluded, disadvantaged groups in the decision-making process. The link between UNCED and the WSSD, with regard to Agenda 21 and women, will also be made.

From UNCED to the WSSD

The World Summit on Sustainable Development has been called to review progress on Agenda 21 and it is expected to "identify measures for further implementation" of Agenda 21 and other UNCED outcomes, including sources of funding (UN, 2000). The WSSD is preceded by a series of preparatory committees (or "prep coms") which are expected to identify areas of debate, and to draft the texts of agreements to be signed. The review of Agenda 21 has been undertaken by the prep coms, the second of which is expected to agree the text of the review, its conclusions and recommendations for action. What is taken to Johannesburg is intended to "reinvigorate at the highest political level, the global commitment to a North/South partnership and a higher level of international solidarity to the accelerated implementation of Agenda 21 and the promotion of sustainable development" (UN, 2000). Agenda 21 is generally held to be the most effective outcome of UNCED. While the Framework Convention on Climate Change, the Convention on Biodiversity and the Statement of Forest Principles were heavily contested and compromised by a lack of agreement between governments of the Global North, and those of the Global South, Agenda 21 emerged with support from a wide range of non-governmental organisations (NGOs) (including women's campaigning groups) which had been involved in its drafting in the preceding prep coms. This NGO involvement was an innovative contribution to a UN conference where policy formulation had previously exclusively been the preserve of member states; arguably this has had a significant impact on its relatively high profile at the local level. Its 40 chapters set out strategies which needed to be developed if environmental sustainability was to be achieved, but also focused on key groups which are commonly excluded from decision making: children and young people (Chapter 25), indigenous peoples and their communities (Chapter 26), non-governmental organisations (Chapter 27), workers and trades unions (Chapter 29), and farmers (Chapter 32), as well as women (Chapter 24). It also referred to the inclusion of business and industry (Chapter 30) and local government (Chapter 28), all of which has led to an increased focus on what has become known as "stakeholder" involvement in environmentally related decision making. This focus on the process of decision making has required governments (both central, and, more particularly, local) to find ways in which all these groups may be more active participants in environmental decision making, to the extent that the UN Economic and Social Council finds that:

"The increase in major group participation has been a key area of success in the post-UNCED period. Agenda 21 follow up has been more participatory, open and accessible to a broad range of state actors" (2001, p. 42).

The same report considers that:

"The strength of LA21 initiatives has been their multi-stakeholder approach to local decision making, the identification of priorities, funding solutions and implementation" (2001, p. 42).

Such claims raise interesting issues about the way in which information about local projects and decision-making processes are communicated to the rather remote offices of the UN. While there has been discussion of Agenda 21 at the local level in much of Europe, North America and Australasia (see discussion by Voisey, *et al.*, 1996; Parker and Selman, 1997; and LASALA, 2001), there is evidence that widespread participation from the marginalised groups identified by the UN is limited (Buckingham-Hatfield, 1999; Buckingham-Hatfield and Matthews, 1999; Evans and Percy, 1999). While some local plans have used innovative and inclusive forms of participation to envisage sustainable futures, many have relied of familiar (and generally limited) forms of pubic participation and consultation. The UN itself has noted that "few countries provide incentives to non-governmental actors to participate, limiting their contributions" (2001, p. 43). Specifically, this report notes that the participation of women at all levels is relatively low. Indeed, the author of this article has observed in Local Agenda 21 setting in the London area that not only are the quantitative contributions of women lower, but that less regard is given to these contributions. Meetings are frequently chaired by and dominated by particular interest groups which value contributions in different ways. For example, it has been common for concerns about dog litter and graffiti, usually expressed by women, to be devalued compared with public transport and cycling provision (Buckingham-Hatfield, 1999). Consequently, there is likely to be a gap between an increasingly familiar discourse of equality, participation and sustainability, and plans which actually get implemented. Likewise, at the global level, the interests of women now appear to be well integrated into UN conventions and, increasingly, into global institutional frameworks (such that the World Bank now considers gender impacts on a number of projects it funds), but there is little evidence that this is changing practice in any fundamental way.

Persistent inequalities

Minu Hemmati (2001) argues that the reasons for a failure to improve women's participation in environmental decision making include a lack of political will, which results in a lack of available resources and a lack of information available to current decision makers. She suggests that the only way in which women are likely to become more meaningfully involved is if more financial resources are committed to raise education levels and awareness levels of both women and men. Both locally initiated and donor-aided development needs to assess its gendered impact, which will require gender-disaggregated data to be made available. Community development needs to create space for women and more opportunities for men and women to enter into dialogue. Finally, the UN, governments and other "stakeholders" need to invest in the participation of women at all levels (UN, 2001, p. 70).

Women in formal decision making

One way in which women's participation in decision making can be measured is to look at the number of women representatives in national Parliaments, and as government ministers. The UK has long held an ambiguous position with regard to equal representation. While paying lip service to equal opportunities, British political parties (with the exception, for a short while, of New Labour in the 1990s) have resisted the affirmative action of all-women shortlists, arguing that good candidates will be selected regardless of gender. However, in a recent Fawcett Society report, 60% of Labour women and 81% of Conservative women agreed that selection committees in their party tended to look more favourably on male than on female candidates (Ashley, 2002). While the number of women MPs in the UK rose when New Labour was elected in 1997, this number fell from 101 at the 1997 election to 96 after the 2001 election, when a number of existing women MPs stepped down, citing women-unfriendly practices as a cause of their resignation. At the same time, in most Western countries the number of women in ministerial posts is rising (for example, in the UK this rose from 9 in 1994 to 24 in 1998). However, 38 countries recorded a drop in women's representation at this level, including a number of countries normally thought to be in the vanguard of equal opportunities (Finland from 39 to 29; Germany from 16 to 8, the Netherlands from 31 to 28 and Norway, from 35 to 20). Eight of these countries dropped to zero representation of women (including Japan and Israel), while 19 countries had no women ministers at either dates (data from World Bank, 2001). While a straightforward numerical rise of women in senior government posts does not necessarily lead to greater gender equality or environmental sustainability, as a proxy, these figures suggest that no significant progress has been made towards greater women's involvement in decision making. On the other hand, it is notable from these World Bank statistics that women are becoming more prominent in government in sub-Saharan Africa and South America.

Labour force participation, education and literacy

In low-income countries women's participation in the labour force remained static between 1990 and 1999 (a 0.6 female:male ratio), while high-income countries experienced a rise from 0.7 to 0.8 (although this obscures regional variations between Europe and Central Asia up to 0.9 and the rise in the Middle East and North Africa to 0.4) (World Bank, 2001). As with several of the indicators under discussion here, labour force participation in itself does not necessarily indicate that women are gaining greater independence and there is a world of difference between participation in the micro-credit schemes referred to earlier, which result in greater empowerment of women, and the virtual slave labour of export processing zones in which mainly women work for little over US$1 a day in intolerable conditions.

The world adult literacy rate is 68% for women (up from 54% in 1980) and 82% for men (up from 72% in 1980). Young women continue to be less literate than young men, with particularly low female literacy in South Asia (41%) and sub-Saharan Africa (51%) (DFID, 2000). Interesting, only one country reported higher literacy rates among young women compared to young men: South Africa has a female:male ratio of 1.1 (World Bank, 2001). Women's relative illiteracy is a product of gender inequality in schools, with girls in developing countries less likely (92% the enrolment level of boys) to be enrolled than boys, even in primary school (DFID, 2000). The UN Economic and Social Council (2001) suggests that a lack of investment in basic education is one of the reasons for the lack of progress on education for sustainable development, although it should be noted that the statistics indicated that for every region in the Global South, women's literacy rates have improved, as have primary school enrolments. (It is also worth noting that these levels have also improved for boys.) Specifically it argues that a focus on education for women and girls translates directly into better nutrition for the whole family, better health care, lower fertility, reduced poverty and better overall economic performance.

Women's health

In its preparation document *Implementing Agenda 21*, the UN Economic and Social Council identified HIV/AIDS as a major issue that has risen to prominence since UNCED in 1992, referring to it as "the fastest growing health threat to development today" (2001, p. 18). While, worldwide, 40% of all adult HIV infections were reported by women, by 1995, 50% of all AIDS cases reported were women (Baden and Wach, 1998). One of the main reasons for this disparity seems to be the lack of women's and girls' sex education. Most societies put a high premium on girls' sexual innocence, which makes it socially unacceptable for girls to seek or be given information on sexually transmitted diseases (UNAIDS, 1999). World Bank data reveals that young women aged between 15 and 24 are almost twice as likely as young men of the same age to be HIV/AIDS infected in low-income countries (particularly in sub-Saharan Africa). This is due to their biological vulnerability[3], unequal power relations in which girls are far more likely than boys to be coerced or forced into having sex, often by older men, and the lack of sex education available to women (particularly of sexually transmitted diseases) (UNAIDS, 1999). This level of infection among young women (even though fewer of them are sexually active than men), has clear implications for childbirth and children's exposure to HIV/AIDS. This is particularly acute considering that significant numbers of women do not receive any prenatal care: 62% of pregnant women in

low-income countries and 77% in middle-income countries (World Bank, 2001). In many respects, this situation reflects health opportunities and care more generally. UNAIDS cites research that shows "a generalised gender-based attitude towards health care that seems to favour boys over girls" (1999, p. 9) and even among the under 5 year olds, boys are more often taken for health care, and taken earlier in their illnesses.

Other health related issues have a more direct link with environmental factors. The World Health Organisation estimates that poor environmental quality contributes to 25% of all preventable ill health in the world today (UN, 2001, p. 17). For example, air pollution is a major contributor to respiratory illnesses, which particularly affect children, the elderly and women. Eighty per cent of deaths resulting from indoor air pollution from the burning of biomass fuels are of women and girls, as they are the main food preparers. This is exacerbated by the use of low quality biomass (such as straw, husks and dung) used when forest trimmings are no longer available because of commercial deforestation.

Conclusions: prospects for greater environmental equality

In addition to HIV/AIDS, the United Nations has identified globalisation and information and communications technology as major developments since 1992, which the World Summit on Sustainable Development will need to address. Both of these "developments" have gendered aspects, which can only briefly be presented here. Globalisation tends to favour the world's wealthy. Many of the world's poorest countries and significant proportions of the population in most other countries have been left behind by globalisation and ICT. As trans-national companies seek out cheaper production costs and sites, it is frequently women who are employed in export processing zones on below-subsistence wages and in intolerable conditions. ICT is available to the relatively wealthy and the educated and this article has already demonstrated that women are less well represented in either of these groups.

International political discourse now includes reference to women and to sustainable development in ways uncommon before 1992, and this is in large part due to the international campaigning activities of women's groups from both the Global North and South. However, the inclusion of women and sustainable development in policies from international institutions obscures a very real lack of movement towards the pre-conditions needed to achieve a form of environmental sustainability that involves all groups in its definition, let alone its achievement. While there have been improvements in women's education and literacy levels (as part of overall improvements), economic inequality has seen little change and gendered health differences remain significant. While some of the indicators addressed in this article are only proxy indicators for improvement (as discussed earlier, for example, more women in government does not necessarily mean more general equality or greater environmental sustainability), research indicates that structural inequalities are currently overwhelming progress made at the micro level. Consequently, the WSSD has a mammoth task ahead of it to attempt to translate equitable decision making in the sustainable development arena into national and international policy and practice.

Notes

1. The UN has a programme of Women's Conferences, the fourth of which was held in Beijing in 1995. A special plenary session of the UN was held in 2000 to review progress made since Beijing (known as "Women 2000" or Beijing +5).

2. Other agreements reached at UNCED were a Statement on Forest Principles, the Framework Convention on Climate Change, a Convention on Biodiversity, and a set of Principles—The Rio Principles on Environment and Development.

3. Women of all ages are more likely than men to become infected with HIV during unprotected vaginal intercourse, but girls, whose genital tract is still not fully mature, are especially vulnerable.

References

Anker, R. (1997) "Theories of occupational segregation by sex: an overview", *International Labour Review*, 136, p. 3.

Ashley, J. (2002) "Revealed: how Labour sees women", *New Statesman*, 4 February.

Baden, S. and Wach, H. (1998) *Gender, HIV/AIDS Transmission and Impacts: A review of issues and evidence*. University of Sussex: Institute of Development Studies.

Black, R. (1998). *Refugees, Environment and Development*. Harlow: Longman.

Buckingham-Hatfield, S. (2000) *Gender and Environment*. London: Routledge.

Buckingham-Hatfield, S. and Matthews, J. (1999) "Including women: addressing gender" in Buckingham-Hatfield, S. and Percy, S. (eds) *Constructing Local Environmental Agendas*. London: Routledge, pp. 194–209.

Buckingham-Hatfield, S. (1999) "Gendering Agenda 21: women's involvement in setting the environmental agenda", *Journal of Environmental Policy and Planning*, 1, pp. 121–32.

Bunch C. with Antrobus, P., Frost, S. and Reilly, N. (2001) "International networking for women's human rights" in Edwards, M. and Gaventa, J. (eds) *Global Citizen Action*. London: Earthscan.

Chant, S. (1997) *Women-Headed Households: Diversity and dynamics in the developing world*. London: Macmillan.

De Groot, G. J. (1999) "A force for change", *The Guardian*, 14 June.

DFID (2000) *Achieving Sustainability, Poverty Elimination and the Environment*. London: Department for International Development.

Evans, R. and Percy, S. (1999) "The opportunities and challenges for local environmental policy and action in the UK" in Buckingham-Hatfield, S. and Percy, S. (eds) *Constructing Local Environmental Agendas*. London: Routledge, pp. 172–85.

Hashemi, S. M. and Morshed, L. (1997) "Grameen Bank: a case study" in Wood, G. D. and Sharif, I. A. (eds) *Who Needs Credit? Poverty and Finance in Bangladesh*. London: Zed Books, pp. 217–30

Hemmati, M. (2001) "Women and sustainable development: from 2000–2002" in Dodds, F. (ed) *Earth Summit 2002, A New Deal*. London: Earthscan, pp. 165–83.

Kempson, F., Bryson, A. and Rowlingson, K. (1994) *Hard Times*. London: Policy Studies Institute.

LASALA Project Team (2001) *Accelerating Local Sustainability—Evaluating European Local Agenda 21 Processes*. Freiburg, Germany: ICLEI.

Mellor, M. (1992) *Breaking the Boundaries, Towards a feminist green socialism*. London: Virago Press.

Osborn, D. and Bigg, T. (1998) *Earth Summit II, Outcomes and Analysis*. London: Earthscan.

Pettman, J. J. (1996) *Worlding Women: A feminist international politics*. London: Zed Books.

Selman, P. and Parker, J. (1997) "Citizenship, civicness and social capital in Local Agenda 21", *Local Environment*, 2, p. 2.

United Nations (2000) "Ten year review of progress achieved in the implementation of the outcome of the United Nations Conference on Environment and Development", www.johannesburgsummit.org (accessed 5 March 2002).

United Nations (1992) *Earth Summit 1992*. Geneva: UN.

UNAIDS (1999) *Gender and HIV/AIDS: Taking stock of research and programmes*. Geneva: UNAIDS.

United Nations Economic and Social Council (2001) *Implementing Agenda 21, Report of the Secretary General*. New York: United Nations.

Voisey, H., Beurmann, C., Sverdrup, L. A. and O'Riordan, T. (1996) "The political significance of Local Agenda 21: the early stages of some European experience", *Local Environment*, 1, p. 1.

Wickramasinghe, A. (1997) *Land and Forestry: Women's local resource based occupations for sustainable survival in South Asia*. Columbo, Sri Lanka: CORRENSA.

World Bank (2001) *World Development Indicators*. New York: World Bank.

Susan Buckingham-Hatfield is a Senior Lecturer in the Department of Geography and Earth Sciences at Brunel University, Uxbridge, Middlesex UB8 3PH (e-mail: susan.buckingham-hatfield@brunel.ac.uk, tel: 01895 274000).

From *Geography*, July 2002, pp. 227-233. © 2002 by The Geographical Association.

Index

Index

Test Your Knowledge Form

We encourage you to photocopy and use this page as a tool to assess how the articles in *Annual Editions* expand on the information in your textbook. By reflecting on the articles you will gain enhanced text information. You can also access this useful form on a product's book support Web site at *http://www.dushkin.com/online/*.

NAME: _____ DATE: _____

TITLE AND NUMBER OF ARTICLE:

_____

BRIEFLY STATE THE MAIN IDEA OF THIS ARTICLE:

_____

LIST THREE IMPORTANT FACTS THAT THE AUTHOR USES TO SUPPORT THE MAIN IDEA:

_____

WHAT INFORMATION OR IDEAS DISCUSSED IN THIS ARTICLE ARE ALSO DISCUSSED IN YOUR TEXTBOOK OR OTHER READINGS THAT YOU HAVE DONE? LIST THE TEXTBOOK CHAPTERS AND PAGE NUMBERS:

_____

LIST ANY EXAMPLES OF BIAS OR FAULTY REASONING THAT YOU FOUND IN THE ARTICLE:

_____

LIST ANY NEW TERMS/CONCEPTS THAT WERE DISCUSSED IN THE ARTICLE, AND WRITE A SHORT DEFINITION:

We Want Your Advice

ANNUAL EDITIONS revisions depend on two major opinion sources: one is our Advisory Board, listed in the front of this volume, which works with us in scanning the thousands of articles published in the public press each year; the other is you—the person actually using the book. Please help us and the users of the next edition by completing the prepaid article rating form on this page and returning it to us. Thank you for your help!

ANNUAL EDITIONS: Developing World 03/04

ARTICLE RATING FORM

Here is an opportunity for you to have direct input into the next revision of this volume.
We would like you to rate each of the articles listed below, using the following scale:

1. **Excellent: should definitely be retained**
2. **Above average: should probably be retained**
3. **Below average: should probably be deleted**
4. **Poor: should definitely be deleted**

Your ratings will play a vital part in the next revision.
Please mail this prepaid form to us as soon as possible.
Thanks for your help!

RATING	ARTICLE	RATING	ARTICLE
	1. The Great Divide in the Global Village		32. The Population Implosion
	2. Prisoners of Geography		33. Saving the Planet: Imperialism in a Green Garb?
	3. The Poor Speak Up		34. Local Difficulties
	4. The Rich Should Not Forget the ROW (Rest of the World)		35. Poverty and Environmental Degradation: Challenges Within the Global Economy
	5. Putting a Human Face on Development		36. Endangered Humans
	6. The Free-Trade Fix		37. A Dirty Dilemma: The Hazardous Waste Trade
	7. Trading for Development: The Poor's Best Hope		38. Death Stalks a Continent
	8. Learning From Doha: A Civil Society Perspective From the South		39. An Epidemic of Neglect: Neglected Diseases and the Health Burden in Poor Countries
	9. Do as We Say, Not as We Do		40. Empowering Women
	10. Spreading the Wealth		41. Women Waging Peace
	11. Surmounting the Challenges of Globalization		42. Women & Development Aid
	12. The Sacking of Argentina		43. Gender Equality: A Prerequisite for Sustainable Development
	13. Safe Deposit: The Case for Foreign Aid		
	14. The Cartel of Good Intentions		
	15. Nongovernmental Organizations (NGOs) and Third World Development: An Alternative Approach to Development		
	16. Fishermen on the Net		
	17. Fight to the Finish		
	18. The Crisis Within Islam		
	19. Mixed Message: The Arab and Muslim Response to 'Terrorism'		
	20. Back to Brinksmanship		
	21. Africa's Great War		
	22. Zimbabwe: The Making of an Autocratic "Democracy"		
	23. Talking Peace, Waging War		
	24. Ending the Death Dance		
	25. Colombia Peace in Tatters		
	26. Democracies: Emerging or Submerging?		
	27. Democracy Inc.		
	28. Rebuilding Afghanistan		
	29. Venezuela's "Civil Society Coup"		
	30. The Many Faces of Africa: Democracy Across a Varied Continent		
	31. Past Successes, Present Challenges: Latin American Politics at the Crossroads		

(Continued on next page)

BUSINESS REPLY MAIL
FIRST-CLASS MAIL PERMIT NO. 84 GUILFORD CT

POSTAGE WILL BE PAID BY ADDRESSEE

McGraw-Hill/Dushkin
530 Old Whitfield Street
Guilford, Ct 06437-9989

NO POSTAGE
NECESSARY
IF MAILED
IN THE
UNITED STATES

IIIₒₒₒIIₒₒIₒIₒIₒIIIₒIIₒₒIIIₒIₒIₒIₒIₒIₒIIₒIₒIIIₒₒIₒIₒI

ABOUT YOU

Name _____ Date _____

Are you a teacher? ☐ A student? ☐
Your school's name _____

Department _____

Address _____ City _____ State _____ Zip _____

School telephone # _____

YOUR COMMENTS ARE IMPORTANT TO US!

Please fill in the following information:
For which course did you use this book?
_____

Did you use a text with this ANNUAL EDITION? ☐ yes ☐ no
What was the title of the text?
_____

What are your general reactions to the *Annual Editions* concept?
_____

Have you read any pertinent articles recently that you think should be included in the next edition? Explain.
_____

Are there any articles that you feel should be replaced in the next edition? Why?
_____

Are there any World Wide Web sites that you feel should be included in the next edition? Please annotate.
_____

May we contact you for editorial input? ☐ yes ☐ no
May we quote your comments? ☐ yes ☐ no